SOLID INTERSTELLAR MATTER: THE ISO REVOLUTION

Les Houches Workshop, February 2-6, 1998

Editors

L. d'HENDECOURT
C. JOBLIN
A. JONES

Springer

Berlin
Heidelberg
New York
Barcelona
Hong Kong
London
Milan
Paris
Singapore
Tokyo

EDP Sciences

7, avenue du Hoggar
Parc d'Activités de Courtabœuf
BP. 112
91944 Les Ulis Cedex A, France

EDP Sciences
875-81 Massachusetts Avenue
Cambridge, MA 02139, U.S.A.

Centre de Physique des Houches

Books already published in this series

1 Porous Silicon Science and Technology
J.-C. VIAL and J. DERRIEN, Eds. 1995

2 Nonlinear Excitations in Biomolecules
M. PEYRARD, Ed. 1995

3 Beyond Quasicrystals
F. AXEL and D. GRATIAS, Eds. 1995

4 Quantum Mechanical Simulation Methods for Studying Biological Systems
D. BICOUT and M. FIELD, Eds. 1996

5 New Tools in Turbulence Modelling
O. MÉTAIS and J. FERZIGER, Eds. 1997

6 Catalysis by Metals
A. J. RENOUPREZ and H. JOBIC, Eds. 1997

7 Scale Invariance and Beyond
B. DUBRULLE, F. GRANER and D. SORNETTE, Eds. 1997

8 New Non-Perturbative Methods and Quantization on the Light Cone
P. GRANGÉ, A. NEVEU, H.C. PAULI, S. PINSKY and E. WERNER, Eds. 1998

9 Starbursts Triggers, Nature, and Evolution
B. GUIDERDONI and A. KEMBHAVI, Eds. 1998

10 Dynamical Networks in Physics and Biology
D. BEYSENS and G. FORGACS, Eds. 1998

Book series coordinated by Michèle LEDUC

Editors of "Solid Interstellar Matter: The ISO Revolution" (No. 11)
L. d'Hendecourt (IAS, Orsay, France)
C. Joblin (CESR, Toulouse, France)
A. Jones (IAS, Orsay, France)

ISBN 3-540-65809-2 Springer-Verlag Berlin Heidelberg New York
ISBN 2-86883-398-5 EDP Sciences Les Ulis

Printed in France

PREFACE

L. d'Hendecourt[1], C. Joblin[2] and A. Jones[1]

[1] *Institut d'Astrophysique Spatiale,*
Université Paris-XI, bâtiment 121,
91405 Orsay Cedex, France
[2] *Centre d'Étude Spatiale des Rayonnements,*
9 avenue du Colonel Roche, BP. 4346,
31028 Toulouse Cedex 04, France

The European Infrared Space Observatory (ISO[1]), launched in November 1995, successfully performed its mission throughout 30 months of observational operations, almost twice its planned lifetime. This European Space Agency mission has been a remarkable collaboration between the 14 European countries which were involved in the management, planning and all the technological developments necessary to achieve the ambitious scientific goals. These successes have now to be followed by a thorough evaluation of the scientific return of this mission and, in the domain of the interstellar medium chemistry, this was the aim of the Workshop "Solid Interstellar Matter: the ISO Revolution" held at Les Houches in early February 1998.

The four instruments on board ISO (the camera CAM, the short- and long-wavelength spectrometers SWS and LWS, and the photometer PHOT) were well-suited to the study of solid interstellar matter and its evolution in the interstellar medium. For the first time, a very large wavelength range has been covered spectroscopically with the SWS and LWS instruments giving access to the chemical composition of dust from its nucleation sites to places of its evolution and destruction. These spectroscopic studies were complemented by the vast mapping capabilities offered by the CAM and PHOT instruments.

During the construction of ISO, various institutes launched efforts to develop experiments to simulate specific interstellar environments in order to create analogues of interstellar dust to aid the interpretation of the ISO astronomical

[1] ISO is an ESA project with instruments funded by ESA Member States (especially the PI countries: France, Germany, the Netherlands and the United Kingdom) and with the participation of ISAS and NASA.

data. These efforts involved laboratory simulations using infrared spectroscopy as a diagnostic tool in the study of the evolution of these dust analogues. This approach proved extremely successful in producing, almost in real time, laboratory spectra closely duplicating the ISO data.

The aim of the Les Houches School, "Solid Interstellar Matter: the ISO Revolution", was to gather together young scientists particularly active in the following fields: ISO observations and data reduction, laboratory spectroscopy of dust analogues and the interpretation of these data within the context of our current knowledge of the physics and chemistry of the interstellar medium. This diversity of subjects was reflected in the wide range of the affiliations of the participants. While about half of the talks referred directly to ISO data, the rest were largely devoted to the presentation and discussion of laboratory approaches and complementary observations for the interpretation of the data. This book, apart from presenting a coherent view of the first ISO results on the subject, promotes a multidisciplinary approach to the problem of the physics and chemistry of the interstellar matter and is therefore presented as a reference textbook in this field through into the next millennium.

The first theme ("General Observations of Dust") focuses on observations of dust in various regions of the interstellar and interplanetary media. The next two themes ("Carbon in Dust" and "Cold Dust: Ices and Silicates") are devoted to the identification of the main components of the interstellar solid matter through observational and laboratory studies. These include refractory solid grains composed of carbon (large aromatic molecules and/or various forms of amorphous carbon) and silicates. In the coldest regions, these refractory grains are covered by mantles of volatile ice. The various contributions cover the physico-chemical evolution of solid matter from diffuse to dense regions, from protostellar environments to the circumstellar shells of evolved stars. The fourth theme ("Dust Processing") more specifically addresses the question of the nucleation and evolution of dust in the interstellar medium. The last theme ("Interstellar Matter in the Laboratory") discusses the link between interplanetary and interstellar dust with special emphasis on the silicate minerals extracted from interplanetary dust grains.

Since the title of the book emphasizes the ISO revolution, one may wonder where this revolution lies. It is firstly a revolution in the method, mostly due to the very interdisciplinary fields that are touched upon. This school, dedicated to the observational results from a large satellite mission, has brought together participants from many different fields, including astronomy, astrophysics, laboratory astrophysics and infrared spectroscopy. As already mentioned, about half of the contributions in this book, are directly related to the ISO observations, whereas the other half present many laboratory investigations that are essential for the interpretation of these data. The various laboratory experimental programs that were specifically planned and designed during the development of the satellite find their full justification here.

Naturally, an obvious revolution comes from the indisputable fact that infrared spectroscopy is the only way to determine the grain chemical and structural composition. In the visible, scattering dominates the extinction of starlight thus rendering the interpretation of the extinction measurements particularly model dependent at these wavelengths. However, in the infrared, pure absorption dominates and the grains generally remain optically thin to their own radiation. Thus the chemical nature of the grains may be completely revealed. Furthermore, the ISO instruments provided unique spectroscopic capabilities that were particularly well-adapted to the study of solids over a very large wavelength range at adequate spectral resolution.

The possible scientific revolutions emerging from the ISO results presented here, may be outlined by looking at the four dominant classes of solid materials present in space: carbonaceous compounds, ices, silicates and refractory oxides.

The notorious Unidentified Infrared Bands (UIBs) are entirely within the range of ISO studies. They are ubiquitously detected in the general diffuse interstellar medium and the overall spectrum of these UIBs appears to be surprisingly constant from region to region, independent of the radiation field. However, one may note that, owing to the very complex nature of the carrier, infrared spectroscopy alone is unable to provide a definite identification of the carrier. This identification must rely on a coherence between models and the constraints of the observations. A true revolution may be hidden in the vast amount of data which still awaits interpretation.

The second major, and expected, advance lies in the field of the chemical composition of ices. Here, the accessibility of a rather large and simultaneous wavelength coverage, allows for the clear detection of ices made of rather simple molecules (H_2O, CO, CO_2, CH_4 ...). Most of them are unobservable using ground-based observatories (and two of them have no transitions in the radio range), yet their abundances prove to be significant and must be taken into account in the models of the chemistry in dense regions. Furthermore, the photochemical evolution of the ices, easily simulated in the laboratory, is largely accessible through the ISO data. The detection of large amounts of CO_2, as well as the ubiquity of this molecule, was clearly a surprise to many observers. This molecule might yet prove to be an important diagnostic of the relationship between grain surfaces and the gas phase chemistry which is driven by classical ion molecule reactions. The presence of significant amounts of solid-state methanol and formaldehyde in protostellar objects and the evolution towards molecular complexity is emphasized by ISO. The connection between these and the chemical composition of cometary volatiles suggests a possible evolutionary link between the two.

Clearly, the most spectacular ISO revolution comes from the observation of crystalline silicate dust around young stellar objects (YSOs) and late-type stars. This largely unexpected result comes primarily from the fact that ISO opened a new spectral window at wavelengths longer than 20 μm where most of the signatures of crystalline silicates lie. For the first time, mineralogical studies of interstellar grains can be undertaken because the crystalline nature of the silicates is revealed through a number of bands that allow the precise determination of the chemical nature of the silicates. Moreover, the possibility of estimating the ratio of annealed to unannealed silicates sheds light on the physical processes involved in their formation and processing. These results will certainly help to understand the fundamental and complex processes of the nucleation and evolution of interstellar grains. The evident similarity between silicates observed in late-type stars and those observed in the comet Hale-Bopp may help us to understand the connection between interstellar matter and comets. Indeed, the structural evolution of the dust from its birth place to its incorporation into comets, passing (or not) through a diffuse interstellar medium stage may be directly studied by observing a variety of astronomical objects in which the dust is found at different evolutionary stages.

Finally, we must mention the effort of various groups to unravel the presence of other grain components, some of them expected from grain models. These components, made of the most refractory materials, such as metallic oxide grains (aluminium, iron, magnesium and titanium oxides) or carbides (silicon and titanium) or other pure forms of carbon, such as diamond or graphite, are often found in primitive meteorites, where they are believed, or proven to be, of extra-solar origin. Their recovery from meteorites shows that such materials must also be present in space where they have yet to be spectroscopically detected. There is no doubt that ISO is the best instrument to provide such detection.

This first presentation of the ISO results, specifically devoted to the nature of solid matter in the interstellar medium, reveals many great leaps forward in our current understanding of this complex subject. This is illustrated by the breadth of the contributions presented here. However, the final ISO database, after 30 months of successful operations, remains largely untouched today. There can be no doubt that the final picture of interstellar and circumstellar dust to emerge from this database will reinforce the true originality provided by the technological success of the Infrared Satellite Observatory.

We wish to thank the following organisations for their financial and material support: le Centre National de la Recherche Scientifique (Formation Permanente – Délégation Midi-Pyrénées), le Ministère des Affaires Étrangères, Délégation Générale à l'Armement, l'Institut d'Astrophysique Spatiale (CNRS), le Centre d'Étude Spatiale des Rayonnements (CNRS), l'Aérospatiale and the Space Research Organisation of the Netherlands. Finally, we wish to acknowledge the helpfulness and efficiency of the management of the Les Houches École de Physique and the University Joseph Fourier in Lyon.

The Infrared Space Observatory Acknowledgement

Many papers in this book are based wholly or partly on Infrared Space Observatory (ISO) data.

ISO is an European Space Agency (ESA) project with instruments funded by ESA Member States (especially the PI countries: France, Germany, The Netherlands and the United Kingdom) and with the participation of ISAS and NASA.

The reference describing the ISO mission is to be found in:

Kessler M.F., *et al.*, *A&A* **315** (1996) L27.

The reference descriptions for the four instruments are:

ISOCAM:	Cesarsky C.J., *et al.*, *A&A* **315** (1996) L32.
LWS:	Clegg P.E., *et al.*, *A&A* **315** (1996) L38.
ISOPHOT:	Lemke D., *et al.*, *A&A* **315** (1996) L64.
SWS:	de Graauw T., *et al.*, *A&A* **315** (1996) L49.

A number of specific data reduction packages have been produced to assist with ISO data reduction. They have been used in many of the papers in this book.

- ISOCAM Interactive Analysis (CIA):

  The ISOCAM Interactive Analysis (CIA) software is a joint development by the ESA Astrophysics Division and the ISOCAM Consortium. The ISOCAM Consortium is led by the ISOCAM PI, C. Cesarsky, Direction des Sciences de la Matière, C.E.A., France.

  The preferred reference to CIA is: S. Ott, *et al.*, "Design and Implementation of CIA, the ISOCAM Interactive Analysis System", ASP Conference Series, Vol. 125, (1997).

- ISO Spectral Analysis Package (ISAP):

  The ISO Spectral Analysis Package (ISAP) is a joint development by the LWS and SWS Instrument Teams and Data Centers. Contributing institutes are CESR, IAS, IPAC, MPE, RAL and SRON.

- ISOPHOT Interactive Analysis (PIA):

  The ISOPHOT Interactive Analysis (PIA) software is a joint development by the ESA Astrophysics Division and the ISOPHOT Consortium.

  The preferred reference to PIA is: Gabriel C., *et al.*, "The ISOPHOT Interactive Analysis PIA, a calibration and scientific analysis tool" in Proc. of the ADASS VI conference, ASP Conf. Ser., Vol. 125, edited by G. Hunt and H.E. Payne (1997) p. 108.

AUTHORS

A. Abergel, Institut d'Astrophysique Spatiale, UMR-CNRS, Université Paris-Sud, bâtiment 120, Campus d'Orsay, 91405 Orsay Cedex, France

P. Ábrahám, Max-Planck-Institut für Astronomie, Königstuhl 17, 69117 Heidelberg, Germany, and, Konkoly Observatory of the Hungarian Academy of Sciences, P.O. Box 67, 1525 Budapest, Hungary

J. Ballet, DSM/DAPNIA/Service d'Astrophysique, CEA/Saclay, 91191 Gif-sur-Yvette, France

J.P. Bernard, Institut d'Astrophysique Spatiale, UMR-CNRS, Université Paris-Sud, bâtiment 120, Campus d'Orsay, 91405 Orsay Cedex, France

F. Boulanger, Institut d'Astrophysique Spatiale, Université Paris XI, 91405 Orsay, France

J.P. Bradley, MVA Inc, 5500/200 Oakbrook Parkway, Norcross, GA 30093, U.S.A., and, School of Materials Science and Engineering, Georgia Institute of Technology, Atlanta, GA 30332-0245, U.S.A.

D.E. Brownlee, Department of Astronomy, University of Washington, Seattle, WA 98195, U.S.A.

E. Bussoletti, Istituto Universitario Navale, Via A. De Gasperi 5, 80133 Napoli, Italy

C.J. Cesarsky, DSM/DAPNIA/Service d'Astrophysique, CEA/Saclay, 91191 Gif-sur-Yvette, France

D. Cesarsky, Institut d'Astrophysique Spatiale, Université Paris XI, 91405 Orsay, France

A. Claret, DSM/DAPNIA/Service d'Astrophysique, CEA/Saclay, 91191 Gif-sur-Yvette, France

L. Colangeli, Osservatorio Astronomico di Capodimonte, Via Moiariello 16, 80131 Napoli, Italy

P. Cox, Institut d'Astrophysique Spatiale, Université Paris XI, 91405 Orsay, France

E. Dartois, Institut d'Astrophysique Spatiale, bâtiment 121, Campus d'Orsay, Université Paris XI, 91405 Orsay Cedex, France, and, IRAM, 300 rue de la Piscine, 38406 Saint-Martin-d'Hères, France

K. Demyk, Institut d'Astrophysique Spatiale, bâtiment 121, Campus d'Orsay, Université Paris XI, 91405 Orsay Cedex, France

L. d'Hendecourt, Institut d'Astrophysique Spatiale, bâtiment 121, Campus d'Orsay, Université Paris XI, 91405 Orsay Cedex, France

C. Dominik, Leiden Observatory, P.O. Box 9513, 2300 RA Leiden, The Netherlands

T. Douvion, DSM/DAPNIA/Service d'Astrophysique, CEA/Saclay, 91191 Gif-sur-Yvette, France

P. Ehrenfreund, Leiden Observatory, P.O. Box 9513, 2300 RA Leiden, The Netherlands

M. Gerin, Radioastronomie Millimétrique, Laboratoire de Physique de l'ENS, 24 rue Lhomond, 75231 Paris Cedex 05, France

M. Giard, Centre d'Étude Spatiale des Rayonnements, UPR-CNRS 8002, Conventionné avec l'Université Paul Sabatier Toulouse, 9 avenue du Colonel Roche, 31029 Toulouse Cedex, France

O. Guillois, Service des Photons, Atomes et Molécules, CEA-Saclay, bâtiment 522, 91191 Gif-sur-Yvette Cedex, France

M.S. Hanner, Jet Propulsion Laboratory, MS 183-501, California Institute of Technology, Pasadena, CA 91109, U.S.A.

Th. Henning, Astrophysical Institute and University Observatory (AIU), Schillergäßchen 3, 07745 Jena, Germany

G. Lagache, Institut d'Astrophysique Spatiale, UMR-CNRS, Université Paris-Sud, bâtiment 120, Campus d'Orsay, 91405 Orsay Cedex, France

P.O. Lagage, DSM/DAPNIA/Service d'Astrophysique, CEA/Saclay, 91191 Gif-sur-Yvette, France

J.M. Lamarre, Institut d'Astrophysique Spatiale, UMR-CNRS, Université Paris-Sud, bâtiment 120, Campus d'Orsay, 91405 Orsay Cedex, France

R.J. Laureijs, ISO Science Operations Centre, Astrophysics Division of ESA, Apartado 50727, 28080 Madrid, Spain

P. Le Coupanec, DESPA, Observatoire de Paris, 92195 Meudon, France

G. Ledoux, Service des Photons, Atomes et Molécules, CEA-Saclay, bâtiment 522, 91191 Gif-sur-Yvette Cedex, France

A. Léger, Institut d'Astrophysique Spatiale, CNRS, 91400 Orsay, France

Ch. Leinert, Max-Planck-Institut für Astronomie, Königstuhl 17, 69117 Heidelberg, Germany

D. Lemke, Max-Planck-Institut für Astronomie, Königstuhl 17, 69117 Heidelberg, Germany

V. Mennella, Osservatorio Astronomico di Capodimonte, Via Moiariello 16, 80131 Napoli, Italy

F.J. Molster, Astronomical Institute Anton Pannekoek, University of Amsterdam, Kruislaan 403, 1098 SJ Amsterdam, The Netherlands

M.H. Moore, NASA Goddard Space Flight Center, Astrochemistry Branch, Greenbelt, MD 20771, U.S.A.

C. Moutou, Observatoire de Haute Provence, CNRS, 04870 Saint-Michel-l'Observatoire, France

I. Nenner, Service des Photons, Atomes et Molécules, CEA-Saclay, bâtiment 522, 91191 Gif-sur-Yvette Cedex, France

F. Pajot, Institut d'Astrophysique Spatiale, UMR-CNRS, Université Paris-Sud, bâtiment 120, Campus d'Orsay, 91405 Orsay Cedex, France

P. Palumbo, Istituto Universitario Navale, Via A. De Gasperi 5, 80133 Napoli, Italy

R. Papoular, Service des Photons, Atomes et Molécules, CEA-Saclay, bâtiment 522, 91191 Gif-sur-Yvette Cedex, France

Y.J. Pendleton, NASA Ames Research Center, Mail Stop 245-3, Moffett Field, CA 94035, U.S.A.

J.L. Puget, Institut d'Astrophysique Spatiale, UMR-CNRS, Université Paris-Sud, bâtiment 120, Campus d'Orsay, 91405 Orsay Cedex, France

C. Reynaud, Service des Photons, Atomes et Molécules, CEA-Saclay, bâtiment 522, 91191 Gif-sur-Yvette Cedex, France

I. Ristorcelli, Centre d'Étude Spatiale des Rayonnements, UPR-CNRS 8002, Conventionné avec l'Université Paul Sabatier Toulouse, 9 avenue du Colonel Roche, 31029 Toulouse Cedex, France

P.R. Roelfsema, SRON Laboratory for Space Research Groningen, P.O. Box 800, 9700 AV Groningen, The Netherlands

A. Rotundi, Istituto Universitario Navale, Via A. De Gasperi 5, 80133 Napoli, Italy

F. Salama, NASA-Ames Research Center, Space Science Division, MS: 245-6, Moffett Field, CA 94035-1000, U.S.A.

W.A. Schutte, Raymond and Beverly Sackler Laboratory for Astrophysics at Leiden Observatory, P.O. Box 9513, 2300 RA Leiden, The Netherlands

K. Sellgren, Astronomy Department, Ohio State University, Columbus, OH 43210, U.S.A.

G. Serra, Centre d'Étude Spatiale des Rayonnements, UPR-CNRS 8002, Conventionné avec l'Université Paul Sabatier Toulouse, 9 avenue du Colonel Roche, 31029 Toulouse Cedex, France

T.P. Snow, Center for Astrophysics and Space Astronomy, University of Colorado, Boulder, CO 80309-0389, U.S.A.

R.J. Sylvester, Department of Physics and Astronomy, University College London, London WC1E 6BT, UK

J.P. Torre, Service d'Aéronomie, UPR-CNRS, Verrières-le-Buisson, France

L. Verstraete, Institut d'Astrophysique Spatiale, CNRS, 91400 Orsay, France

C. Waelkens, Instituut voor Sterrenkunde, Katholic University Leuven, Celestijnenlaan 200B, 3001 Heverlee, Belgium

L.B.F.M. Waters, Astronomical Institute Anton Pannekoek, University of Amsterdam, Kruislaan 403, 1098 SJ Amsterdam, The Netherlands, and, SRON Laboratory for Space Research Groningen, P.O. Box 800, 9700 AV Groningen, The Netherlands

CONTENTS

GENERAL OBSERVATIONS OF DUST

LECTURE 1

Interplanetary Dust as Seen in the Zodiacal Light with ISO

by P. Ábrahám, Ch. Leinert and D. Lemke

LECTURE 2

Small Dust Particles in the ISM

by F. Boulanger

LECTURE 3

Properties of Dust in High Latitude Clouds

by R.J. Laureijs

LECTURE 4

Very Cold Dust in Star Forming Regions

by I. Ristorcelli, G. Serra, J.M. Lamarre, M. Giard, J.P. Bernard, F. Pajot, J.P. Torre, A. Abergel, G. Lagache and J.L. Puget

CARBON IN DUST

LECTURE 5

Polycyclic Aromatic Hydrocarbons in the Interstellar Medium: A Review

by F. Salama

LECTURE 6

PAHs in Reflection Nebulae, Fullerenes in the ISM

by C. Moutou, K. Sellgren, A. Léger, L. Verstraete and P. Le Coupanec

LECTURE 7

Present Situation of the Coal Model for Interstellar Carbon Dust

by O. Guillois, G. Ledoux, I. Nenner, R. Papoular and C. Reynaud

LECTURE 8

Organics in the ISM: Where Do They Come from?

by Y.J. Pendleton

LECTURE 9

Laboratory Analogues for Interstellar Carbon Dust

by L. Colangeli, V. Mennella, E. Bussoletti, P. Palumbo and A. Rotundi

COLD DUST: ICES AND SILICATES

LECTURE 10

Combined SWS-LWS Spectra of Solid State Features

by P. Cox and P.R. Roelfsema

LECTURE 11

Molecules in the Gas and Solid State Towards the Young Massive Protostar RAFGL7009S

by E. Dartois, K. Demyk, M. Gerin and L. d'Hendecourt

LECTURE 12

Ice Evolution in the Interstellar Medium

by W.A. Schutte

LECTURE 13

The Physics and Chemistry of Ices in the Interstellar Medium

by M.H. Moore

LECTURE 14

Crystalline Silicates in Circumstellar Shells

by L.B.F.M. Waters, F.J. Molster and C. Waelkens

LECTURE 15

Comparison of Ices and Silicates in Comets and in the Interstellar Medium

by P. Ehrenfreund

DUST PROCESSING

LECTURE 16

Grain Formation and Evolution in the Interstellar Medium

by Th. Henning

LECTURE 17

Emission from Carbon-Rich Stellar Envelopes

by R.J. Sylvester

LECTURE 18

ISO Observations of Vega-Like Stars

by C. Dominik

LECTURE 19

Dust Formation in Supernovae

by P.O. Lagage, T. Douvion, J. Ballet, F. Boulanger, C.J. Cesarsky, D. Cesarsky and A. Claret

INTERSTELLAR MATTER IN THE LABORATORY?

LECTURE 20

Mg-Rich Olivine and Pyroxene Grains in Primitive Meteoritic Materials: Comparison with Crystalline Silicate Data from ISO

by J.P. Bradley, T.P. Snow, D.E. Brownlee and M.S. Hanner

General Observations of Dust

LECTURE 1

Interplanetary Dust as Seen in the Zodiacal Light with ISO

P. Ábrahám[1,2], Ch. Leinert[1] and D. Lemke[1]

[1] *Max–Planck–Institut für Astronomie, Königstuhl 17, 69117 Heidelberg, Germany*
[2] *Konkoly Observatory of the Hungarian Academy of Sciences, P.O. Box 67, 1525 Budapest, Hungary*

Abstract. ISO performed an extensive observing programme on the zodiacal light. New and interesting results are presented from multi-filter photometry of the interplanetary dust cloud, observations of the asteroidal bands and of a cometary dust trail, and investigation of the small scale brightness fluctuations. The pioneering measurements on the mid-infrared spectrum of the zodiacal light can tightly constrain the nature of the constituents and the size distribution of the interplanetary grains. We review the initial results of ISOPHOT and ISOCAM.

1. INTRODUCTION

In June 1991 P. Lamy summarized, in this lecture room, the basic knowledge that an ISO observer should have of the zodiacal light [1]. He emphasised that, although for many observers the thermal emission of the interplanetary dust cloud is just a strong foreground radiation which has to be removed, the zodiacal light is also a rich source of information on the interplanetary dust particles, and in general on cosmic dust. Some seven years later we already know that ISO has successfully performed an extensive observing programme on the zodiacal light. Since the evaluation of these measurements requires precise calibration as well as good understanding of instrumental effects, the publication of these results may take several years. In this paper we review those fields of zodiacal light research where important contributions can be expected from ISO, and summarise the published as well as present some previously unpublished results. We focus on zodiacal light measurements from the ISOPHOT Guaranteed Time programme, but also cite results from ISOCAM observations.

Historically the name "zodiacal light" refers to a dim glow visible with the naked eye after sunset or before sunrise. This conspicuous phenomenon is caused by the scattering of sunlight by dust particles orbiting in interplanetary space. For infrared astronomers, however, zodiacal light is the thermal reradiation of sunlight absorbed by the same particles. The cloud formed by these particles, called the Interplanetary Dust Cloud (IDC), occupies a region covering the inner solar system and extending out to at least the asteroidal belt. It contains only $10^{16} - 10^{17}$ kg of dust, equivalent to the mass of a large comet [2]. In spite of the low density of the IDC ($\approx 10^{-19}$ kg m^{-3}), the sunlight absorbed and re-radiated by the dust within the cloud dominates the infrared sky in the $3 - 70$ μm wavelength range. The absorption and scattering of solar radiation decreases the orbital velocities of the particles due to the Poynting-Robertson effect, limiting the lifetime of individual grains to $10^4 - 10^5$ years. The most likely source to replenish the evaporated dust is the destruction of comets and asteroids. A comprehensive review on the basic observational facts on the Interplanetary Dust Cloud is presented in [3].

Since the Earth is orbiting within the IDC, there are many ways to study the dust particles, including optical and radio observations of meteors, collection of particles in the upper atmosphere or in the Antarctic, analysis of the zodiacal light (both the scattered and thermal components), *in situ* capture of dust by space probes, and the study of lunar micro-craters. The different methods, however, are sensitive to different particle sizes, therefore their results are complementary. For a review on the properties of the interplanetary dust particles derived from the mentioned observing methods we refer to [2].

The zodiacal light observed in the infrared is always an integral along the line-of-sight over regions of different physical conditions, therefore derivation of local dust properties requires careful modelling or the usage of inversion methods. There are, however, several fields where the best available tool to study the IDC is infrared photometry:

- Due to the orbital motion of the Earth, the surface brightness of the zodiacal light towards a given sky position exhibits an annual variation. Extensive mapping of the zodiacal light over the year can be used to deduce the 3-dimensional structure of the IDC as performed by the infrared satellites IRAS and COBE.
- Substructures in the brightness distribution of the cloud discovered by IRAS (asteroidal bands, cometary trails) seem to carry important information on the replenishment mechanism, *i.e.*, on the origin of dust particles.

- If the dust grains are replenished by discrete sources, then the smoothness of the zodiacal emission on smaller scales provides information on the efficiency of mixing within the cloud.

- The mid-infrared spectrum of the zodiacal light has a special importance because the shape of the spectrum in the $5-15$ μm range is very sensitive to dust properties, and in case individual spectral features were identified even the chemical composition can be determined. From differential measurements within the cloud it is also possible to derive the local dust emissivity factors.

- One should also keep in mind that the majority of analysed particles were collected in the neighbourhood of the Earth. If the IDC is not well mixed then it may turn out that all these particles originate from the same set of parent bodies, and that the derived properties may not be representative of the whole cloud. The zodiacal light observations can provide information on more distant parts of the cloud.

- In order to test our understanding of the IDC, one can model and predict the brightness and the spectrum of the zodiacal light in different directions by combining the 3-dimensional density and temperature distribution of the cloud with the size distribution and optical properties of the particles. Infrared observations can efficiently constrain these models and serve as final check on the model assumptions.

Zodiacal light observations with ISO have both advantages and disadvantages. ISO was designed for point source photometry, and neither its small field-of-view ($\approx 3'$) nor the observing strategy made it possible to repeat the systematic all-sky surveys of the IRAS and COBE satellites. In addition, ISO is calibrated on point sources and the conversion to extended source calibration is not always straightforward. On the other hand, with the small ISO beam can efficiently avoid point sources and cirrus structure, and the good filter coverage in the $3-200$ μm range as well as the possibility of performing mid-infrared spectrophotometry provides better spectral coverage than the previous satellites had. The straylight rejection of ISO has been proved to be excellent [4]. Since there is an unresolved zero point difference between the IRAS and DIRBE absolute calibrations in the far-infrared [5], ISO can also help to fix the absolute brightness level of the zodiacal light.

In this contribution we review zodiacal light observations from the Guaranteed Time programme of ISOPHOT by following the items listed above. We put a special emphasis on the mid-infrared spectrum, because it was unknown before ISO, and it thus constitutes one of the most important results of the whole programme.

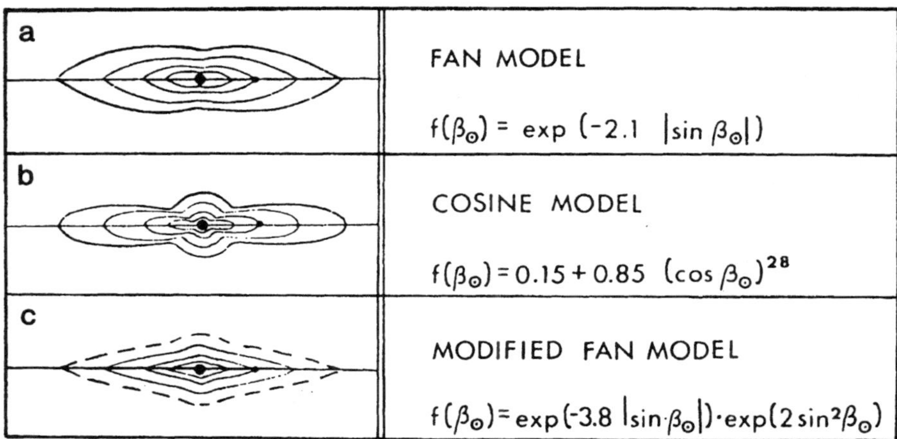

Fig. 1. — Three basic types of models for the density distribution of the interplanetary dust cloud [2]. $\beta_\odot$ is the angle above the ecliptic plane.

2. LARGE SCALE STRUCTURE OF THE INTERPLANETARY DUST CLOUD

Optical and infrared studies result in a picture of a flattened lenticular cloud. Cylindrical symmetry seems to be a good approximation, but the symmetry plane of the cloud is tilted to the Ecliptic by $1 - 3°$. Figure 1 shows three basic types of models for the density distribution of the cloud [2]. The IRAS and COBE satellite teams derived their own zodiacal light models in order to subtract the interplanetary contribution from the measured signals [5,6]. These satellites, however, did not observe the inner part of the cloud, therefore the relevant models only describe the outer part of the IDC. The COBE model includes, in addition to a general fan shape, the asteroidal bands and also the Earth's resonant dust rings.

The derivation of a zodiacal light model for all ISOPHOT and ISOCAM filters is a calibration task. With ISOPHOT we build up our model in three steps:

- we observed 18 dark areas distributed on the sky accessible by ISO ($60° < D < 120°$, where D is the distance from the sun), using at least 7 and at most 13 filters in the $3.6 - 200$ μm wavelength range. The positions were carefully examined for low cirrus levels in the IRAS 100 μm maps, for locations far from any known infrared point sources, and we tried to avoid any stars in the beam visible on the Palomar Observatory Sky Survey. Absolute photometry mode was used, in many cases including dark current measurements and checks for the zero level in the faint calibration source measurements.

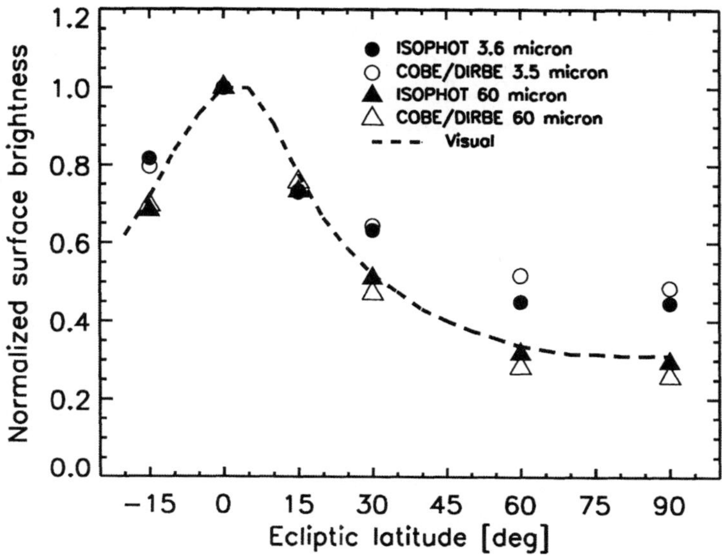

Fig. 2. — Surface brightness of the zodiacal light measured by ISOPHOT and COBE/DIRBE at $\lambda - \lambda_{\odot} = 90°$. The brightness values are normalised to the value in the Ecliptic.

- We adopt the COBE model [6] and adapt it to our filters by fitting it to our measurements.
- We will set up a database of all ISOPHOT background measurements, and check in detail – or if necessary, improve – the adopted COBE models.

Figure 2 shows, as a preliminary result, the brightness variation along the main circle at $\lambda - \lambda_{\odot} = 90°$, $\beta = [-15°, 0°, +15°, +30°, +60°, +90°]$, together with the variation measured in the visible as well as that observed by COBE. The ISOPHOT and COBE results show the same trend, indicating that the adaptation of the COBE model to ISOPHOT can be done with high accuracy. Comparing directly ISOPHOT and COBE observations, however, one should keep in mind the large beam size difference between the two instruments.

3. ORIGIN OF THE ZODIACAL DUST: ASTEROIDAL BANDS AND COMETARY TRAILS

The main contribution to the zodiacal light comes from dust grains in the $20 - 200\ \mu$m size range [7]. The lifetime of individual particles of this size at 1 AU is about 10^4 yrs, and to maintain the zodiacal cloud in a steady state an average dust input of $\approx 9 \times 10^6$ g s^{-1} is required [8]. Possible dust sources are collisions within the asteroidal belt, active comets, satellites and planetary rings, as well as a stream of interstellar dust particles. The relative contributions of these sources are not well determined, and may depend on the position of the observer within the IDC. The new measurements with ISO may help to quantify the fraction of dust particles of asteroidal and cometary origin, the two most important sources, by constraining the models of dust ejection from the asteroidal belts and from cometary trails. Another way to evaluate the relative contributions is to separate them on the basis of spectral features which will be described in Section 5.

3.1. Asteroidal bands

The asteroidal bands were discovered as two local maxima in the IRAS scans across the Ecliptic at $\beta \approx 0°$ and as shoulders at $\beta \approx \pm 10°$ [9]. The amplitude of the bands is $1 - 3\%$ of the zodiacal light brightness, their width at half maximum is $1° - 3°$ [10]. The bands are interpreted as the result of collisions in the asteroidal belt. The Themis and Koronis asteroidal families produce the double bands at $\beta \approx 0°$ [11]. The higher latitude bands are attributed to the Eos family [11], although this was later questioned because the proper inclination of the Eos family seems to be too large [12]. The discovery of the dust bands showed unambiguously that there is an important asteroidal contribution to the IDC. Modelling of the IRAS dust bands indicates that their dust content constitutes $5 - 10\%$ of the zodiacal cloud [12]. Another estimate of the fraction of asteroidal dust comes from modelling of the Earth's resonant circumsolar ring and leads to a value of about 30% [13].

With ISOPHOT we observed the asteroidal bands by performing multi-filter scans across the Ecliptic. Details of the observations are given in Table I. We present here some preliminary results (these results are expected to be improved in the near future using a better data reduction procedure). In Figure 3 we show the brightness profiles derived from two low-resolution scans (No. 1 and No. 5, both are at 25 μm). Since the observed ecliptic latitudes of the bands depend on the annual motion of the observer as well as on the distance from the sun, we predict and indicate in Figure 3 the latitudes of the band peaks using the formula of Reach [10]. The plots demonstrate that, in spite of their very low surface brightness, ISOPHOT detected the bands close to the expected positions and with approximately the expected widths. In the future these studies will be extended; we now detail the planned studies. The measurements will be used to study the brightness and temperature profiles of the bands. The derived average temperatures will be compared with the blackbody temperature at the same distance from the sun, while their colours will be compared with the colours of the asteroids and with those of the general interplanetary dust. We will also look for a systematic effect which is expected if the dust in the bands is strongly affected by the Poynting-Robertson effect: the bands extend over a range of heliocentric distances with the inner (warmer) parts appearing at progressively higher ecliptic latitudes. Such information could give

Table I. — ISOPHOT scans across the Ecliptic. At $\lambda \leq 25$ μm the 180″ circular aperture was selected; at longer wavelengths the C100 (3×3 pixels, 46″/pix) and the C200 (2×2 pixels, 92″/pix) cameras were used.

No.	Filters [μm]	λ	$\lambda - \lambda_\odot$	β range	Step	Date
1	25	182°	+98°	−5°, +5°	20′	Jun. 15, 96
2	12,25,60	18°	+90°	−13°, +5°	3′	Jan. 8, 97
3	7.3,12,60,150,180	18°	−90°	−0.7°, −3.7°	30′	Jul. 11, 97
4	7.3,12,60,150,180	18°	−90°	−13°, −10°	30′	Jul. 11, 97
5	25	18°	+90°	−14°, +3.8°	30′	Jan. 8, 98

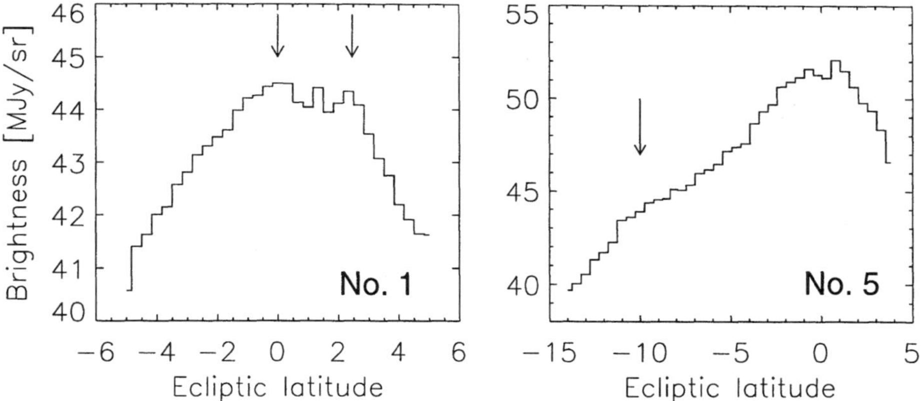

Fig. 3. — Scans across the Ecliptic taken by ISOPHOT at 25 μm with the 180″ circular aperture. Details of the scans are given in Table I.

a clue to the age of these features, and could be used to estimate the asteroidal mass input into interplanetary space as well as to ascertain the relative importance of the different dynamical effects on this dust.

3.2. Cometary trails

Crossing the orbits of a few periodic comets (Tempel 2, Encke, Kopff, Tempel 1, Gunn, Schwassmann-Wachmann 1, Churyumov-Gerasimenko, and Pons-Winnecke) IRAS detected long narrow streams [14]. These dust "trails" typically extend 10° behind and 1° ahead of the comet, and they should not be confused with the visually more prominent "tails" formed by submicrometer-sized particles blown away by the solar radiation pressure. Rather they correspond to the anti-tails observed in the visible for particular geometric arrangements of the Sun, Earth and comet, and seemingly extend towards the Sun. The trails consist of mm-sized dust particles ejected from the comet during active times extending over many years [15]. Theoretical investigations of dust ejection from comets predict that the common dust production of 85 short-period and 101 long-period comets is of the order of only 3×10^4 g s^{-1} [16]. This production rate is by far too low to maintain the zodiacal cloud. Recent infrared observations indicate, however, that due to the very low albedo of the cometary dust, the mass input rate may be higher than suspected [17]. In addition, the dust injection rate is not necessarily constant in time, and the contribution from active comets in the past is not known. A higher dust supply rate, in the form of comet showers at 37 and 50 million years ago, is indicated by measurements of the ^{3}He content in interplanetary dust particles from oceanic core samples [18].

A key parameter in the calculations of the dust supply rate from comets is the timescale on which the ejected grains are mixed with the general interplanetary medium. Davies *et al.* [17] observed the dust trail of comet P/Kopff at positions 0.5° and 1.0° behind the nucleus, and compared the results with measurements from 13 years earlier obtained with IRAS [17]. The new observations were performed at 12 μm with ISOCAM, the infrared camera on board ISO. Figure 4 shows the raster images. The trail is clearly detected towards both positions with a surface brightness of the order of 0.33 MJy sr^{-1}. This value is about a factor of 2 lower than that predicted from the IRAS data, indicating different levels of activity between perihelion passage before and after the IRAS observations. Comparing the sizes of the P/Kopff trail observed by the two satellites, a significant broadening was observed during this 13 yr period.

With ISOPHOT we also observed the trail of comet P/Kopff at 12, 25, and 60 μm in October 1996. Figure 5 shows both the observed positions and the derived 25 μm brightness profiles taken perpendicular to the orbit at different distances behind and ahead the comet. Note that there will be future improvements in the calibration of this data and that we have made no attempt to correct for the curved baselines caused by detector transients. The figure shows that the trail was clearly observed at 0.25° behind the comet but only marginally detected at larger distances behind and ahead of the nucleus. The temperature determined from IRAS measurements is systematically higher than expected for blackbodies at the same heliocentric distance. ISOPHOT 12/25/60 μm photometry will be used in the future to determine the temperature of the trail particles and to confirm the expected temperature increase towards the comet.

4. ARCMINUTE STRUCTURE OF THE ZODIACAL LIGHT

One of the main characteristics of the zodiacal light is the large-scale smoothness of its brightness distribution over the sky. Little is known of the

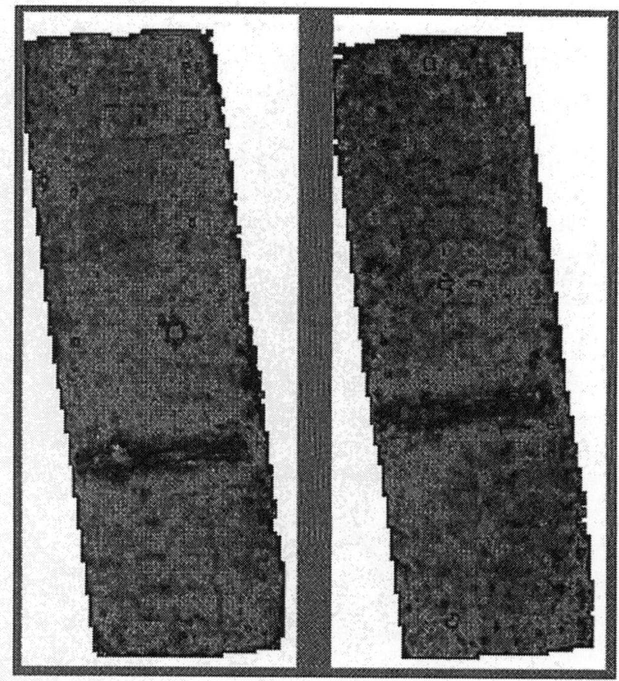

Fig. 4. — ISOCAM observations of the dust trail of comet P/Kopff at 0.5° and 1° behind the nucleus (from [17]).

inhomogeneities in the zodiacal light brightness at small spatial scales. Structure must exist at some level because the zodiacal light is produced by particles which are replenished from localised sources, comets and asteroids.

We mapped a few fields of $\approx 0.5° \times 0.5°$ at 25 μm with ISOPHOT in order to look for fluctuations in the zodiacal light at low, intermediate and high ecliptic latitudes. These fields were selected for low cirrus emission and avoid any bright infrared point sources. Figure 6 shows one of the maps with pixel size of $3'$. The image smoothness is quantified in Figure 6 which shows the histogram of the map. The width of the almost perfectly Gaussian distribution – after removing the instrumental noise contribution – is about 0.2% of the total brightness level. This very low fluctuation supports the concept of a generally smooth zodiacal light distribution. More details of this study can be found in [19].

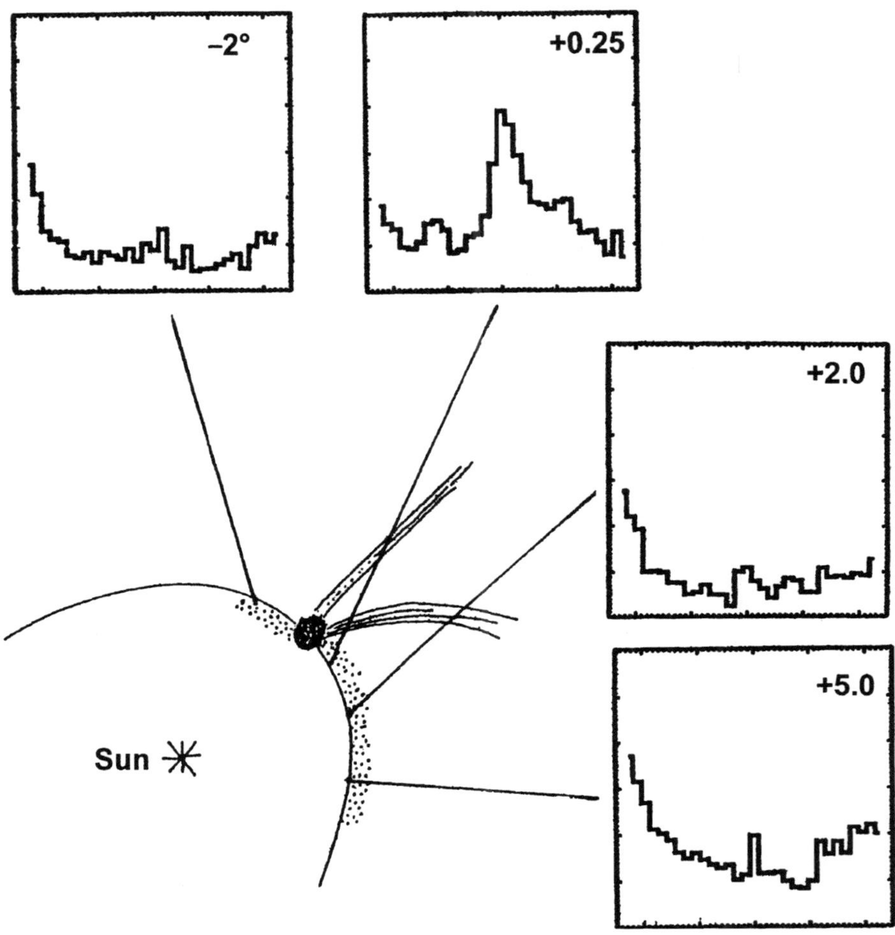

Fig. 5. — ISOPHOT observations of comet P/Kopff at 25 μm.

5. SPECTRUM OF THE ZODIACAL LIGHT

5.1. Observational results

The spectral energy distribution of the zodiacal light is dominated by scattered sunlight at $\lambda \leq 3.5$ μm and by thermal emission from dust grains at longer wavelengths. The thermal spectrum, which is of primary interest for ISO, cannot be observed from the ground. Starting in 1971 [20], broad-band photometry from rockets and balloons were performed in order to define

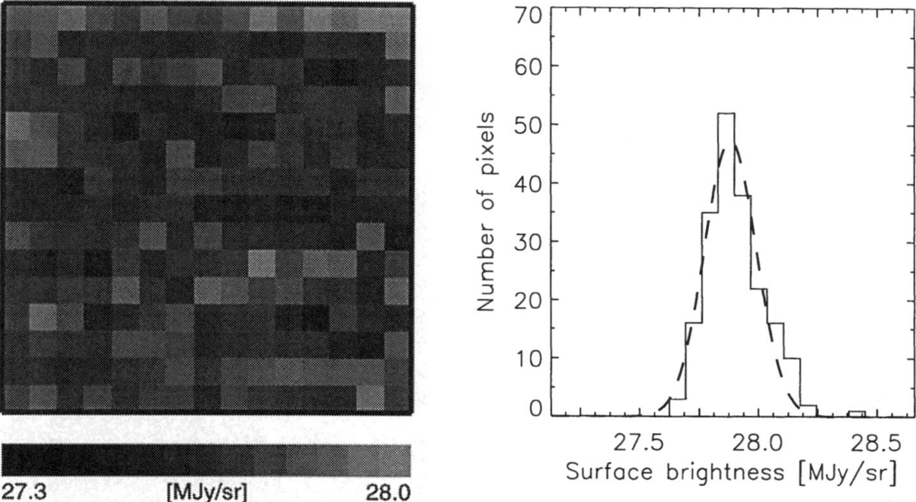

Fig. 6. — Map of $45' \times 45'$ obtained at the North Galactic Pole at 25 μm (left) and the histogram of the brightnesses measured in the central 13×13 pixels (right).

the general shape of the zodiacal light in the thermal infrared [21–23]. The most extensive data sets are derived from the satellite missions IRAS [24] and COBE [6]. The COBE/DIRBE results are estimated to be accurate to $\approx 10\%$ at $\lambda \leq 25$ μm and to $\approx 20\%$ at longer wavelengths.

ISO observed the spectrum of the zodiacal light by performing multi-filter photometry with ISOCAM and ISOPHOT at $\lambda \leq 16$ μm and by ISOPHOT at wavelengths up to 200 μm. In Figure 7 we plot the broad band spectrum measured towards a dark position at $\lambda - \lambda_\odot = 90°$ and $\beta = 0°$ using the absolute photometric mode of ISOPHOT. Note that while the calibration at $\lambda \geq 60$ μm is almost finalised, the values at shorter wavelengths should be still considered as preliminary. It is also likely that despite of the careful selection of the dark field, weak galactic and extragalactic emission is present in the long wavelength surface brightness values. The general agreement between the ISOPHOT and COBE/DIRBE results is, however, already quite good, demonstrating that by now not only the spectral shape but also the absolute level of the zodiacal emission is well determined.

One of the most important contributions of ISO to zodiacal light studies is the mid-infrared ($6 - 16$ μm) spectrophotometry, where the brightness of the zodiacal light increases by two orders of magnitude. Before the ISO mission only one rocket measurement of the mid-infrared spectrum had been attempted [25]. Reach *et al.* [26] observed the spectrum at one sky position with the ISOCAM circular variable filters from $5 - 16$ μm (Fig. 8). They found that: (1) the spectrum is remarkably well-fit by a Planck curve at 270 K; (2) there are no indications of sharp lines; but (3) there is a broad hump in the $9 - 11$ μm region detected at the 10% level, with more emission on the longer

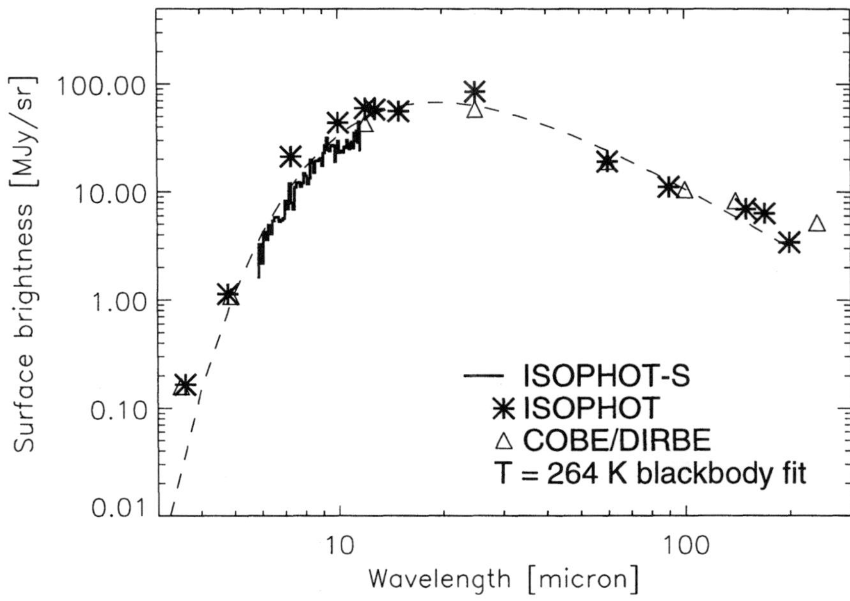

Fig. 7. — ISOPHOT photometry of the zodiacal light at $\lambda - \lambda_\odot = 90°$ and $\beta = 0°$

wavelength side. There is a good agreement with the COBE/DIRBE broadband photometry at 4.8 μm and 12 μm.

With ISOPHOT we also observed the mid-infrared spectrum, although the spectral coverage of the spectrophotometer ISOPHOT-S is limited to the slightly smaller $5 - 11.7$ μm range. We collected 18 measurements at different sky positions, some of these are identical to the positions where ISOPHOT multi-filter photometry was performed. The average spectrum [27] can be fitted with a 264 K blackbody with no obvious spectral features at any wavelengths (details of the data reduction of the spectra are also given in [27]). The individual spectrum measured at $\lambda - \lambda_\odot = 90°$ and $\beta = 0°$ is overplotted in Figure 7 with solid line. The discrepancy between the spectrum and the photometry reflects the unfinalised status of ISOPHOT extended source calibration at $\lambda \leq 25$ μm. Using the COBE/DIRBE 12 μm photometry to scale together the ISOCAM and ISOPHOT-S spectra, the agreement between the two measurements is satisfactory. Due to the limited spectral coverage of ISOPHOT-S, we cannot comment on the existence of the broad $9 - 11$ μm spectral feature detected by ISOCAM. Very recently a third measurement of the mid-infrared spectrum of the zodiacal light was observed by the Japanese IRTS satellite [28] The agreement among the three instruments is within the error bars.

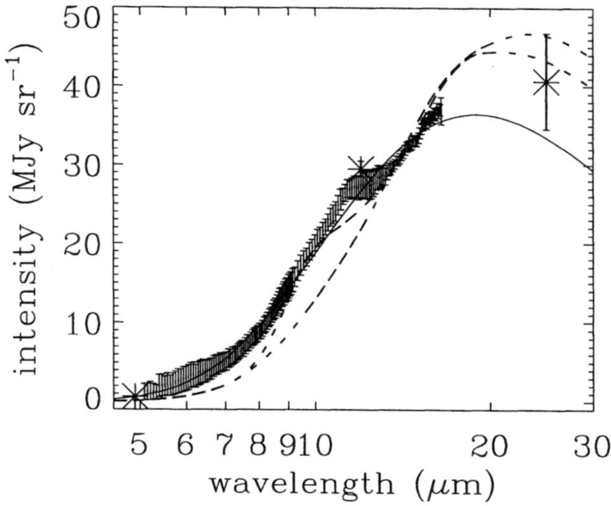

Fig. 8. — Mid-infrared spectrum of the zodiacal light measured by ISOCAM from Reach *et al.* [26]. The asterisks represent DIRBE measurements.

5.2. Interpretation of the spectrum

The observed spectrum of the zodiacal light is an integral of dust emission along the line of sight. The volume emissivity depends on the optical properties of the grains, on their size distribution, and on the temperature. Though the infrared observations cannot be used to derive directly these parameters, the spectrum of the zodiacal light provides a very strong constraint on models which use different optical constants and size distributions. The mid-infrared spectral range is very sensitive to dust temperature, because it corresponds to the Wien part of a Planck curve for large particles at 1 AU. The temperature, on the other hand, depends strongly on the wavelength dependent absorption properties of the dust grains.

Synthetic spectra of the zodiacal light were calculated for typical materials like obsidian, olivine, andesite, graphite, magnetite, water ice, and different size distributions [29, 30]. Measured refractive indices were used to calculate the scattering and absorbtion properties of dust particles using the Mie theory, assuming spherical particles. Reach *et al.* [26] compared the mid-infrared spectrum of the zodiacal light observed by ISOCAM with model calculations (Fig. 9). They found that none of the assumed constituents or size distributions could reproduce the observed spectrum. The best results were achieved for andesite with a "lunar" size distribution (derived from lunar micro-crater studies by [7]), and for "astronomical silicates" [31] with an "interplanetary"

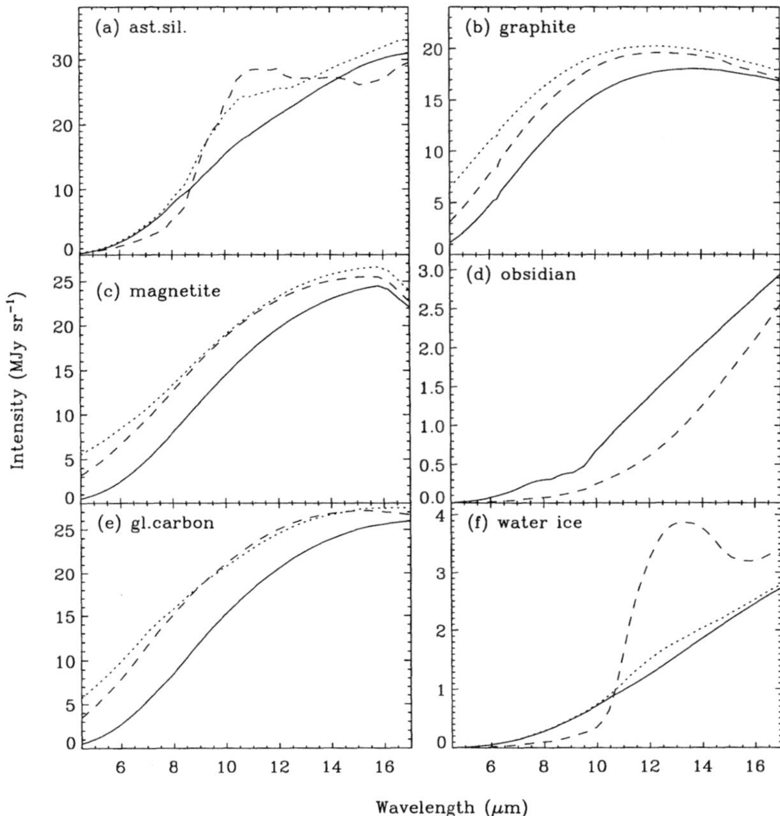

Fig. 9. — A suite of models of the zodiacal light spectrum. From [26]. The three curves correspond to different size distributions. Solid: interplanetary, dotted: lunar, dashed: P/Halley.

size distribution [7]. The other models are clearly ruled out, since the strongly absorbing materials (graphite, magnetite, glassy carbon) turned out to be too hot, and the more transparent constituents (obsidian, water ice) are too cold. The lack of a strong 10 μm feature in the observed spectrum, similar to that observed in comets, rules out the presence of an important population of very small grains.

6. TOWARDS A CONSISTENT PICTURE OF THE ZODIACAL DUST

The new measurements of the infrared zodiacal light provided by the satellites COBE/DIRBE and ISO raise the possibility of addressing two important questions. The first question is whether our knowledge of the properties of interplanetary dust particles derived from the analysis of captured particles is consistent with the measurements of the zodiacal light. Having a good model for the three-dimensional distribution of the interplanetary dust cloud, provided by COBE/DIRBE, and recent laboratory measurements of the optical properties of many different dust constituents, we should be able to reproduce the observed ISO and DIRBE zodiacal light spectra over the entire infrared spectral range. This modelling will be an important step towards a consistent picture of the zodiacal dust.

The second question is related to the origin of interplanetary particles: are the particles of different origins distinguishable by their observed properties? It was expected that the spectrum of dust at high ecliptic latitudes should show similarities to the spectrum of comets at high inclination, while dust in the ecliptic should resemble the asteroids. Our ISOPHOT-S data taken at 18 different positions over the sky, however, do not show any significant variation in the spectra other than temperature effects. It appears that the origin of the particles cannot be determined from their mid-infrared spectrum, because of lack of spectral features, and because the present IDC appears well mixed.

REFERENCES

[1] Lamy P., Infrared Astronomy with ISO, edited by Th. Encrenaz and M.F. Kessler (Nova Science Publishers Inc., New York, 1992) p. 251.

[2] Leinert Ch. and Grün E., Physics and Chemistry in Space, edited by R. Schwenn and E. Marsch (Springer-Verlag, Berlin, 1990) p. 207.

[3] Leinert Ch., Bowyer S., Haikala L.K., *et al.*, *A&AS* **127** (1998) 1.

[4] Klaas U., Lemke D., Kranz T., *et al.*, Proc. of the SPIE Infrared Astronomical Instrumentation Conference, edited by A.M. Fowler (SPIE, Vol. 3354, 1998) p. 996.

[5] Wheelock S.L., Gautier T.N., Chillemi J., *et al.*, IRAS ISSA Explanatory Supplement, JPL, Pasadena (1994).

[6] Reach W.T., Franz B.A., Kelsall T. and Weiland J.L., Unveiling the cosmic infrared background, AIP Conf. Proc. 348, edited by E. Dwek (AIP, New York, 1996) p. 37.

[7] Grün E., Zook H.A., Fechig H. and Giese R.H., *Icarus* **62** (1985) 244.

[8] Flynn G.J., *Nature* **376** (1995) 114.

[9] Low F., Beintema D., Gautier F.N., *et al.*, *ApJ* **278** (1984) L19.

[10] Reach W.T., *ApJ* **392** (1992) 289.

[11] Dermott S.F., Nicholson P.D., Burns J.A., Houck J.R., *Nature* **312** (1984) 505.

[12] Dermott S.F., Durda D.D., Gustafson B.A.S., *et al.*, Asteroids, Comets, Meteors, edited by A. Milani, M. Martini and A. Cellino (Kluwer, Dordrecht, 1993) p. 127.

[13] Dermott S.F., Jayaraman S., Xu Y.L., Gustafson B.A.S. and Liou L.C., *Nature* **369** (1994) 719.

[14] Sykes M.V., Lebofsky L.A., Hunten D.M. and Low F., *Science* **232** (1986) 1115.

[15] Sykes M.V., Lien D.J. and Walker R.G., *Icarus* **86** (1990) 236.

[16] Mukai T., *Highlights Astron.* **8** (1989) 305.

[17] Davies J.K., Sykes M.V., Reach W.T., *et al.*, *Icarus* **127** (1997) 251.

[18] Farley K.A., *Nature* **376** (1995) 153.

[19] Ábrahám P., Leinert Ch. and Lemke D., *A&A* **328** (1997) 702.

[20] Soifer T.B., Houck J.R. and Harwit M., *ApJ* **168** (1971) 73.

[21] Price S.D., Murdock T.L. and Marcotte L.P., *AJ* **85** (1980) 765.

[22] Murdock T.L. and Price S.D., *AJ* **90** (1985) 375.

[23] Salama A., Andreani P., Dall'Oglio G., *et al.*, *AJ* **93** (1987) 467.

[24] Hauser M.G., Gillett F.C., Low F.J., *et al.*, *ApJ* **278** (1984) 15.

[25] Briotta D.A. Jr., Ph D. Thesis (Cornell Univ., Cornell, 1976) p. 1.

[26] Reach W.T., Abergel A., Boulanger F., *et al.*, *A&A* **315** (1996) L381.

[27] Ábrahám P., Acosta-Pulido J.A., Laureijs R.J., *et al.*, Proceedings of the First ISO Workshop on Analytical Spectroscopy, ESA SP-419, edited by A.M. Heras, K. Leech, N.R. Trams and M. Perry (ESA, Paris, 1998) p. 119.

[28] Ootsubo T., Onaka T., Yamamura I., *et al.*, *Earth Planets Space* **50** (1998) 507.

[29] Röser S. and Staude H.J., *A&A* **67** (1978) 381.

[30] Reach W.T., *ApJ* **335** (1988) 468.

[31] Draine B.T. and Lee H., *ApJ* **285** (1984) 89.

LECTURE 2

Small Dust Particles in the ISM

F. Boulanger

Institut d'Astrophysique Spatiale,
Université Paris XI, 91405 Orsay, France

Abstract. I review new perspectives opened by ISO for the study of small dust particles, and more particularly, the analysis of the interstellar medium mid-infrared spectra obtained with the ISO camera (ISOCAM). ISO observations of interstellar matter have allowed the detection of the emission features attributed to Polycylic Aromatic Hydrocarbons (PAHs) over a wide range of physical conditions including cirrus clouds. From translucent clouds to bright reflexion nebulae, the mid-infrared spectra of photo-dissociation regions are observed to scale with the intensity of the radiation field with little spectral change. This is a remarkable result confirming what is expected for the emission of small particles transiently heated by the absorption of individual photons. Across photo-dissociation regions, the variations of the band intensities, reproduceable from object to object, fit with what is expected for PAH ionization. The constancy of the feature positions and widths and their Lorentzian profiles argue that the bands are to a large extent intrinsic to interstellar PAHs rather than a blend of narrower features shifted with respect to one another. We suggest that the bulk of interstellar PAHs are molecules and or hydrogenated aromatic carbon clusters in the size range of 100 to 1000 atoms. The abundance of carbon in these particles is estimated to be $\sim$20% of the solar abundance. The mid-infrared spectra of H II regions are dominated by the continuum and low contrast features of a population of very small grains intermediate in size between PAHs and large dust grains.

1. INTRODUCTION

The all-sky surveys carried out by the Infrared Astronomical Satellite (IRAS) and the Diffuse Infrared Background Experiment (DIRBE) have shown that a large fraction of the power radiated by interstellar dust comes from small particles transiently heated by the absorption of individual photons. A wealth of ISO observations have been devoted to the study of the near and mid-infrared emission of these particles in a wide range of objects and environments. These observations are giving new momentum to investigations aimed at understanding the nature of small dust particles and their role in the physical and chemical evolution of matter in space.

This paper reports on the analysis of mid-infrared spectra from interstellar matter, with a special emphasis on results obtained with the circular variable filter of the ISO camera (ISOCAM). ISO achieved one of its important goals in demonstrating the presence of the near and mid-infrared carbon dust features in the spectra of diffuse interstellar matter heated by the mean solar neighborhood radiation field (Sect. 2). In brighter regions, the ISO instruments provided the necessary spatial resolution to map variations of the emission properties with radiation field intensity and gas density (Sect. 3). Some remarkable observational properties are coming out of the data analysis. They provide important constraints on the nature of small dust particles (Sect. 4). Outstanding questions and perspectives are outlined in the last section.

2. CIRRUS SPECTRUM

An important goal of spectroscopic observations of the mid-IR emission from interstellar clouds was to verify that the 12 μm diffuse emission measured by IRAS was associated with the well known set of dust emission features at 3.3, 6.2, 7.7, 8.6, and 11.3 μm often referred to as the Unidentified Infrared emission Bands (UIBs). These bands are considered to be characteristic of Polycyclic Aromatic Hydrocarbons (PAHs) [1, 2] (also see the contribution of Salama in this volume). The presence of the 3.3 and 6.2 μm emission features in the spectrum of the diffuse emission from the Galaxy was first observed by the AROME balloon experiment [3, 4]. But it is only recently that Galactic plane spectra obtained with ISO [5] and IRTS [6,7] show the full set of dust features.

An important milestone for ISO was the detection of emission bands in a cloud heated by the solar neighborhood Interstellar Radiation Field (ISRF). A low resolution mid-infrared spectrum of the edge of a translucent cloud in Chamaeleon, DC300-17, is presented in Figure 1. In the measured spectrum the PAH features appear on top of the brighter continuum emission from interplanetary dust particles. The zodiacal emission and the Galactic background have been removed with a reference spectrum measured on a nearby position during the same observation. Some of the dust features highlighted by the ISOCAM spectrum have also been detected with narrow band filters using the

ISO photometer [8]. The overall opacity of DC300-17 at its center is $A_v = 2$ [9]; at the position of the ISOCAM observation the extinction is $A_v = 1$. The cloud has a molecular core but no CO emission is detected at this position [10] where the mean gas density is estimated to be a few 100 H cm^{-3} from the modelling of the far-infrared emission profile [11]. The observed position of DC300-17 may be considered as a low density Photo-Dissociation Region (PDR) heated by the solar neighborhood ISRF.

The Chamaeleon spectrum is remarkably similar to that of the diffuse Galactic emission measured at Galactic longitude and latitude, $l = 30°$ and $b = 3°$, and that of brighter PDRs (Fig. 2). Two important facts come from this comparison of spectra. First, the feature positions and widths are remarkably constant from object to object. Second, the intensity ratios among the various bands show little variations over three orders of magnitude in their absolute intensity. The good spectral match between these spectra is a remarkable result confirming a strong prediction of the excitation of small particles by single photons: the spectrum shape should be independent of radiation field intensity. Relative to the large grain emission in the far-IR, the PAH features are twice as bright in DC300-17 than along the Galactic line of sight $l = 30°$, $b = 3°$. We consider this as due to an abundance enhancement of PAHs in the halo of DC300-17.

3. EMISSION FROM PHOTO-DISSOCIATION REGIONS

3.1. Emission features

In Figure 2, a set of spectra covering five orders of magnitude in intensity are presented. In addition to the Chamaeleon and the Galactic plane spectra discussed in the previous section, spectra obtained for several PDRs are shown. From low to high radiation field, the first spectrum, Oph N, corresponds to the northern edge of the Ophiucus molecular cloud heated by the ρ Oph star [12]. The second PDR, Oph W, corresponds to the long filament to the west of the Ophiucus cloud in the large mosaic of ISO images presented by Abergel *et al.* [16]. This PDR is heated by the B2IV star HD 147889 located 10′ to the South-West of the interface. The last two PDRs are the well known NGC 7023 reflexion nebula and M17-SW molecular cloud-H II region interface. For each PDR, the selected line of sight corresponds to the position where the mid-infrared emission peaks. The ISOCAM spectra can be very accurately fit by a set of five Lorentz curves centered at the position of the 6.2, 7.7, 8.6, 11.3 and 12.7 μm features ([17] and Fig. 3). We believe that the emission peak at 16.5 μm at the end of the NGC 7023 spectrum in Figure 3 is also real. After subtraction of the main bands, the residual emission shows minor features at 5.2, 5.8, 13.5 and 14.2 μm which are also present in higher resolution spectra obtained with the Short Wavelength Spectrometer (SWS) [18,19] (also see the contribution of Moutou in this volume). The intensity enhancements at 6.9 and 9.7 μm coincide with the S(5) and S(3) lines of H_2.

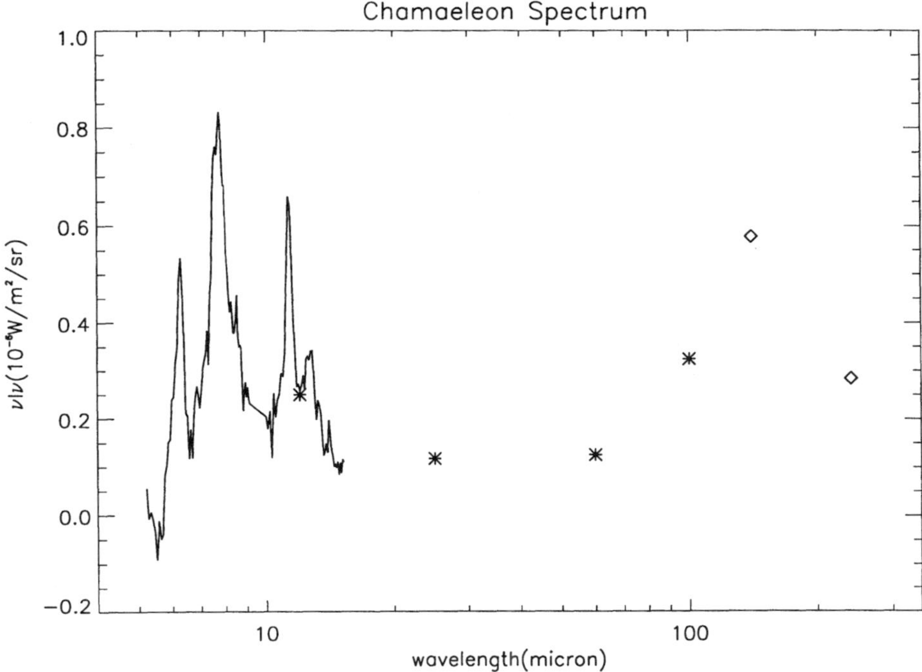

Fig. 1. — The ISOCAM spectrum obtained for the cloud DC300-17 in Chamaeleon corresponds to the mean emission within the $3' \times 3'$ field of view of the camera. Laureijs *et al.* [9] have presented a detailed investigation of the IRAS emission of this cloud. Here, the ISOCAM data is combined with the broad band IRAS brightnesses at the same position (stars). The diamonds represents the 100 μm IRAS brightness scaled with the average colors $I_\nu(140\ \mu\text{m})/I_\nu(100\ \mu\text{m})$ and $I_\nu(240\ \mu\text{m})/I_\nu(100\ \mu\text{m})$ measured for the whole cloud with the $40'$ DIRBE data [10].

The Lorentz fits provide a valuable mean of quantifying and comparing the spectra. Feature intensities measured with such fits scale with estimates of the stellar radiation density from Chamaeleon to the Ophiucus West interface [13]. At higher radiation fields, the NGC 7023 and the M17 data points are somewhat below the linear extrapolation of the other PDRs. This could be an indication of a decreased abundance of small dust particles at high radiation fields. However, NGC 7023 and M17 are also more distant objects than Ophiucus and Chamaeleon and we do not exclude the possibility that some of the observed decrease in intensity could be accounted for by beam dilution.

The fact that the bands are well described by Lorentz profiles can be related to the physics of the emission of large molecules at high temperature [21, 22]. Boulanger *et al.* [17] proposed that the feature widths are intrinsic to the

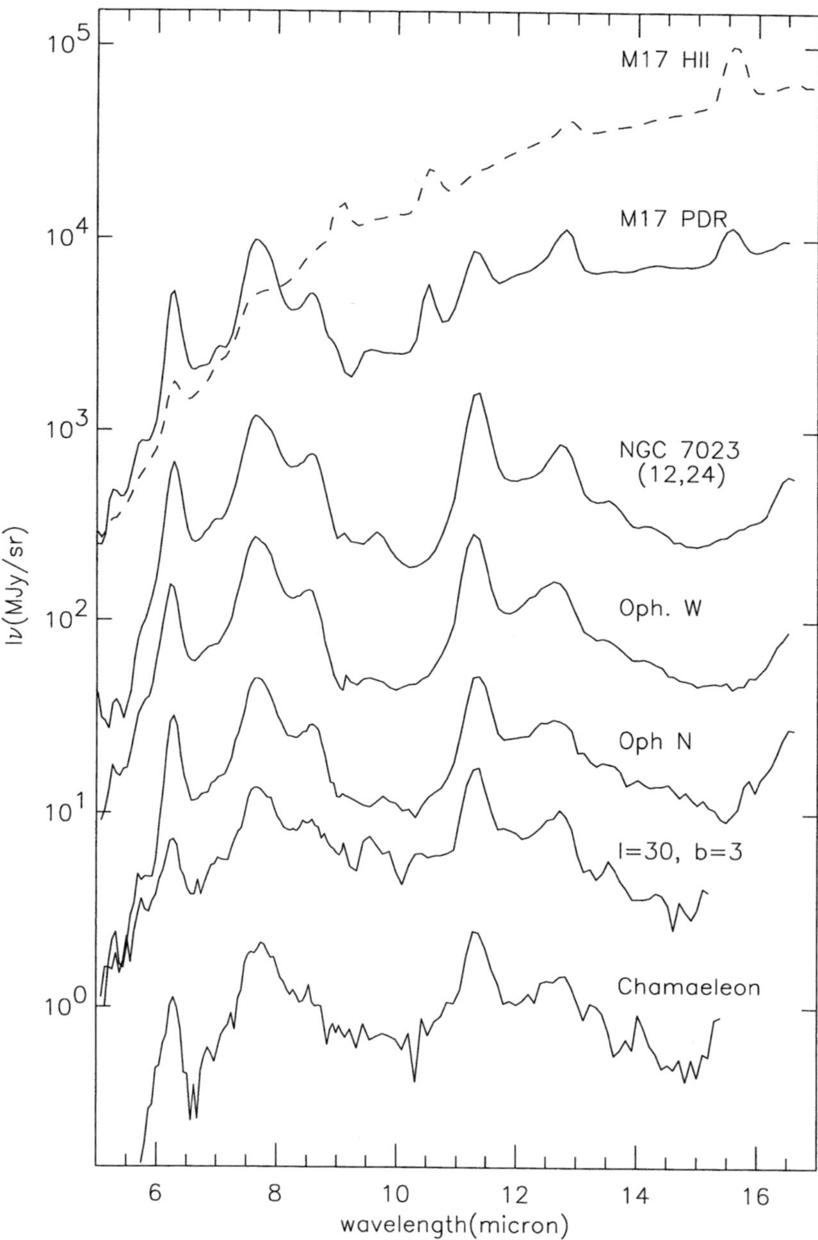

Fig. 2. — ISOCAM spectra obtained on five objects covering five orders of magnitude in brightness intensity. Original references for these spectra are Boulanger *et al.* [12, 13] (Ophiuchus) and Cesarsky *et al.* [14, 15] (NGC 7023 and M17). The M17 spectra include the Ar III (9.0 μm), S IV (10.5 μm), Ne II (12.8 μm), Ne III (15.6 μm) lines from ionized gas.

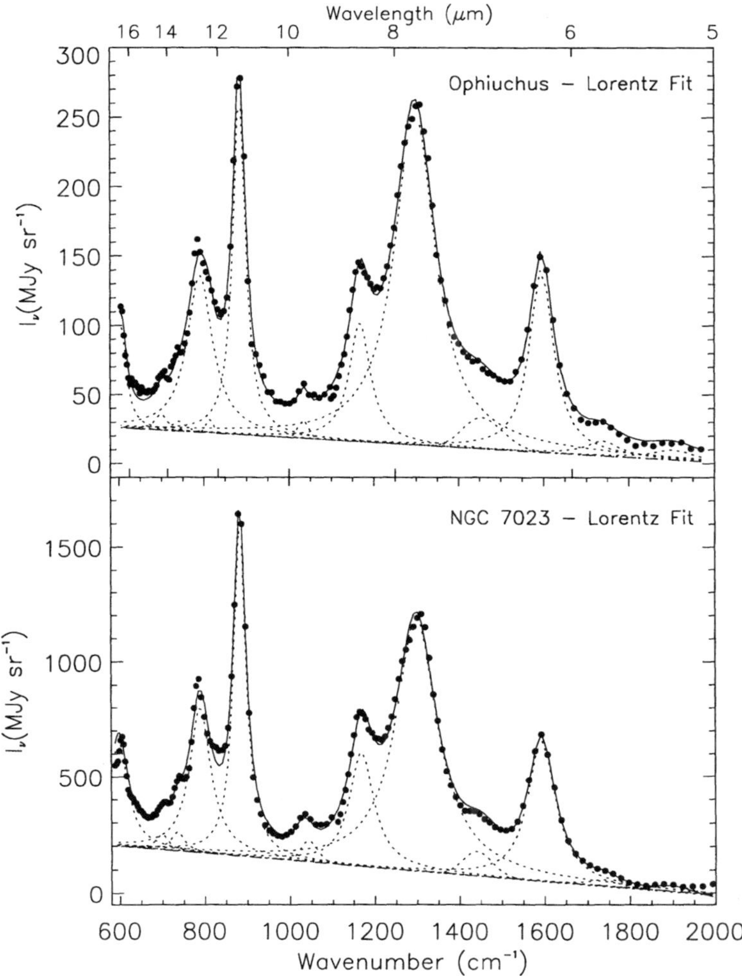

Fig. 3. — Lorentz decomposition of two typical ISOCAM Spectra. The NGC 7023 spectrum corresponds to that presented in Figures 2 and 4 at the camera pixel (12,14). Parameters of the Lorentzian decomposition can be found in Boulanger *et al.* [17]. The individual Lorentz profiles are represented with a dotted line.

emitting particles and not the result of a blend of narrower features shifted with respect to one another. The Lorentzian shape would be associated with the very short lifetime (a few 10^{-14} s) of the excitation of a given vibrational state before the energy is transferred to another vibrational state. This process known as Internal Vibrational Redistribution (IVR) has been demonstrated by laboratory experiments [23]. Note that the very short lifetime of the excitation of a given mode is incommensurate with the radiative cooling time which is of the order of seconds [1]. The Lorentzian decomposition presented for the ISOCAM data glosses over some aspects of the emission revealed by the higher resolution spectra obtained with the SWS. These data (*e.g.* the NGC 7023 spectrum [19], also see the contribution of Moutou in this volume) show that the true band shapes must be the combination of a larger set of Lorentz profiles. For example, the 6.2 and 11.3 μm are clearly non-symmetric in the SWS spectra with a broader wing towards longer wavelengths. This could be due to a shift of the central wavelength with temperature observed in laboratory for emission of hot PAH molecules in the gas phase [20]. The SWS spectra also show that the 7.7 μm has substructure and that it is a blend of at least two independent bands. The combination of profiles is important in fitting the emission peaks but has little effect on the width and intensity of the band wings which are seen in the SWS as well as in the ISOCAM spectra. In particular, it does not affect the fact that most of the pseudo-continuum underlying the features is accounted for by the wings of the main emission bands.

3.2. Spectral variations

In Figure 4, we show how the emission spectrum varies across the NGC 7023 PDR [14]. The variations in the relative band intensities illustrated by this figure seem to represent a general trend observed in other objects of the same type such as the M17 PDR [24] where, however, the effect is weaker. A systematic increase of the 11.3 and 12.7 μm features with respect to the 6.2, 7.7 and 8.6 μm group, is observed from the star to the molecular cloud. The ratio between the integrated intensity in these two groups of bands varies by a factor 4.5 between the top and bottom spectra. This variation fits with the predicted effect of PAH ionization (*e.g.* [25,26], and the contribution of Salama this volume) from cations near the star to neutrals in the shielded molecular gas. However, other physical parameters such as the size distribution and the reddening of the stellar radiation can also affect the emission spectrum. Characterisation of the physical conditions in the observed regions is necessary in order to assess plausible interpretations. Since the effect of ionization is the strongest on the 3.3 μm band, the existing data on the 3.3 μm emission in NGC 7023 [27] offer a key test. The variations in relative intensities observed across PDRs contrast with the remarkable similarity among spectra in Figure 2 other than that of the M17 HII region. For these spectra the ratio between the 6.2, 7.7, 8.6 μm on one hand and the 11.3 and 12.7 μm features on the other hand is the same within 20% of the mean value. Any explanation of the spectral changes across PDRs must also explain the remarkable constancy of relative band intensities from the Chamaeleon cloud to the brighter PDRs. The ionization state depends on both the radiation field and the electron density [28]. For the lines of sight of Figure 2 the ionization state of PAHs may

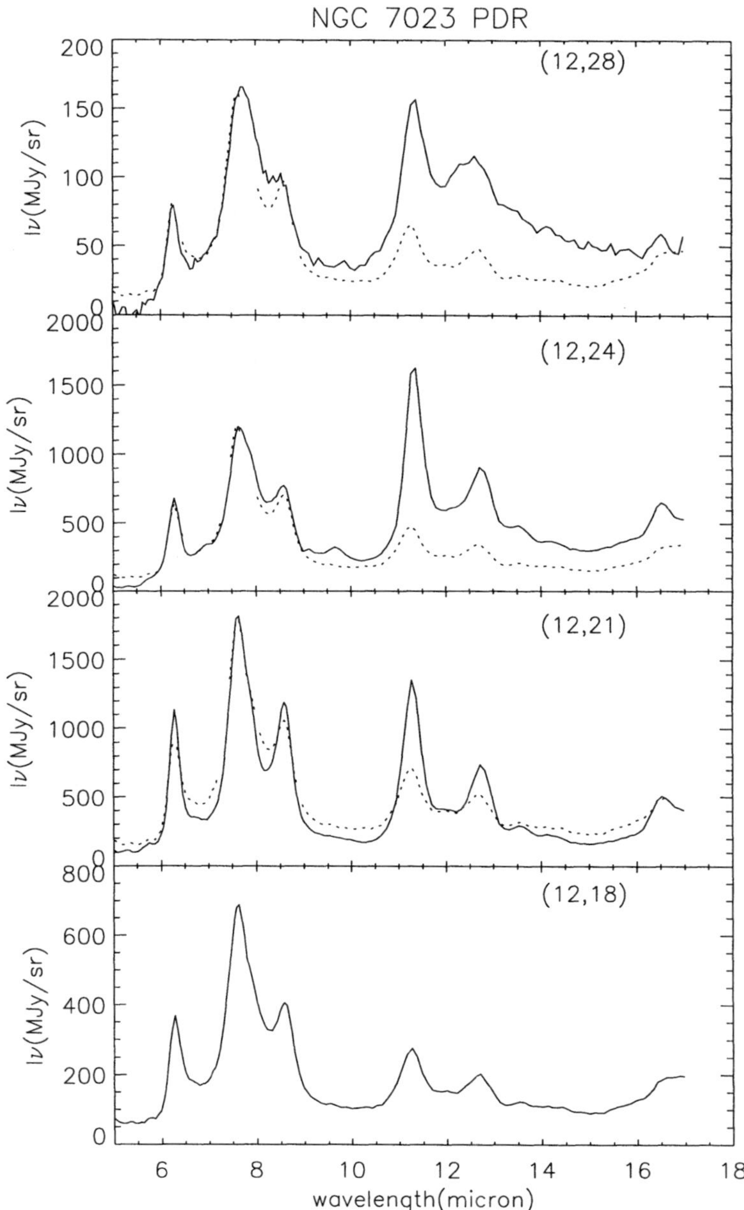

Fig. 4. — Variations of the emission spectrum across the NGC 7023 PDR. The spectra, identified by the corresponding pixel numbers, are spaced by 18″. The selected positions go from photo-dissociated matter close to the exciting star (pixel 12,18 at the bottom) to the shielded molecular gas (pixel 12,28 to the top) through the bright photo-dissociation front (pixels 12,21 and 12,24). The dotted spectrum represents the (12,18) spectrum scaled to match the peak of the 7.7 μm feature. Near the star the short wavelength continuum may be contaminated by stellar stray light.

not vary much if the increase in radiation field intensity is compensated by an increase in gas density.

The spectra in Figure 4, from bottom to top, also show a shift towards longer wavelength and a widening of the 7.7 μm feature while the central wavelength and width of the other features remain constant. An SWS spectrum shows that the widening of the 12.7 μm feature in the (12,28) spectrum is due to the S(2) H_2 line at 12.3 μm. Since the 7.7 μm feature is the combination of at least two bands, the variations in its width and position are likely to be due to a change in the relative strength of these bands.

3.3. Emission from very small grains

Particles in the intermediate size range between PAHs and large dust grains are often referred to as very small grains (VSGs). Nothing is firmly known about the nature of these grains and their relation with PAHs but much is expected on this question from the analysis and modelling of ISO spectra of warm sources. In the M17 PDR and H II region, the VSGs are warm enough to contribute to the emission over the ISOCAM wavelength range (Fig. 1). The PAHs are also probably to a large extent destroyed. In this context, we believe that the VSGs account for the rising continuum emission towards longer wavelengths that is seen in these two sources; the VSGs are also present in regions with less intense radiation fields but they are too cold to emit within the ISOCAM wavelength range. Low contrast spectral features are present in the M17 H II region spectrum [24] and those of compact H II regions ([29], see also the contribution of Cox and Roelfsema in this volume). These features differ significantly from the standard *PAH features* and share similarities with those observed for anthracite and HACs [30]. It is thus speculated that the VSGs could be carbon clusters with variable degrees of hydrogenation and mixed hybridisation which can be thought of as aggregates of PAHs. The silicate feature around 9.6 μm is seen in emission only very close to very luminous stars (*e.g.* Trapezium stars in Orion) where the equilibrium temperature of large dust grains is sufficiently high for large grains to emit in the mid-IR.

4. INTERSTELLAR PAHs

4.1. Emission properties

The results presented in the previous sections support the existence of small aromatic hydrogenated particles in the general interstellar medium in two fundamental ways. First, the 12 μm emission observed by IRAS in nearby interstellar matter heated by the general Galactic ISRF has been shown to be associated with the set of emission features considered to be characteristic of C–C and C–H bonds in aromatic hydrocarbons. Second, the shape of the spectrum has been shown to be remarkably constant for PDRs over three orders of magnitude in radiation field intensity. This is a strong prediction of the emission from small particles transiently heated by the absorption of single photons. Further, the observed scaling of the feature intensity with radiation field intensity establishes, on observational grounds, the fact that the particles are excited by stellar light.

The excitation by single photons puts strong constraints on the size of the emitting particles. After absorption of a photon, a small particle is heated to a high temperature and emits most of its energy near that peak temperature, because the heat capacity is a steeply increasing function of temperature. There is thus a rough correspondence between particle size and emission wavelength [31]. Schutte *et al.* [32] computed that most of emission in the $5-16$ μm range comes from particles of a few hundred atoms. For the 11.3 and 12.7 μm features, particles with up to 1000 carbon atoms contribute. Such large particles might be three-dimensional aromatic hydrocarbon clusters rather than very large planar molecules. Molecules with less than 100 carbon atoms are necessary to account for the 3.3 μm feature but they represent a minor fraction of the carbon in aromatic hydrocarbon particles. From the energy emitted in the $3-4$ μm range (Sect. 3), we estimate that the abundance of carbon in these molecules is only 1% (5% of the total abundance of carbon in aromatic hydrocarbons with less than 1000 C atoms). The interstellar aromatic hydrocarbons are thus on average larger than the molecules studied in the laboratory (*e.g.* [26,33]) and modelled by quantum mechanics calculations (*e.g.* [25]). This difference may explain why none of the published laboratory spectra provide a satisfactory match to the interstellar spectra.

The constancy of the central wavelengths and widths of the interstellar features is a remarkable characteristic of interstellar PAHs which contrasts with the great variations observed for small PAHs (molecules with less than a few tens of atoms) studied in the laboratory. To match the Orion bar mid-infrared spectrum Allamandola *et al.* [26] had to propose a very specific mixture of selected PAH cations (see Fig. 5 in the contribution of Salama in this volume). In view of the remarkable stability of the interstellar features, such a specific solution is not satisfactory. We think that the feature widths are intrinsic to the emitting particles rather than the blend of narrower features shifted with respect to one another. The Lorentzian shape of the bands strongly argues for this interpretation and is also supported by the physics of the emission from large molecules. An important implication of this view is that the vibrational modes of interstellar PAHs do not depend much on the exact shape of the particle. In the absence of theoretical or experimental work on PAHs with 100 or more atoms, one can only speculate that for large particles, band shifts are greatly reduced because boundary and specific symmetries become less important. Interstellar PAHs also need to have wider features than those observed for small PAHs. The widths of 6.2 and 8.6 μm interstellar features are 2 to 3 times larger than those measured in the laboratory for coronene at 700 K [20].

4.2. Carbon abundance in PAHs

Current estimates of the abundance of carbon in small aromatic hydrocarbons are derived from the mean solar neighborhood 12 μm emission per hydrogen derived from the IRAS data [34]. The ISO spectra validate the bolometric correction, used at that time, to estimate the emission in the $2-15$ μm wavelength range from the 12 μm IRAS flux. Based on the ISOCAM spectra in Figure 2 and the L band ($3-4$ μm) measurements of DIRBE [35–37], we find that $I(2-15\ \mu\text{m}) = 1.5\,\nu I_\nu(12\ \mu\text{m}) = 1.6 \times 10^{-31}\,\text{W/(H atom)}$, very close to the value of $1.5 \times 10^{-31}\,\text{W/(H atom)}$ in [34]. The emission measured by DIRBE in the $3-4$ μm wavelength range accounts for only 5% of this integrated emission. From laboratory data, Joblin *et al.* [38] computed the mean specific absorption of PAHs to be $2.3 \times 10^{-27}\,\text{W/(C atom)}$ for the solar neighborhood ISRF. Combining these numbers, we get an abundance of carbon in small aromatic hydrocarbons of 7.0×10^{-5}, or 19% of the solar abundance.

ISO has also allowed the detection of the 6.2 μm aromatic feature band in absorption in a few sources [39]. Based on the lack of correlation with the H_2O, CO and CO_2 ice features, the 6.2 μm absorption feature is thought to be associated with the diffuse medium rather than with molecular clouds. Since this band is observed to be much weaker for neutral PAHs, its detection implies that PAHs should be mostly cations in the diffuse medium. In this case, an abundance similar to that derived from the emission would account for the observed absorption band. The particles seen in absorption should thus be the same as those seen in emission. At any moment, only a small fraction of these particles are hot enough to emit in the mid-infrared. The absence of absorption bands at the position of the C=O carbonyl group at 5.95 μm and the C-H aromatic features would suggest that interstellar PAHs are poor in oxygen and little hydrogenated.

5. PERSPECTIVES

In this paper we have presented only a few aspects of the mid-IR observations of the interstellar medium. Many more spectroscopic results than those described here need to be analysed and put together. In particular, no measurements of the 3.3 and 3.4 μm bands have been presented while they would be important to discuss the charge state of interstellar PAHs and study the connection between interstellar PAHs and the carbon particles responsible for the 3.4 μm absorption feature (see the contribution of Pendleton in this volume). These features were not detected by ISO instruments as easily as the longer wavelengths features but some data do exist [27]. With all the available data, we can expect to have a complete view of the evolution of the emission properties of carbon dust within the interstellar medium from the particles generated in circumstellar shells to protostellar condensations and proto-planetary disks. In addition, it is essential to gather data on the gas emission in order to correlate dust properties with physical conditions. Even with this large data base at hand, the key question of the nature of interstellar PAHs will not be answered without obtaining laboratory spectra matching the astrophysical dust features. Some experiments of synthesis and spectroscopy on large PAHs and carbon clusters have been initiated [40, 41]. An important modelling effort is also required to compare laboratory data with astrophysical spectra. Many of the statements made in the previous sections remain preliminary until quantitative models allow an assessment of the respective influences of the size distribution, the intensity and spectral distribution of the heating radiation and the emission properties of interstellar particles on the observed spectra. The contribution of interstellar PAHs to the extinction curve is also an open question. The detection of the aromatic features in environments with low UV radiation in M31 [42] and in the reflexion nebula vdB 133 [43] argue for excitation of interstellar PAHs by visible photons and not uniquely UV photons. While the vdB 133 spectrum looks very much like the spectra presented here, the emission spectra in the nucleus and disk of M31 are odd since they are completely dominated by the 11.3 μm feature.

In addition to spectroscopy, imaging of clouds with ISOCAM and ISOPHOT will provide important information in understanding the evolution of dust within interstellar clouds with respect to gas density, radiation field and velocity field (*e.g.* [16, 44]). Also, one can hope that some aspects of the evolution of dust can be sufficiently understood in order that one can use the spectral energy distribution of the dust emission as a diagnostic of physical conditions within interstellar clouds.

REFERENCES

[1] Léger A. and Puget J.L., *A&A* **137** (1984) L5.

[2] Allamandola L.J., Tielens A.G.G.M. and Barker J.R., *ApJ* **290** (1985) L25.

[3] Giard M., Lamarre J.M., Pajot F. and Serra G., *A&A* **286** (1994) 203.

[4] Ristorcelli I., Giard M., Meny C., *et al.*, *A&A* **286** (1994) L23.

[5] Mattila K., Lemke D., Haikala L.K., *et al.*, *A&A* **315** (1996) L353.

[6] Onaka T., Yamamura I., Tanabe T., Roellig T. and Yuen L., *PASJ* **48** (1996) L41.

[7] Tanaka M., Matsumoto T., Murakami H., *et al.*, *PASJ* **48** (1996) L53.

[8] Lemke D., Mattila K., Lehtinen K., *et al.*, *A&A* **331** (1998) 742.

[9] Laureijs R.J., Chlewicki G., Clark F.O. and Wesselius P.R., *A&A* **220** (1989) 226.

[10] Boulanger F., Bronfman L., Dame T.M. and Thaddeus P., *A&A* **332** (1998) 273.

[11] Bernard J.P., Boulanger F. and Puget J.L., *A&A* **277** (1993) 609.

[12] Boulanger F., Reach W.T., Abergel A., *et al.*, *A&A* **315** (1996) L325.

[13] Boulanger F., Abergel A., Bernard J.P., *et al.*, Star formation with ISO, ASP Conf. 132, edited by J.L. Yun and R. Liseau (ASP, San Francisco, 1998) p. 15.
[14] Cesarsky D., Lequeux J., Abergel A., *et al.*, *A&A* **315** (1996) L305.
[15] Cesarsky D., Lequeux J., Abergel A., *et al.*, *A&A* **315** (1996) L309.
[16] Abergel A., Bernard J.P., Boulanger F., *et al.*, *A&A* **315** (1996) L329.
[17] Boulanger F., Boissel P., Cesarsky D. and Ryter C., *A&A* **339** (1998) 194.
[18] Beintema D., van den Ancker M.E., Molster F.J., *et al.*, *A&A* **315** (1996) L369.
[19] Moutou C., Seelgren K., Léger A., *et al.*, Star formation with ISO, ASP Conf. 132, edited by J.L. Yun and R. Liseau (ASP, San Francisco, 1998) p. 47.
[20] Joblin C., Boissel P., Léger A., d'Hendecourt L. and Défourneau D., *A&A* **299** (1995) 835.
[21] Barker R.J., Allamandola L.J. and Tielens A.G.G.M., *ApJ* **315** (1987) L61.
[22] Cook D.J. and Saykally R.J., *ApJ* **493** (1998) 793.
[23] Ionov S.I., Stuchebryukhov A.A., Bagratashvili V.N., *et al.*, *Appl. Phys. B* **47** (1988) 229.
[24] Verstraete L., Puget J.L., Falgarone E., *et al.*, *A&A* **315** (1996) L337.
[25] Pauzat F., Talbi D. and Ellinger Y., *A&A* **293** (1995) 263.
[26] Allamandola L.J., Hudgins D.M. and Sandford S.A., *ApJ* (1999) in press.
[27] Rouan D., LeCoupanec P., Lacombe F., *et al.*, The Universe as seen by ISO, edited by P. Cox and M. Kessler (ESA Special Publications Series (SP-427), Paris, 1999) in press.
[28] Salama F., Bakes E.L.O., Allamandola L.J. and Tielens A.G.G.M., *ApJ* **458** (1996) 621.
[29] Roelfsema P., Cox P., Tielens A.G.G.M., *et al.*, *A&A* **315** (1996) L289.
[30] Papoular R., Connard J., Guillois O., *et al.*, *A&A* **315** (1996) 222.
[31] Puget J.L. and Léger A., *ARA&A* **27** (1989) 161.
[32] Schutte W.A., Tielens A.G.G.M. and Allamandola L.J., *ApJ* **415** (1993) 397.
[33] Moutou C., Léger A. and d'Hendecourt L., *A&A* **310** (1996) 297.
[34] Boulanger F. and Pérault M., *ApJ* **330** (1988) 964.
[35] Bernard J.P., Boulanger F., Désert F.X., *et al.*, *A&A* **291** (1994) L5.
[36] Dwek E., Arendt R.G., Fixsen D.J., *et al.*, *ApJ* **475** (1997) 565.
[37] Arendt R.G., Odegard N., Weiland J.L., *et al.*, *ApJ* **508** (1998) 74.
[38] Joblin C., Léger A. and Martin P., *ApJ* **393** (1992) L79.
[39] Schutte W.A., van der Hucht K.A., Whittet D.C.B., *et al.*, *A&A* **337** (1998) 261.
[40] Joblin C., Masselon C., Boissel P., *et al.*, *Rapid Comm. Mass Spectrom.* **11** (1997) 1619.
[41] Herlin N., Bohn I., Reynaud C., *et al.*, *A&A* **330** (1998) 1127.
[42] Cesarsky D., Lequeux J., Pagani L., *et al.*, *A&A* **337** (1998) L35.
[43] Uchida K.I., Sellgren K. and Werner M., *ApJ* **493** (1998) L109.
[44] Laureijs R.J., Haikala L., Burgdorf M., *et al.*, *A&A* **315** (1996) L317.

LECTURE 3

Properties of Dust in High Latitude Clouds

R.J. Laureijs

ISO Science Operations Centre, Astrophysics Division of ESA,
Apartado 50727, 28080 Madrid, Spain

Abstract. The properties of dust as inferred from observations of high latitude clouds are discussed. A general overview is given of the extinction and emission by dust in the diffuse ISM with special attention to the emission properties in the infrared. The first observations with the *Infrared Space Observatory* (ISO) are reviewed.

1. INTRODUCTION

The IRAS survey at 60 and 100 μm has shown that the sky at high galactic latitudes is dominated by emission from mostly irregularly shaped clouds [1]. The first studies indicated that these structures are associated with HI filaments, optical reflection [2], red emission [3], and Lynds dark and bright nebulae [4]. The bright regions in a large number of these clouds were detected in CO surveys suggesting the presence of molecular material [5]. Analysis of the infrared emission quickly pointed towards some unexpected infrared properties of these clouds: firstly, the colour temperature derived from the 60 and 100 μm surface brightness ratio is too high for "classical" dust grains in the general interstellar radiation field (ISRF) [6], and secondly, the clouds also emit in the 12 and 25 μm IRAS bands which cannot be explained by emission from grains in thermal equilibrium with the ISRF [7]. These results triggered further studies of high latitude clouds and have led to an improved picture of the composition of dust in the diffuse interstellar medium (ISM).

In this review I discuss the properties of dust in the diffuse ISM as derived from observations of quiescent high latitude dust clouds. These clouds are well suited for such a study because: (1) they are situated in the local solar neighbourhood ($d <$ 500 pc) which provides a handle on the external radiation field and the distance, (2) the clouds are in many cases isolated due to the fact that confusion along the line of sight is minimized, (3) it is possible to study clouds with a wide range of opacities which can be used to test grain models under varying radiation field conditions, and (4) from an observational point of view, high latitude clouds cover many angular sizes so that smaller telescopes can be used to observe them. However, due to the absence of internal radiation sources, the extended emission in the optical and infrared is generally weak. Only the brightest cirrus structures can be used to measure emission spectra with reasonable signal to noise ratios.

The study of high latitude clouds in the infrared has the additional advantage of the availability of all sky HI, CO and infrared (COBE and IRAS) surveys which can be used to obtain additional information, such as the gas column density from HI and CO and the infrared spectral energy distribution. Moreover, many extinction observations cover high latitude lines of sight, which can be used as an additional source of information.

In Section 2 an overview is presented of the general properties of dust in the light of the parameters that can be observed in cirrus. Section 3 gives the status of the general extinction properties of dust ranging from the UV to the far IR (FIR). Observations of extended red emission (ERE) in high latitude clouds are discussed in detail in Section 4. The observations that have led to the refinement of the infrared dust properties are presented in Section 5. Section 6 reviews the ISO achievements so far. Finally, in Section 7 an outlook is given of what can still be done with ISO observations.

2. GENERAL PHYSICAL PROPERTIES

Before discussing the properties of dust in high latitude clouds, an overview is given of the physical parameters involved. The starting point is the balance equation where the incident radiation field J heats dust particles with size a which in turn emit light at different wavelengths:

$$\sum_i \int \int \Phi_i(a,\lambda) 4\pi J(\lambda)\, \mathrm{d}a\, \mathrm{d}\lambda = \int 4\pi I(\lambda) \mathrm{d}\lambda, \tag{1}$$

in which the $\Phi_i(a,\lambda)$ term characterizes the dust absorption, size distribution, composition and shape:

$$\Phi_i(a,\lambda) = f_i(a)\, C^i_{\mathrm{abs}}(a,\lambda), \tag{2}$$

where $f_i(a)$ gives the dust abundance and size distribution, $C^i_{\mathrm{abs}}(a,\lambda)$ is the absorption cross-section for a dust particle with size a and composition i. $C^i_{\mathrm{abs}}(a,\lambda)$ can be derived from the optical properties (usually measured in laboratory) of proposed grain materials. For spherical particles Mie and Rayleigh theory are applied in the derivation of $C^i_{\rm abs}(a,\lambda)$. For more complex shaped particles, such as composite or porous grains, additional computational methods are necessary [8].

The sum in equation (1) is performed over all grain components. The emission term $I(\lambda)$ can be expressed in terms of hydrogen column density $N_{\rm H}$ by:

$$I(\lambda) = \sum_i \frac{\mu m_{\rm H} Z^i_{\rm d}}{\langle m^i_{\rm d}\rangle} N_{\rm H} \int {\rm d}a\ \Phi_i(a,\lambda) \int {\rm d}T\ G_i(a,T) B(\lambda,T), \qquad (3)$$

where μ, $m_{\rm H}$, $Z_{\rm d}$, and $\langle m_{\rm d}\rangle$ are the mean atomic mass, the hydrogen mass, dust abundance, and mean dust mass, respectively. In the above expression it is assumed that when observed over a period of time, a given dust particle emits at different temperatures due to the discrete nature (photons!) of the incident radiation. The temperature distribution as a function of size is included in the function $G(a,T)$. For particles larger than about 10 nm, the function $G(a,T)$ is well approximated by a delta function in T implying a constant ("equilibrium") temperature. However, for smaller sizes, the particle heat capacity is of the same order as the UV photon energy so that the absorbed photon can cause a significant rise in temperature. In a dilute radiation field, such as the general ISRF, where the mean time between the absorption of UV photons is long, the emission of photons by the particle itself causes the temperature to drop significantly before the next photon is absorbed. Expressions for $G(a,T)$ are derived in [9,10].

The FIR part of the absorption cross section $C_{\rm abs}$ in the emission term (Eq. (2)) can be approximated by a power law λ^{-b} (also referred to as an "emissivity" law) with b typically between 1 and 2 for spherical particles. From the above relationships it can immediately be inferred that both Φ and G tend to broaden the dust emission spectrum to one that is wider than that expected from a modified black body (*i.e.*, $\lambda^{-b}B(\lambda)$).

Dust models are strongly constrained by the abundance factor $Z_{\rm d}$. Recent abundance determinations of gas phase oxygen in the diffuse ISM using the Hubble space telescope and ground based observations of O, C, and N abundances in early type stars and HII regions provided growing evidence that the overall abundance of the heavy elements in the ISM is significantly less than solar with $Z_{\rm ISM}/Z_\odot$ as low as 0.7 [11,12]. The lower abundances are incompatible with nearly all earlier dust models, *e.g.*, the MRN model [13] which requires 85% of the solar carbon abundance in grains. To overcome the problem, non-compact ("fluffy") dust particles are proposed and the implications of these models have been extensively discussed [14,15].

In the UV and visible, the extinction is the sum of absorption and scattering and computational modelling is necessary to separate the two. In equation (1) the scattering terms have been omitted. Although scattered light is easily observed, measuring the scattering properties of dust is very difficult due to uncertainties in a cloud's illumination geometry as well as uncertainties in the properties of the radiation field [16]. The albedo and scattering asymmetry factor at blue and visible wavelengths ($\lambda > 200$ nm) are found to be high (0.6 – 0.8 in both cases) implying that the grains are strongly forward scattering [4]. At shorter wavelengths both scattering parameters are smaller.

Scattering is negligible in the FIR and the absorption cross section is inferred from the spectral energy distribution of the dust emission. In the regime where non-equilibrium emission is negligible, a useful observable quantity is the dust optical depth τ_λ:

$$\tau_\lambda = \frac{I(\lambda)}{B(\lambda,T)} \,. \tag{4}$$

To determine T, the FIR emission must be measured at at least two wavelengths. The value of τ_λ (Eqs. (3) and (4)) is proportional to the total column density $N_{\rm H}$ and $\Phi(a,\lambda)$. Once the column density has been derived, from *e.g.*, A_V, HI or CO, the FIR dust properties can be inferred *via* $\Phi(a,\lambda)$.

In the remainder of this review I discuss the properties of dust derived from observations in the optical and infrared. Due to the many poorly-determined quantities entering equation (1), the derivation of a consistent picture can only be obtained by combining different observations with different targets, telescopes and wavelengths.

3. DUST EXTINCTION/ABSORPTION

The classical properties of dust are inferred from extinction observations in the UV and visible. The spectrum of a reddened star of known spectral type is compared to that of an unreddened star of the same spectral type and the ratio of the two spectra gives the extinction curve. An excellent discussion of all the observational aspects can be found in Krelowski and Papaj [17].

The ratio between the total visual extinction A_V and selective extinction $E(B-V) = A_V - A_B$ along the line of sight is denoted by R_V. R_V has a constant value around 3.1 for the diffuse ISM. For denser clouds the value for R_V can be more than 5. Typical extinction curves from the UV to the FIR are given in Figures 1 and 2.

The extinction curve in the visible and UV range shows three distinct domains which have been the subject of much discussion [19]:

- a rise in the optical regime proportional to λ^{-1}. This part of the curve is dominated by a population of grains with radii of the order of 0.1 – 0.3 μm;
- an absorption bump at 217.5 nm. Due to its strength, only materials composed of the most abundant elements can be considered as serious candidates for its origin. Up to now a population of small graphitic-type of particles (of the order of 5 nm diameter) gives the closest and most consistent match. Pure graphite exhibits a strong resonance at 220 nm, but its centre wavelength shifts to longer wavelengths for particles > 5 nm;

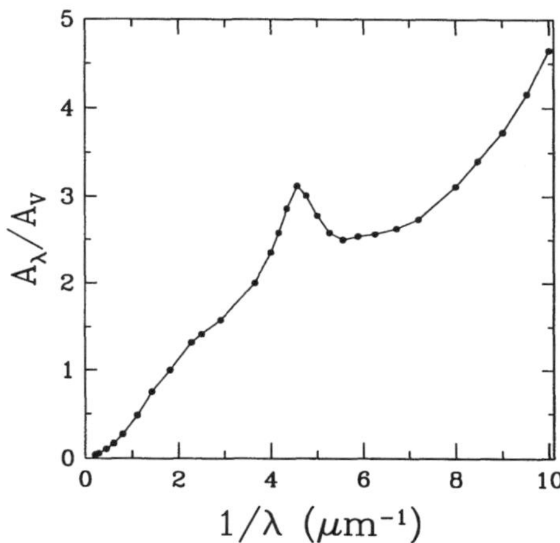

Fig. 1. — A "typical" extinction curve from the near IR to the UV assuming $R_V = 3.1$, the data are from Whittet [18]. Three trends can be distinguished: (1) linear rise in the visible range ($1\ \mu m < \lambda^{-1} < 3\ \mu m$), (2) the "bump" ($3\ \mu m < \lambda^{-1} < 6\ \mu m$), and (3) the FUV rise ($\lambda^{-1} > 6\ \mu m$).

- a steady far-UV rise 170 – 100 nm. The rise at these wavelengths can only be due to a population of very small particles with radii of the order of 10 nm or smaller.

Cardelli *et al.* [20] found that for $\lambda < 1\ \mu m$ all extinction curves can be represented by one functional form and that the variations among curves can be matched by adjusting only one parameter, R_V. Apparently, changes in the extinction curve involve variations in all contributing dust populations and sizes according to a strict law. This argues in favor of a general size distribution for dust particles. Mathis [15] interprets the R_V dependency as the result of an efficient coagolation process where small grains (responsible for the UV part of the extinction curve) stick onto the larger ones in dense regions thereby increasing R_V. However – as pointed out by Krelowski and Papaj [17] – when considering individual extinction curves one should be aware that large deviations from the Cardelli *et al.* law can occur.

The direct comparison between extinction and infrared emission may allow us to study the extinction properties of the dust emitting in the infrared. The visual or blue extinction shows a good correlation with the far-infared emission at 100 μm, see Section 5.2. Unfortunately, correlations between *e.g.*, the 220 nm bump or the Cardelli *et al.* parameters and the FIR emission are generally inconclusive due to the difficulties in identifying the emission regions responsible for the extinction and due to uncertainties in the infrared background subtraction [21].

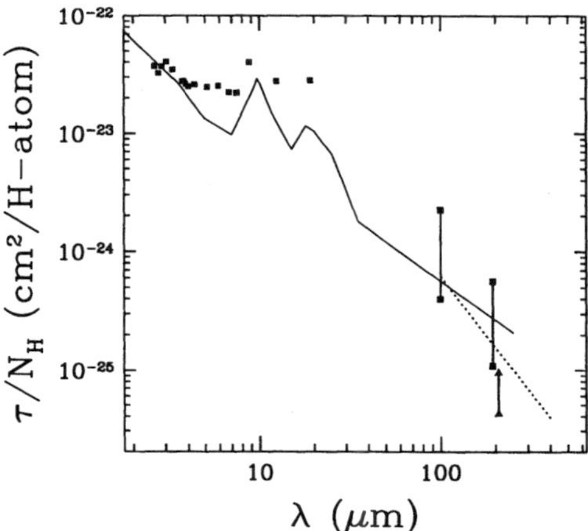

Fig. 2. — The extinction curve in the infrared ($\lambda > 2.5$ μm). The solid curve is the standard extinction curve given by Mathis [19]. The FIR part ($\lambda > 25$ μm) matches the opacities given by Hildebrand [23]. The dotted line gives the opacity found by Boulanger *et al.* [24] from COBE. Included are the values derived by Lutz *et al.* [25] from H recombination line data with SWS (filled squares), FIR opacities from Lehtinen *et al.* ([26], connected squares) and Laureijs *et al.* ([27], connected triangles) obtained from ISOPHOT observations.

At wavelengths beyond 1 μm the extinction curve becomes more uniform with no significant variations between diffuse and dense regions [22]. The curve in the $1-5$ μm regime can be represented by a power law with index of $b \approx 1.8$. The constancy of the slope for different lines of sight suggests that the coagulation of large grains which must alter the near infrared extinction curve does not occur in denser environments. In the $8-30$ μm wavelength range the extinction curve is dominated by two absorption features at 9.7 and 18 μm. Amorphous silicate particles exhibit similar features and give the best evidence that an important fraction of the interstellar dust consists of silicates.

The FIR part of the extinction curve ($\lambda >$ 25 μm) can only be derived from the analysis of dust emission (*cf.* Eq. (4)). Since the derivation of τ_λ involves both the determination of dust temperature and column density, there are therefore large uncertainties. The values obtained from KAO observations summarized by Hildebrand [23] are still commonly used. Boulanger *et al.* [24] derived τ_λ for the diffuse ISM using the correlation between COBE and HI data, see also Section 5.2.

4. EXTENDED RED EMISSION (ERE)

Studies of a variety of extended dusty objects (such as reflection nebulae, circumstellar emission around evolved stars, planetary nebulae, HII regions, *etc.*) have shown the existence of a red emission feature which cannot be attributed to scattered light from dust grains (see *e.g.*, [16]). Initially the extended red emission (ERE) was attributed to H_α emission, but improved spectral resolution indicated that the emission feature: a) peaks between 600 and 800 nm, b) is very broad with a FWHM of 120 – 190 nm, and c) does not seem to require very specific radiation field conditions. Relatively intense red emission inferred from images in the B, R, and I bands indicates that the ERE is also present in high latitude cirrus clouds [3]. Whittet [18] suggested that the very broad structure (VBS) seen in extinction curves [17] might be an artifact due to ERE in the spectra. By analyzing blue and red measurements collected by the Pioneer 10 and 11 spacecraft outside the zodiacal dust cloud (thereby avoiding zodiacal light contamination) Gordon *et al.* [28] concluded that the ERE is a general emision component of the diffuse ISM.

Only recently the spectral energy distribution of the ERE in a number of high latitude clouds was measured for the first time by Szomoru and Guhathakurta [29]. These observations confirm that ERE is a general emission component with spectral properties similar to that found in the brighter objects. The ERE intensity is about one third of the scattered light intensity.

The ERE properties indicate that the phenomenon may be caused by photoluminescence from a dust component. The observations show that the photon conversion efficiency of the process is high: more than 10% of all the energy absorbed in the UV and the blue appears in the ERE. This implies that the agent causing ERE should be an abundant material which is capable of efficient photoluminescence – only carbon or silicate based materials are serious candidates.

Laboratory emission spectra of amorphous hydrogenated carbons (HACs) provide a good match to the observed ERE spectra. The plausibility of carbon based carriers like HACs was supported by the observation that no O-rich PN shows ERE while most C-rich PNe do. However, recent ISO observations have challenged the reasoning (see contribution by Waters this volume). Szomoru and Guhathakurta [29] found that the ERE in cirrus peaks around 600 nm which is bluer than the ERE peak wavelengths associated with more intense UV irradiated environments. For example, the ERE in HII regions peaks around 800 nm. This systematic shift with inceasing UV hardness is also seen in laboratory measurements on HAC films. Despite the matching of the emission spectra, the most serious problem with HAC is that the luminescence photon efficiency is low [30, 31].

The absence of a clear correlation between ERE and the 3.3 μm emission feature in a number of objects [32] argues against the hypothesis that the ERE and the unidentified infrared bands (UIBs) have a common carrier.

An alternative interpretation has been provided recently by Ledoux *et al.* [31] and Witt *et al.* [30] based on laboratory observations of very porous crystalline silicon. This material shows strong photoluminescence with a high quantum yield in the red. The spectral energy distribution matches the observations and the wavelength of maximum quantum yield corresponds to the observed ERE wavelengths. The high photoluminescence efficiency is due to the confinement of excited electrons in the nanometre-sized quantum wells present in porous silicon. Ledoux *et al.* propose crystalline particles of pure silicon as the carrier of ERE. Their laboratory data show that a particle diameter of $3.5 - 4$ nm is required to match the astronomical observations. The particle temperature as well as the excitation wavelength have little effect on the emission spectrum. Only the particle size affects the spectrum: smaller particles shift the emission peak to the blue. Similarly, Witt *et al.* [30] proposed crystalline silicon cores of $1 - 3$ nm in diameter surrounded by a mantle of SiO_2.

Witt *et al.* pointed out that the nanoparticles must be responsible for 20% of the UV/visible absorption which implies that 36% of all the silicon is in these particles. It would be extremely interesting to know the optical properties of these silicon particles. Applying bulk optical properties of "astronomical" silicate Draine and Anderson [9] found that 3 nm silicate particles cannot explain the high infrared continuum in the $12 - 60$ μm range in the diffuse ISM.

5. PROPERTIES OF DUST IN THE INFRARED

5.1. Small grain emission

IRAS has clearly shown that cirrus emission can be detected in all 4 survey bands at 12, 25, 60 and 100 μm. The high emission shortward of 100 μm could only be explained by introducing new dust components not predicted from UV and optical extinction [7,33].

Puget *et al.* [34] attribute the excess 12 μm emission to polycyclic aromatic hydrocarbons (PAHs) which mainly emit in discrete molecular bands at 3.3, 6.2, 7.7, 8.6, 11.3, and 12.7 μm. There might be bands at longer wavelengths, but the energy involved is much lower [35]. The PAH hypothesis is strengthened by the fact that the emission features are seen in nearly all dusty sources with prominent UV heating [36]. In addition, balloon borne experiments showed that the aromatic 3.3 μm feature is ubiquitous in the galactic plane [37]. However, so far it has not been possible to identify a single molecule or a mixture of PAH molecules which can match in detail the observed UIB spectra and associated extinction curves.

To account for the emission in the 25 and 60 μm bands a power law size distribution is proposed starting at the PAH molecules (80 carbon atoms) up to carbonaceous particles with sizes up to 15 nm [35,38,39]. Only very small grains (VSGs) of carbonaceous material can attain high enough temperatures through transient heating for continuum emission in the $15 - 60$ μm range, very small silicates remain too cold even at sub-nm sizes [9]. In some models *e.g.*, [38] the carbonaceous VSGs are assumed to be also responsible for the 2175 nm extinction bump.

By analysing several cloud complexes, Boulanger *et al.* [40] showed that the spatial variations in 12/100 μm emission ratio cannot entirely be due to variations in the radiation field but must be due to genuine abundance variations of the 12 μm carriers. This is in strong contrast to the behaviour of the 60 μm emission which is constant with respect to the 100 μm emission in the diffuse ISM.

5.2. Large grain emission

The emission in the IRAS 100 μm band is found to be consistent with emission from the grains causing the extinction in the optical. Using the ratio $I_\nu(100\ \mu\text{m})/N(\text{HI})$ obtained from the correlation of the 100 μm emission with HI emission [6,41] the inferred dust temperature that can be derived assuming classical dust particles (*cf.* Eq. (3)) agrees well with what has been predicted for such particles assuming a standard ISRF.

The COBE observations confirmed this result by directly measuring the spectral energy distribution of the diffuse emission between 3 and 1000 μm. To separate the diffuse infrared emission associated with the HI emission, Boulanger *et al.* [24] correlated the COBE/DIRBE band maps with all sky HI maps. The slopes of the pixel-to-pixel correlations yield the spectral energy distribution normalized to HI. They found that the FIR maximum can be fit with a 17.5 K component in combination with a λ^{-2} emissivity law. The resulting value for the FIR opacity (*cf.* Fig. 2) is in good agreement with the MRN prediction for classical compact grains. A similar analysis was performed by Dwek *et al.* [39] by correlating all COBE band maps with the map at 100 μm.

The 60 to 100 μm surface brightness ratio is found to be very constant throughout the Galaxy [42] and in low opacity high latitude clouds [43]. However, the ratio suddenly drops in relatively small isolated regions of clouds where the opacity is high enough to show detectable ^{13}CO emission [44–46]. By subtracting the 60 μm maps scaled with the inverse 60/100 μm ratio from the 100 μm maps it is possible to separate the diffuse component of clouds from the denser regions associated with ^{13}CO.

Using the all-sky DIRBE and FIRAS maps, Lagache *et al.* [47] decomposed the high latitude FIR emission in regions with different temperature components. The "cirrus" or warm component was separated from the cold component by subtracting scaled 60 μm DIRBE maps from the maps at the longer wavelengths. They found that the cirrus component has a temperature distribution between 16 and 20 K, peaking at 17.5 K. Regions with a significant cold component cover only about 3% of the high latitude sky and have temperatures ranging between 13 and 16 K peaking at 15 K. These colder regions can be associated with large molecular complexes with low mass star forming activity and have column densities $N_{\text{H}} > 2.5 \times 10^{20}$ H-atoms/cm^2. No

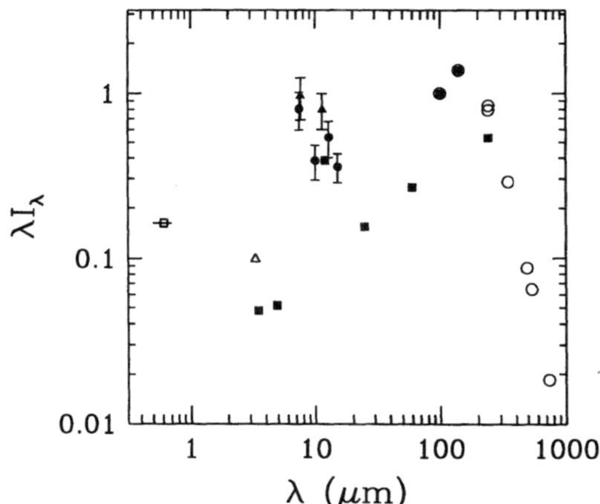

Fig. 3. — Dust thermal emission and fluorescence from the diffuse ISM. The data are normalized to the emission at 100 μm. This composite diagram contains the following datasets: COBE data from Boulanger *et al.* ([24], open circles) and Dwek *et al.* ([39], filled squares); ISOPHOT data from Lemke *et al.* [48] scaled to the 12 μm COBE measurement, the filled circles are the continuum filters, the filled triangles the feature filters at 7.7 and 11.3 μm; 3.3 μm AROME point from Giard *et al.* ([37], open triangle), the 3.3 μm continuum is estimated from the COBE values; ERE emission from Szomoru and Guhathakurta ([29], open square).

evidence was found for the presence of a very cold dust component of less than 10 K which can be associated with discrete regions. Interestingly, it is found that the temperature variations in high latitude diffuse emission corresponds to variations in the ISRF intensity of less than 30%.

The overall infrared spectral energy distribution obtained from the various observations is presented in Figure 3.

6. ISO OBSERVATIONS OF HIGH LATITUDE CLOUDS

6.1. ISO capabilities

By design ISO was ideal for observing faint extended structures with angular sizes of a few arcminutes to a few degrees. The ISOCAM and ISOPHOT instruments were capable of detecting extended emission of less than 1 MJy/sr in the ISOCAM range down to better than 0.1 MJy/sr with ISOPHOT at 200 μm, the longest wavelengths covered.

ISOCAM provided 11 narrow and wide band filters in the range 2 – 18 μm together with a circular variable filter (CVF) operating between 5 and 18 μm. ISOPHOT was composed of 25 filters in the 2.5 – 200 μm range covering most of the spectral distribution of cirrus emission. Long integration grating LWS observations in the range 50 – 200 μm have measured the spectral energy distribution of the brightest cirrus regions which are close to the LWS detection limit of a few tens of MJy/sr.

At the time of writing, the ISOPHOT calibration of extended emission is ongoing. Analyses of dedicated calibration observations that relate the point source calibration with extended source calibration are not yet complete. In the following subsections the first results are discussed.

6.2. Infrared extinction

The first CAM images of the Milky Way in the LW3 (12 – 18 μm) filter revealed the presence of regions where the diffuse background emission is absorbed by very dense clouds with estimated extinctions of $A_V > 25$ mag [49]. These infrared "dark clouds" have also been observed in the ρ Ophiuchi main cloud in the LW2 (7 – 9 μm) and LW3 filters [50] and in the densest regions of molecular cloud complexes at high latitude. These observations offer the possibility to map out the densest regions in molecular clouds. Because the extinction is proportional to the column density, one can in principle directly derive the column density (and mass) of the cloud cores with a resolution several times better than can be achieved from dust emission observations at longer wavelengths. The largest uncertainties in the column density estimate are caused by the determination of the level of diffuse background emisson, usually on top of the much brighter zodiacal emission component, and by the assumed extinction in the infrared [50].

Measurements of the zodiacal emission component with CAM CVF and PHT-S indicate that it has no significant spectral features and that it can be well-modelled by a blackbody at 260 K [51, 52]. In addition PHT-P measurements at 25 μm showed no spatial structure on a 0.5 degree scale [53]; see the contribution by Ábrahám in this volume.

The fact that our knowledge of the infrared extinction law in 5 – 9 μm range is incomplete was been pointed out by Lutz *et al.* [25] who measured the extinction of the ISM towards the Galactic centre using SWS observations of hydrogen recombination lines. Their findings are included in Figure 2 and show that the observations do not confirm the predicted minimum around 8 μm.

Whittet *et al.* [54] measured the diffuse ISM towards Cygnus OB2 No. 12 and confirmed the presence of a hydrated silicate feature around 2.75 μm.

ISOPHOT observations of isolated clouds in the Chamaeleon complex of clouds by Laureijs *et al.* [27] and Lehtinen *et al.* [26] allowed the determination of the FIR opacities based on equation (4). These calculated opacities are included in Figure 2. Although involving only denser clouds, these opacity determinations confirm the values found by Boulanger *et al.* [24] from the COBE–HI correlation. However, the uncertainties are too large to impose constraints on the proposed grain models. It is expected that an improved calibration will decrease the uncertainties on the infrared observations.

6.3. Dust temperatures

The wavelength coverage of ISO up to 200 μm opens up the possibility of directly measuring the temperature of the large grains in isolated cirrus clouds. However, it is found that the emission in the 90 μm ISOPHOT band still contains a significant contribution from the small grain component prominant in emission at 60 μm [27]. Consequently, temperature estimates from colour ratios involving the 90 μm band are likely to be biased towards a higher colour temperature due to the additional emission component.

FIR observations of isolated dense clouds indicate dust temperatures of $12 - 15$ K [27] and $14 - 16$ K [26] assuming a λ^{-2} power law for the dust emissivity. The ranges in temperature are predominantly due to uncertainties in the calibration. By analyzing the surface brightness ratio between the 150 μm and 200 μm PHT filter bands of a number of high latitude clouds, Laureijs and Haikala [43] found that the ratio is constant for different positions within the clouds but varies from cloud to cloud. This suggests that the dust temperature does not vary dramatically inside clouds, but that global temperature variations exist between clouds.

So far, evidence for a very cold dust component ($T < 10$ K) has not been found in high latitude clouds. Detection of such a component in inactive galaxies has been reported by Krügel *et al.* [55] implying that the amount of interstellar matter in such galaxies is much higher than the values estimated from IRAS flux densities.

6.4. Infrared emission

IRAS and COBE provided only broad band observations of the UIBs in the diffuse ISM. With ISO it has become possible to resolve these features in cirrus clouds. Broad and narrow band ISOPHOT observations in the wavelength range between 3.3 and 16 μm of a cirrus cloud with strong 12 μm IRAS emission were carried out by Lemke *et al.* [48]. A positive detection was found for the 7.7 and 11.3 μm emission features, and enhanced emission was found in the $6 - 9$ μm and $12 - 14$ μm ranges suggesting the presence of the 6.2, 8.6, 12.7 μm emission bands. ISOCAM CVF observations of the same cloud is reported elsewhere in this volume by Boulanger. It should be pointed out that the UIBs have been detected by ISO in a variety of extragalactic targets at different redshifts including normal galaxies similar to ours.

6.5. Cirrus structure

By analyzing four high latitude fields, Herbstmeier *et al.* [56] found that the cirrus fluctuation spectrum at 90 μm has the same power and spectral index as the one measured at 180 μm down to a resolution of about $3'$. The fluctuation spectrum of cirrus obtained with ISOPHOT is in good agreement with predictions based on IRAS measurements. In addition, the FIR fluctuation spectra are similar to the spectra found for the HI gas in the ISM suggesting a good mixing between gas and dust on the smallest scales.

7. OUTLOOK AND CONCLUSIONS

Many high latitude regions have been observed with ISO and the analysis of these data is ongoing. Due to the relatively low surface brightness of these clouds many difficult data processing operations, such as the surface brightness calibration (including flatfielding and dark current subtraction) as well as accurate background subtraction still have to be overcome. Figures 2 and 3 illustrate that ISO has the potential to put tight constraints on the proposed dust models. In particular, the ISO database contains information to address the following main topics:

- an improved determination of the spectral energy distribution in the $50 - 200$ μm range. This interval is only covered by 3 DIRBE bands as opposed to 10 by ISOPHOT. The transition between emission from small grains and the large classical grains can be studied in detail. Grating spectroscopy of cirrus with LWS is unprecedented;
- accurate determination of the spectral energy distribution with ISOCAM and ISOPHOT in the $5 - 18$ μm range will give information on the properties of UIBs as well as the underlying continuum which will enable a better determination of the VSG properties;
- the structure and FIR colours of cirrus from raster maps and from the ISOPHOT serendipity survey [57] which provides narrow $3'$ scans at 170 μm at a spatial resolution better than 5 arcmin.

ACKNOWLEDGEMENT

Dr. Dieter Lutz is acknowledged for kindly sending the IR extinction values derived from improved SWS calibration data.

REFERENCES

[1] Low F., Beintema D., Gautier F.N., *et al.*, *ApJ* **278** (1984) L19.
[2] Vries C.P. and le Poole R.S., *A&A* **145** (1985) L7.
[3] Guhathakurta P. and Tyson J.A., *ApJ* **346** (1989) 773.
[4] Laureijs R.J., Mattila K. and Schnur G., *A&A* **184** (1987) 269.
[5] Weiland J.L., Blitz L., Dwek E., *et al.*, *ApJ* **306** (1986) L101.
[6] Terebey S. and Fich M., *ApJ* **309** (1986) L73.
[7] Boulanger F., van Albada G.D. and Baud B., *A&A* **144** (1985) L9.
[8] Absorption and Scattering of Light by Small Particles, edited by C.F. Bohren and D.R. Huffman (Wiley, New York, 1983) p. 1.
[9] Draine B.T. and Anderson N., *ApJ* **292** (1985) 494.
[10] Aanenstadt P.A., Evolution of Interstellar Dust and Related Topics, edited by A. Bonetti, J.M. Greenberg and S. Aiello (North-Holland, New York, 1989) p. 121.
[11] Savage B.D and Sembach K., *ARA&A* **34** (1996) 279.
[12] Mathis J.S., From Stardust to Planetesimals, ASP Conf. 122, edited by Y.J. Pendleton and A.G.G.M. Tielens (ASP, San Francisco, 1997) p. 87.
[13] Mathis J.S., Rumpl W. and Nordsiek K.H., *ApJ* **217** (1977) 425.
[14] Dwek E., *ApJ* **484** (1997) 779.
[15] Mathis J.S., *ApJ* **497** (1998) 824.
[16] Witt A.N., Interstellar Dust, edited by L.J. Allamandola and A.G.G.M. Tielens (Kluwer, Dordrecht, 1989) p. I87.
[17] Krelowski J. and Papaj J., *PASP* **105** (1993) 1209.
[18] Whittet D.C.B., Dust in the Galactic Environment (IOP Publishing Ltd., Bristol, 1992) p. 1.
[19] Mathis J.S., *ARA&A* **28** (1990) 37.
[20] Cardelli J., Clayton G.C. and Mathis J.S., *ApJ* **345** (1989) 245.
[21] Jenniskens P. and Désert F.X., *A&A* **275** (1993) 549.
[22] Martin P.G. and Whittet D.C.B., *ApJ* **357** (1990) 113.
[23] Hildebrand R.H., *QJRAS* **24** (1983) 267.
[24] Boulanger F., Abergel A., Bernard J.-P., *et al.*, *A&A* **312** (1996) 256.
[25] Lutz D., Feuchtgrüber H. and Genzel R., *A&A* **315** (1996) L269.
[26] Lehtinen K., Lemke D., Mattila K. and Haikala L.K., *A&A* **333** (1998) 702.
[27] Laureijs R.J., Haikala L., Burgdorf M., *et al.*, *A&A* **315** (1996) L317.
[28] Gordon K.D., Witt A.N. and Friedmann B.C., *ApJ* **498** (1998) 522.
[29] Szomoru A. and Guhathakurtha P., *ApJ* **494** (1998) L93.
[30] Witt A.N., Gordon K.D. and Furton D.G., *ApJ* **501** (1998) L111.
[31] Ledoux G., Ehbrecht M., Guillois O., *et al.*, *A&A* **333** (1998) L39.
[32] Perrin J.M. and Sivan J.P., *A&A* **255** (1992) 271.
[33] Laureijs R.J., Chlewicki G., Clark F.O. and Wesselius P.R., *A&A* **220** (1991) 226.
[34] Puget J.-L., Léger A. and Boulanger F., *A&A* **142** (1986) L19.
[35] Schutte W.A., Tielens A.G.G.M. and Allamandola L.J., *ApJ* **415** (1993) 397.

[36] Puget J.-L. and Léger A., *ARA&A* **27** (1989) 161.
[37] Giard M., Lamarre J.M., Pajot F. and Serra G., *A&A* **286** (1994) 210.
[38] Désert F.X., Boulanger F. and Puget J.-L., *A&A* **237** (1990) 215.
[39] Dwek E., Arendt R.G., Fixsen D.J., *et al.*, *ApJ* **475** (1997) 565.
[40] Boulanger F., Falgarone E., Puget J.-L. and Helou G., *ApJ* **364** (1990) 136.
[41] Boulanger F. and Pérault M., *ApJ* **330** (1988) 964.
[42] Bloemen J.B.G.M., Deul E.R. and Thaddeus P., *A&A* **233** (1990) 437.
[43] Laureijs R.J. and Haikala L., Star formation with ISO, ASP Conf. 132, edited by J.L. Yun and R. Liseau (ASP, San Francisco, 1998) p. 153.
[44] Laureijs R.J., Clark F.O. and Prusti T., *ApJ* **372** (1991) 185.
[45] Abergel A., Boulanger F., Mizuno A. and Fukui Y., *ApJ* **423** (1994) L59.
[46] Laureijs R.J., Fukui Y., Helou G., *et al.*, *ApJS* **101** (1995) 87.
[47] Lagache G., Abergel A., Boulanger F. and Puget J.-L., *A&A* **333** (1998) 709.
[48] Lemke D., Mattila K., Lehtinen K., *et al.*, *A&A* **331** (1998) 742.
[49] Pérault M., Omont A., Simon G., *et al.*, *A&A* **315** (1996) L165.
[50] Abergel A., Bernard J.P., Boulanger F., *et al.*, Star formation with ISO, ASP Conf. 132, edited by J.L. Yun and R. Liseau (ASP, San Francisco, 1998) p. 220.
[51] Reach W.T., Abergel A., Boulanger F., *et al.*, *A&A* **315** (1996) L381.
[52] Ábrahám P., Acosta Pulido J.A., Klaas U., *et al.*, First ISO workshop on analytical spectroscopy, edited by A.M. Heras, K. Leech, N.R. Trams and M. Perry (ESA Pub., Noordwijk, 1997) p. 119.
[53] Ábrahám P., Leinert C. and Lemke D., *A&A* **328** (1998) 702.
[54] Whittet D.C.B, Boogert A.C.A., Gerakines P.A. *et al.*, *ApJ* **490** (1998) 729.
[55] Krügel E., Siebenmorgen R., Zota V. and Chini R., *A&A* **331** (1998) L9.
[56] Herbstmeier U., Ábrahám P., Lemke D., *et al.*, *A&A* **332** (1998) 793.
[57] Stickel M., Bogun S., Lemke D., *et al.*, *A&A* **336** (1998) 116.

LECTURE 4

Very Cold Dust in Star Forming Regions

I. Ristorcelli[1], G. Serra[1], J.M. Lamarre[2], M. Giard[1], J.P. Bernard[2], F. Pajot[2], J.P. Torre[3], A. Abergel[2], G. Lagache[2] and J.L. Puget[2]

[1] *Centre d'Étude Spatiale des Rayonnenents, UPR-CNRS 8002, Conventionné avec l'Université Paul Sabatier Toulouse, 9 avenue du Colonel Roche, 31029 Toulouse Cedex, France*
[2] *Institut d'Astrophysique Spatiale, UMR-CNRS, Université Paris Sud, Bât. 120, Campus d'Orsay, 91405 Orsay Cedex, France*
[3] *Service d'Aéronomie, UPR-CNRS, Verrières-le-Buisson, France*

Abstract. The high sensitivity submillimeter observations (180 to 1100 μm) performed with the French balloon-borne experiment PRONAOS-SPM (Sept. 96) have shown the existence of an abundant cold component within different regions of the interstellar medium, in particular within the star forming regions of Orion, ρ Ophiuchi, and M17. The very cold condensations discovered (typically $T = 10 - 15$ K, $\emptyset = 0.1 - 1$ pc in size, and $M = 5 - 30\ M_{\odot}$) could be a new type of interstellar object, at a very early phase of star formation, located between the class 0 protostellar cores (≤ 0.05 pc) and the molecular clouds traced by CO. Moreover, the grain emissivity is found to vary between the cold core and the surrounding warm halo, probably due to a different grain and/or icy mantle composition. ISO spectral observations are essential to study the energy balance and the opacity of the condensations, two key parameters for the understanding of star formation.

1. INTRODUCTION

Despite much recent progress in observations and studies of star forming regions, the early phase preceding the protostar formation is still poorly understood. In particular, there is a crucial need for constraining the initial steps of the collapse process and fragmentation mechanism. This pre-stellar phase is supposed to be associated with dense and cold cloud cores initially supported against gravity by thermal, magnetic or turbulent effects, and devoid of any embedded objects. This phase is just prior to the first very brief and isothermal collapse (designated as "class −I" in the evolutionary sequence of Boss [1]) that leads to the formation of an opaque, hydrostatic stellar object.

In these early evolutionary phases, it is now very clear that dust grains play a key role in the different processes involved in star formation. The cloud opacity can be considered as a crucial parameter because its gradients may activate instabilities within the cloud. The opacity is mainly determined by the grain absorption/emission properties: over a broad wavelength range, the opacity due to dust is several orders of magnitudes higher than the molecular and atomic contributions. In addition, with a strong involvement in the chemical processes that determine the degree of ionisation of the cloud, dust grains may play an important role in the ambipolar diffusion (*i.e.*, the process in which neutral species drift with respect to the magnetic field), which is widely accepted as the mechanism regulating star formation [2,3]. In the dense cores, most of the grains are charged and will couple to the magnetic field, leading to ambipolar diffusion which will affect the rate of formation and contraction of fragments [4]. In this scenario, smaller grains are better attached to the magnetic field and are left behind by the neutral grains. This induces a grain size stratification within the core, the larger ones being found towards the centre of the core.

The importance of dust in the different stages of star formation makes the knowledge of its physical and chemical properties decisive for constraining the dynamical evolution and interpreting the observational data (for a review see [5]). This investigation is all the more difficult because the grain properties are expected to depend both on the local stellar/protostellar environment (chemical composition, temperature, density and radiation field) and on its history. In addition, due mainly to the formation of molecular ice mantles and the coagulation of grains, the dust properties in cold and dense pre-stellar cores can be very different from those in the diffuse interstellar medium.

Dust at low equilibrium temperatures within the pre-stellar condensations emits essentially in the far-infrared (FIR) and submillimeter range. In this wavelength domain, the Earth's atmosphere offers reasonable transmission only within a few spectral windows in which the residual atmospheric emission drastically limits the ability to detect very low surface brightness gradients at angular scales of a few arcminutes. The first continuum submillimeter maps of pre-stellar cores were the ground-based observations of Ward-Thomson *et al.* [6]. Most of these observations targeted the dense and starless cores detected in molecular lines by Myers [7]. This allowed the characterisation of the structure and density distribution of the pre-stellar cores, which are less compact than the envelopes surrounding the class 0 sources, *i.e.*, the youngest observed stellar objects [8]. More recently, high resolution and wide-field mapping at 1.3 mm of the ρ Oph main core by Motte *et al.* [9] has revealed a large number of small-scale dust structures, often starless. Most of these fragments are expected to have very low temperatures (10 – 20 K), giving a dust emission peak in the range 100 to 300 μm. To determine and constrain the dust temperature and emissivity in these regions, space-borne observations with high sensitivity are needed. Due to its limited wavelength coverage, the IRAS survey did not allow a direct determination of the grain temperature in dense clouds. However, a study of the colour ratio $I(100\,\mu\mathrm{m})/I(60\,\mu\mathrm{m})$ variation provided signatures of cold dust within the densest parts of molecular clouds [10, 11]. With the DIRBE and FIRAS instruments, COBE has provided emission spectra from near-IR to mm wavelengths, but with a very low angular resolution (0.7° for λ shorter than 240 μm, 7° above). Lagache *et al.* [12] have deduced the presence of a cold emission component with a temperature around 15 K in the direction of known molecular complexes, in particular those with low star forming activity. The ISOPHOT measurements from 40 to 200 μm have allowed the detection of two cold isolated clouds in the Chameleon complex [13,14] and one globule in the Thrumbprint Nebula [15]. Assuming a ν^2 dust emissivity, a 12 – 16 K dust temperature range has been derived.

Covering the wavelength range 180 – 1100 μm with high sensitivity and a resolution from 2′ to 3.5′, the PRONAOS (Programme National d'Astronomie Submillimétrique) balloon-borne experiment is particularly suited to the detection of cold extended condensations and the determination of dust temperature and emissivity. In the following we present some of the results which have been obtained on star forming regions during the first two flights of PRONAOS (1994 and 1996), and later complemented by ISO observations.

2. PRONAOS-SPM INSTRUMENTATION AND OBSERVATIONS

PRONAOS is a stratospheric, balloon-borne, French project devoted to submillimeter astronomy. It uses a 2 metre telescope, pointed with a relative accuracy of 5″. The SPM focal instrument (see [16]) has been designed to perform simultaneous measurements in four broad filters from 180 to 1100 μm. The bolometers, cooled to 0.3 K using ^{3}He refrigerators have an intrinsic noise-equivalent power of about 10^{-15} W Hz$^{-1.5}$, close to the photon noise. The warm optics ensures the beam switching on the sky with an amplitude of 5′ and the in-flight calibration by means of two blackbodies. The accuracy of this calibration is better than 10% in absolute values, and 5% in relative values (between the bands). The main instrument characteristics and sensitivities achieved during the flights are given in Table I. PRONAOS was launched for the first time on 17th September 1994 from Fort Sumner (New Mexico) as part of a NSBF-NASA launch campaign. Unfortunately, the failure of the star sensor during daytime together with too low a ceiling altitude at night, did not permit nominal observations. Only a few were possible, allowing however the discovery of a very cold and extended condensation in Orion [17, 18].

Table I. — PRONAOS instrument characteristics.

Primary mirror diameter and aperture	2045 mm f/10			
Wobbling mirror frequency and throw	20 Hz, $\delta\alpha = 5'$			
Azimuthal pointing accuracy	$20''$ absolute, $5''$relative			
Band	1	2	3	4
$\Delta\lambda$ (μm)	180 – 240	240 – 340	340 – 540	540 – 1100
$\lambda_{\rm equiv}$ (μm)	200	260	360	570
Field of View (arcmin)	2	2	2.5	3.5
Noise Equivalent Brightness (MJy sr^{-1} Hz$^{-1/2}$)	85	50	8	5

The experiment was launched for the second time on 22$^{\rm nd}$ September 1996. The 30 hours flight was successful and allowed 50 observational sequences. One of the main objectives was to map the submillimeter emission arising from different sites of the interstellar medium (star forming regions, quiescent molecular clouds, filaments and cirrus). A mapping of Saturn during the flight enabled a measurement of the FWHM beam sizes and a check of the in-flight calibration. In the following sections, we present the results obtained from the extended mappings ($1^\circ \times 1^\circ$) made in the star forming regions of Orion M42, ρ Ophiuchi and M17.

3. RESULTS

From the measurements in the four photometric bands of PRONAOS, we have deduced emission spectra which can be fitted by a dust emission law:

$$I_\nu{}^{\rm fit} = \epsilon_\nu B_\nu(T), \tag{1}$$

where T is the averaged dust temperature and ϵ_ν is the dust emissivity. At long wavelengths, this emissivity can be assumed to be a power law $\epsilon \propto \nu^\beta$, β being the emissivity spectral index. β and T are deduced from a fit to the

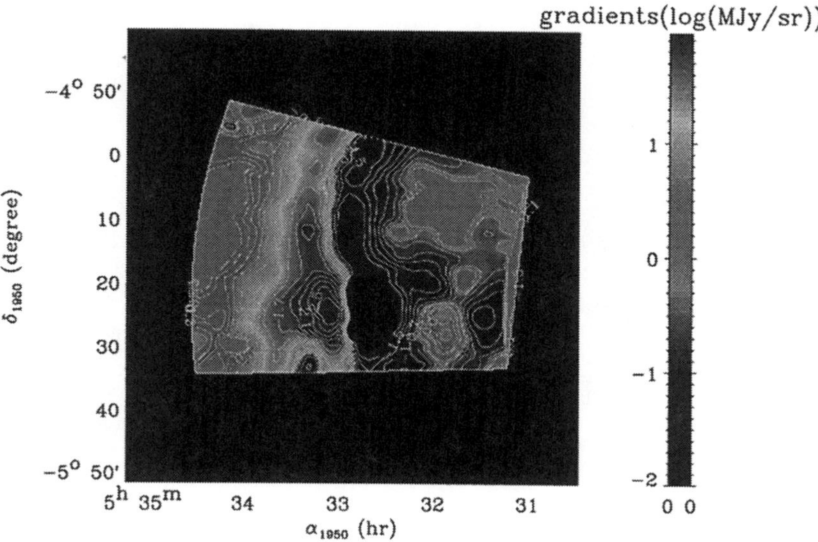

Fig. 1. — Emission measured in the first band of PRONAOS ($\lambda_{\rm equiv} = 200\,\mu$m) in the surrounding of the M42 nebula. The differential signal is due to the modulation (5.35 arcmin amplitude). The cold and extended condensation M42-NF1 is $\sim 30'$ North-West from the brightest peak (BN/KL direction).

measurements using a least squares method. The best fit is determined taking into account the 5% uncertainty in the relative values between the bands of SPM. For each band, the flux is given at an equivalent wavelength taking into account the transmission profile and the source spectrum. In order to avoid differential dilution effects between the four bands, the fits are computed on emission spectra corresponding to the fluxes integrated over the same angular beam (*i.e.*, the largest beam, approximately gaussian, with a 3.5′ FWHM).

3.1. Orion - M42

Figure 1 shows the extended mapping ($1° \times 45'$) performed in the direction of the M42 nebula during the second flight. The signal is presented as measured with the modulated beam in the first photometric band ($180 - 240\,\mu$m). Extended emission is visible over the entire map and has been detected with a sensitivity better than 40, 25, 4, 2 MJy/sr (at 3σ) for bands 1 to 4, respectively. The brightest peak corresponds to the emission surrounding the embedded young sources in the BN/KL direction. The other main sources highlighted on this map are: i) an extended emission located 15′ southeast of BN/KL, visible

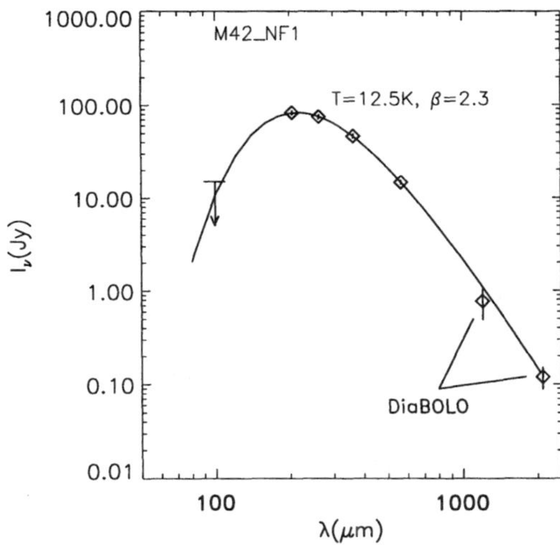

Fig. 2. — Emission spectrum and best fit obtained in the direction of the cold condensation M42-NF1. The 1.2 and 2.1 mm observations were performed with the DiaBOLO experiment at IRAM. All fluxes have been averaged over the 3.5′ beam.

on the ^{13}CO molecular map [19] and likely to be associated with a shocked gas region, ii) the long molecular bridge lying between M42 and M43 and iii) an extended condensation on the West part of the map. The Northern condensation was discovered during the first flight of PRONAOS: it shows very cold dust emission, with an average temperature of 12 K, extended with a 5′ FWHM ($\sim$ 0.7 pc [17, 18]). The mapping of this region performed during the second flight confirmed the existence of this condensation and allowed a determination of its physical parameters. The fluxes integrated over 3.5′ are plotted in Figure 2 together with the measurements at 1.2 and 2.1 mm, obtained with the ground based instrument DiaBOLO [20] located at the focus of the IRAM 30 m telescope.

The cold condensation is not visible on the extended IRAS map at 100 μm which is dominated by the warmer halo with a flux level of $I_\nu \sim 250$ MJy/sr. We have searched for an underlying cold residual component by subtracting that part of the 100 μm emission which correlates with the 60 μm emission (using the method described by Abergel *et al.* [11]). The cold condensation is still not apparent but an upper limit of $I_\nu < 15$ MJy/sr was determined.

Applying equation (1) to the measured fluxes, we deduce an optical depth $\tau_\lambda = I_\lambda / B_\lambda(T)$ in the optically thin approximation which is valid in the longest wavelength band. We obtain: $\tau(570\ \mu\text{m}) = 5.3 \times 10^{-4}$ averaged over the 3.5′ beam.

In order to deduce the corresponding column density and total mass, we have adopted the dust model developed by Ossenkopf and Henning [21] for cold and dense protostellar cores. Taking into account coagulation processes and icy mantle accretion, the model gives a mass absorption coefficient (per gramme of gas and dust): $\kappa(570\ \mu\text{m}) \sim 0.035$ cm/g. Then, we obtain in average over the 3.5′ beam: $N_{\text{H}} = \tau/(\kappa \times m_{\text{H}}) \sim 9.1 \times 10^{21}$ cm^{-2}, $n_{\text{H}} \sim 10^4$ cm^{-3}, and $M_{\text{H}} \sim 12\ M_{\odot}$. These results are linearly dependent on the dust model parameters. Other dust models give values of κ larger by a factor of 1.5 to 4 (*e.g.*, [22,23]), and therefore the values that we report here are lower limit density and mass values.

The total mass derived can be compared with the Jeans critical mass, M_{J}, [24]:

$$\frac{M_{\text{J}}}{M_{\odot}} \approx 10^{-18} R(\text{cm})\, T(\text{K}) \tag{2}$$

corresponding to an equilibrium between the thermal and gravitational forces. The cold condensation characteristics lead to $M_{\text{J}} = 9\, M_{\odot}$.

Considering that the total mass estimated previously is probably a lower limit, this comparison suggests that the condensation is gravitationally unstable, in a very early stage of fragmentation and star formation, or that it is supported by turbulence or a magnetic field.

The relatively high value obtained for the cold extended condensation M42-NF1 (Fig. 2) for the emissivity power law index ($\beta = 2.3 \pm 0.3$) contrasts with the low value deduced from the best fit for the bright region surrounding the BN/KL object ($T = 83 \pm 20$ K, $\beta = 1 \pm 0.1$, [18]). High resolution observations have shown that the BN/KL region is very complex, with a number of massive compact sources embedded in the dense molecular cloud [25]. However, these sources are concentrated within a 30″ cavity and their contribution to the sub-mm fluxes measured over the 3.5′ beam accounts for only 1/4 of the total flux. The emission then mainly arises from the dust in the diffuse surrounding halo. In addition, we have shown that the rather low value for the average spectral index cannot be explained by a dust temperature distribution, nor by radiative transfer effects, but is probably due to the properties of the dominant dust components.

3.2. ρ Ophiuchi main cloud

Figure 3 shows the emission mapped with PRONAOS in the second photometric band ($\lambda \sim 260\,\mu$m) in the ρ Oph main cloud. This region is known to harbour efficient star formation activity [26]. Molecular ^{13}CO, C^{18}O, and DCO$^+$ observations have revealed the presence of several condensations named ρ Oph A to F. The recent high resolution continuum map at 1.3 mm by Motte *et al.* [9] has shown a highly fragmented cloud, with numerous starless cores, which are thought to be pre-stellar objects. With its 2′ beam, PRONAOS cannot resolve the fragments and the detailed structure within the cores. However its high sensitivity has allowed the detection of the extended very low brightness emission gradient down to 2 MJy/sr at 600$\,\mu$m, and to determine for the

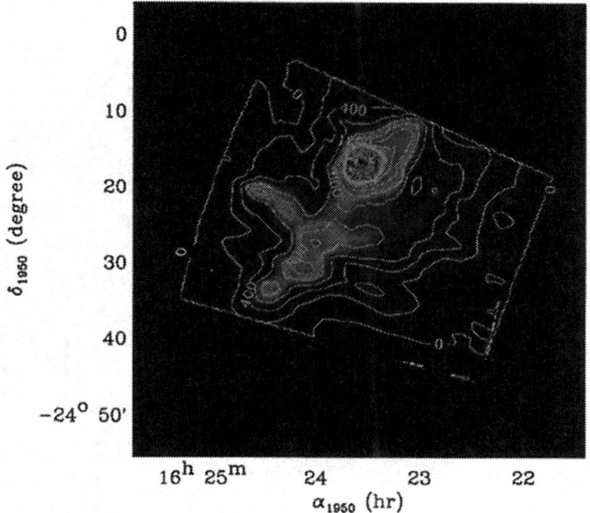

Fig. 3. — Emission mapping of the ρ Oph main cloud in the PRONAOS band 2 ($\lambda \sim 260\,\mu$m, contour step = 200 MJy/sr). The signal has been de-convolved using an inverse Fourier transform.

first time the averaged dust temperature of the condensations [27]. The obtained emission spectra of the main sources are plotted in Figure 4. To better constrain the spectrum of the cold condensation ρ Oph B2, we made additional measurements at 1.2 and 2.1 mm with DiaBOLO and in the $43 - 180\,\mu$m range with ISO-LWS. In order to subtract the halo contribution at IR-FIR wavelengths, the observations with LWS were performed toward two lines of sight: the central core and a reference position taken at $\Delta\delta = 3'$ offset (see Fig. 5). The resulting difference has been overplotted on the ρ Oph B2 spectrum. A scaling factor was used to take into account the beam dilution effect, in agreement with the best fit obtained (Fig. 4). Adopting the same dust absorption coefficient as before for M42-NF1, we deduce for ρ Oph B2, in average over the 3.5′ beam: $\tau(570\ \mu\text{m}) = 4.21 \times 10^{-3}$, $N_\text{H} = \tau/(\kappa \times m_\text{H}) \sim 7.2 \times 10^{22}$ cm^{-2}, $n_\text{H} \sim 2.2 \times 10^5$ cm^{-3}, and $M_\text{H} \sim 11\ M_\odot$.

As for M42-NF1, the total mass is greater than the critical Jeans mass: $M_\text{J} = 2.5\ M_\odot$.

Figure 6 shows the comparison between the emission measured with PRONAOS at $200\,\mu$m (contours) and the ISOCAM mapping in the $5 - 8.5\,\mu$m band [28]. The ρ Oph B2 and C condensations are associated with deep absorption features against the diffuse mid-IR background arising from the cloud envelope. For ρ Oph C, the spatial coincidence could be interpreted as a

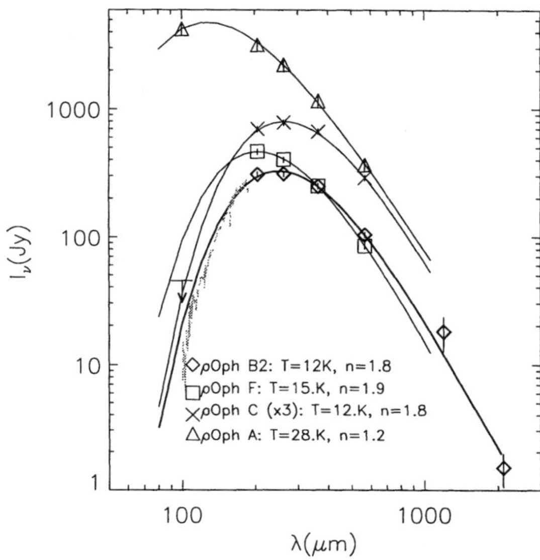

Fig. 4. — Emission spectra averaged over the 3.5′ beam in the direction of the ρ Oph A, F, C, and B2 cores. The 100 μm fluxes and upper limits are taken from IRAS. For ρ Oph B2, the 1.2 and 2.1 mm fluxes have been obtained with the ground-based experiment DiaBOLO, and the $100-180\,\mu$m spectrum is deduced from LWS gradient measurement (bold).

direct absorption due to the condensation along the line of sight of the mid-IR emission. From this comparison, we deduce $\tau_{6\,\mu\mathrm{m}}/\tau_{600\,\mu\mathrm{m}} \sim 300$ averaged over the 3.5′ beam. This value is within the range of the ratios deduced from the extinction curves for various dust models (ranging typically from 300 to 500).

We have shown [27] that the large scale sub-mm emission could be interpreted as arising from a two temperature dust distribution: a warm component at $T = 30$ K, $\beta = 1.3$, correlated with the 100 μm IRAS emission, and a cold component at $T = 12$ K, $\beta \sim 2$. The resulting column densities actually trace two different parts of the ρ Oph main cloud, leading to a cold to warm component mass ratio of 2 in this region.

3.3. M17 Nebula

Together with the M42 nebula in Orion, M17 is one of the best studied sites of high mass star formation in the Galaxy. Numerous observations have been performed in the central part of the nebula, surrounding the HII region and its interface with the bright molecular cloud [29]. A recent complete spectral

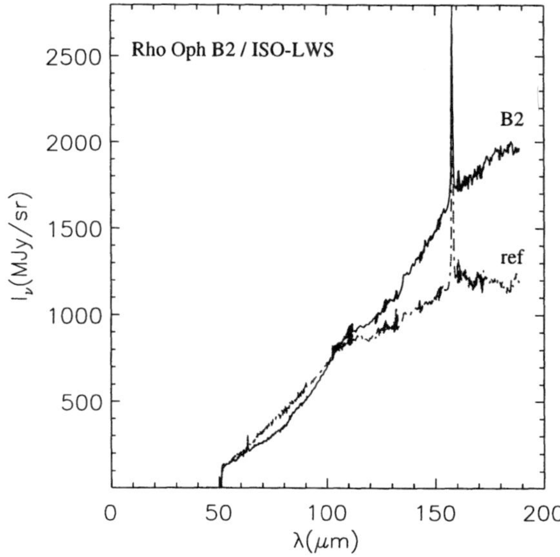

Fig. 5. — LWS emission spectra obtained toward the ρ Oph B2 core and along a reference line of sight with a $\Delta\delta = 3'$ offset.

ISO study has been done by Henning *et al.* [30] in the star forming region lying in the Northern part of the nebula. An extended map ($1° \times 1°$) surrounding M17 was obtained with PRONAOS [31]. Two cold and extended condensations have been found at about $30'$ west of the central M17 brightest IR peak. The corresponding emission spectra are plotted in Figure 7 and indicate unusually large values for the spectral index, *i.e.*, $\beta = 2.6$ and 3. To further constrain β and the temperature, observations with ISO-PHOT and ISO-LWS have been obtained, and the data processing is now in progress.

3.4. Discussion and conclusion

The recent PRONAOS and ISO-PHOT results, together with the COBE spectral large scale survey analysis [12] show that very cold dust ($T \sim 10-14$ K) is widely present in the ISM even inside highly active regions. The cold condensations observed with PRONAOS in the star forming regions of Orion, ρ Ophiuchi and M17 regions could be pre-stellar objects in very early phases, at the gravitational stability limit. The dust emissivity deduced from the sub-mm spectra for most of these objects corresponds to unusually high values for the spectral index, β, ranging $2-3$. If a unique average dust temperature is adopted to fit the observed spectra, the steep rise of the emission cannot be

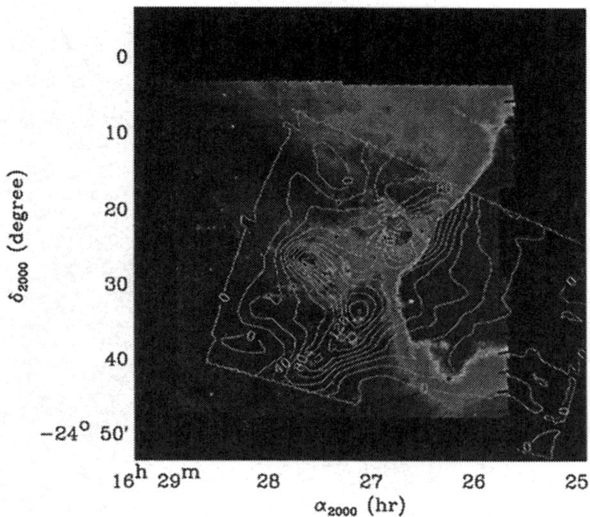

Fig. 6. — Comparison between the mid-IR emission of the ρ Oph main cloud mapped with ISOCAM (greyscale, Abergel *et al.* [28]) and the sub-mm emission at 600 μm measured with PRONAOS (contour, Ristorcelli *et al.* [18]). The cold condensations ρ Oph B2 and ρ Oph C observed in the east of the PRONAOS map are associated with deep absorption features in the diffuse mid-IR background arising from the cloud envelope.

explained by a temperature distribution, nor by radiative transfer effects. In the ρ Ophiuchi and M42 mapped clouds, these high indices contrast with the lower values obtained in the warmer regions (halo and/or embedded sources).

To characterize dust optical properties within dense and cold regions, many effects have to be considered, which can lead to a very different grain compositions compared to the diffuse medium (see reviews by Henning and Stognienko [32] and Jones *et al.* [33]). The main processes to be taken into account are molecular accretion leading to the formation of ice mantles and grain coagulation (inducing modifications in size, shape and fluffiness), as modelled for example by Ossenkopf and Henning [21]. These authors deduce a mass absorption coefficient 5 times higher than in the diffuse medium at $\lambda = 1$ mm, but without significant changes in the dust spectral index ($\beta \sim 1.8$). With complete IR coverage, ISO provided a wealth of data to improve our knowledge of the chemical composition of ice mantles which can considerably influence

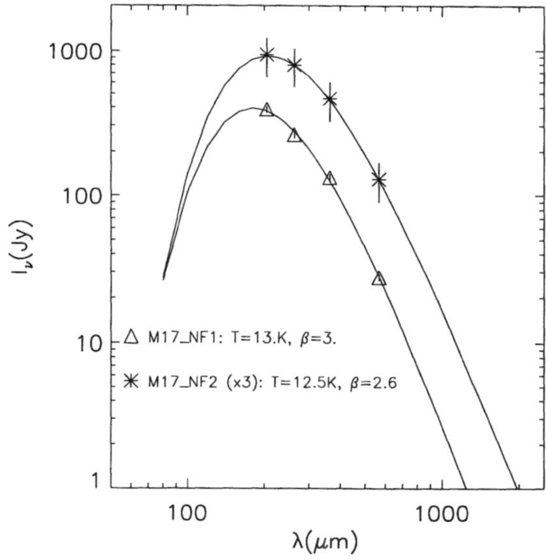

Fig. 7. — Emission spectra of the two cold extended condensations observed in the M17 nebula.

the grain opacity. The main molecular species identified on ice mantles are H_2O, CO, CH_3OH, CO_2, CH_4 OCS [34, 35] and contributions in this book). In particular the ISO results emphasise the presence of abundant CO_2 rich ice mantles [36]. Polar and apolar ices seem to co-exist on grains, arranged in layered, "onion" structures [37]. The physical structure of the grain is also a determining factor in its optical properties. Laboratory measurements on various carbon and graphite grains [38] have revealed a correlation between the FIR spectral index and the degree of crystallization of the grains. The index values range from 0.6 for amorphous carbon up to 2.75 for highly crystallized graphite. More recently, the results of Agladze *et al.* [39] and Mennella *et al.* [40] have shown a strong temperature dependence, due to multiphonon processes, on the FIR-mm absorption, both in terms of absolute values and spectral index. The measurements performed by Agladze *et al.* on amorphous silicates (forsterite and enstatite), over the temperature range 1.2 – 30 K, show an increase of β up to 2.5 at 10 K, and values in the range 1.5 – 2 at very low temperatures and above 20 K. They interpret this variation as due to two level tunnelling systems governing the absorption in glasses. The results obtained by Mennella *et al.* on different kinds of amorphous carbon grains and silicates with various compositions and degrees of structural order (crystalline and amorphous) also show a significant temperature dependance, characterised by higher emissivity indices values at low temperatures. Current progress with observations will provide important insights into the physical-chemistry of cold condensations and their evolution within star forming regions. For example, new results are expected from the IR spectro-photometric ISO observations and the sub-mm and mm observations from ground-based instruments and from PRONAOS. In addition, in the framework of future surveys and observations planned with the PLANCK and FIRST missions, the laboratory and observational characterization of the optical properties of cold matter will be crucial in properly separating the extragalactic from the galactic contributions.

ACKNOWLEDGMENTS

We are indebted to the French space agency CNES which supports the PRONAOS project, manufactured the gondola and managed the systems (including the telescope, which was developed by MMS-Toulouse). We are very grateful to the PRONAOS technical team at the CNRS (led by G. Guyot) and at the CNES (led by F. Buisson). We thank the NASA-NSBF balloon launching facilities group in Fort-Sumner (New Mexico). We also thank F. Vivares for her help in processing the ISO-LWS data.

REFERENCES

[1] Boss A.P. and York H.W., *A&A* **439** (1995) L55.

[2] Mouschovias T.C., The role of dust in the formation of stars, edited by H.U. Käufl and R. Siebenmorgen (Springer, Berlin, 1995) p. 382.

[3] Egan M.P. and Charnley S.B., The role of dust in the formation of stars, edited by H.U. Käufl and R. Siebenmorgen (Springer, Berlin, 1995) p. 373.

[4] Ciolek G.E. and Mouschovias T.C., *ApJ* **418** (1993) 774.

[5] Henning T., The role of dust in the formation of stars, edited by H.U. Käufl and R. Siebenmorgen (Springer, Berlin, 1995) p. 250.

[6] Ward-Thompson D., Scott P.F., Hills R.E. and André P., *MNRAS* **268** (1994) 276.

[7] Myers P.C., *ApJ* **270** (1983) 105.

[8] André P., Ward-Thompson D. and Barsony M., *ApJ* **406** (1993) 122.

[9] Motte F., André P. and Néri R., *A&A* **336** (1998) 150.

[10] Laureijs R.J., Clark F.O. and Prusti T., *ApJ* **372** (1991) 185.

[11] Abergel A., Boulanger F., Mizuno A. and Fukui Y., *ApJ* **423** (1994) L59.

[12] Lagache G., Abergel A., Boulanger F. and Puget J.L., *A&A* **333** (1998) 709.

[13] Laureijs R.J., Haikala L., Burgdorf M., *et al.*, *A&A* **315** (1996) L317.

[14] Ward-Thompson D., André P., Scott P.F. and Hills R.E., Low mass star formation from Infall to Outflow, IAU Symp. 182, edited by F. Malbet and A. Castets (Kluwer, Dordrecht, 1997) p. 257.
[15] Lehtinen K., Lemke D., Mattila K. and Haikala L.K., *A&A* **333** (1998) 702.
[16] Lamarre J.M., Pajot F., Torre J.-P., *et al.*, *Infrared Phys. Technol.* **35** (1994) 277.
[17] Ristorcelli I., De Luca A. and Giard M., The role of dust in the formation of stars, edited by H.U. Käufl and R. Siebenmorgen (Springer, Berlin, 1995) p. 121.
[18] Ristorcelli I., Serra G., Lamarre J.M., *et al.*, *ApJ* **496** (1998) 267.
[19] Castets A., Duvert G., Dutrey A., *et al.*, *A&A* **234** (1990) 469.
[20] Gaertner S., Benoit A., Lamarre J.-M., *et al.*, *A&AS* **126** (1997) 151.
[21] Ossenkopf V. and Henning T., *A&A* **291** (1994) 943.
[22] Preibisch T., Ossenkopf V., Yorke H.W. and Henning T., *A&A* **279** (1993) 577.
[23] Draine B.T. and Lee H.M., *ApJ* **285** (1984) 89.
[24] Larson R., *MNRAS* **145** (1969) 271.
[25] Genzel R. and Stutzki J., *ARA&A* **27** (1989) 41.
[26] Wilking B.A., Low Mass Star Formation in Southern Molecular Cloud, edited by B. Reipurth (ESO Sci. Rep. V11, Garching, 1992) p. 159.
[27] Ristorcelli I., *et al.* (1999) in preparation.
[28] Abergel A., Bernard J.P., Boulanger F., *et al.*, *A&A* **315** (1996) L329.
[29] Stutzki J., Stacey G.J., Genzel R., *et al.*, *ApJ* **332** (1988) 379.
[30] Henning T., Klein R., Launhardt R., Lemke D. and Pfau W., *A&A* **332** (1998) 1035.
[31] Giard M., *et al.* (1999) in preparation.
[32] Henning T., Michel B. and Stognienko R., *Planet. Sp. Sci.* **10-11** (1995) 1333.
[33] Jones A.P., From Stardust to Planetesimals, ASP Conference Series Vol. 122, edited by Y. Pendleton and A.G.G.M. Tielens (ASP, San Francisco, 1997) p. 97.
[34] Whittet D.C.B., *A&A* **315** (1996) L357.
[35] d'Hendecourt L., Jourdain de Muizon M., Dartois E., *et al.*, *A&A* **315** (1996) L365.
[36] Dartois E., d'Hendecourt L., Boulanger F., *et al.*, Star Formation with ISO, ASP Conf. 132, edited by J. Yun and R. Liseau (ASP, San Francisco, 1997) p. 66.
[37] Ehrenfreund P., Boogert A.C.A., Gerakines P.A., Tielens A.G.G.M. and Van Dishoeck E.F., *A&A* **328** (1997) 649.
[38] Koike C., Kimura S., Kaito C., *et al.*, *ApJ* **446** (1995) 902.
[39] Agladze N.I., Sievers A.J., Jones S.A., Burlitch J.M. and Beckwith S.V.W., *ApJ* **462** (1996) 1026.
[40] Mennella V., Brucato J.R., Colangeli L., *et al.*, *ApJ* **496** (1998) 1058.

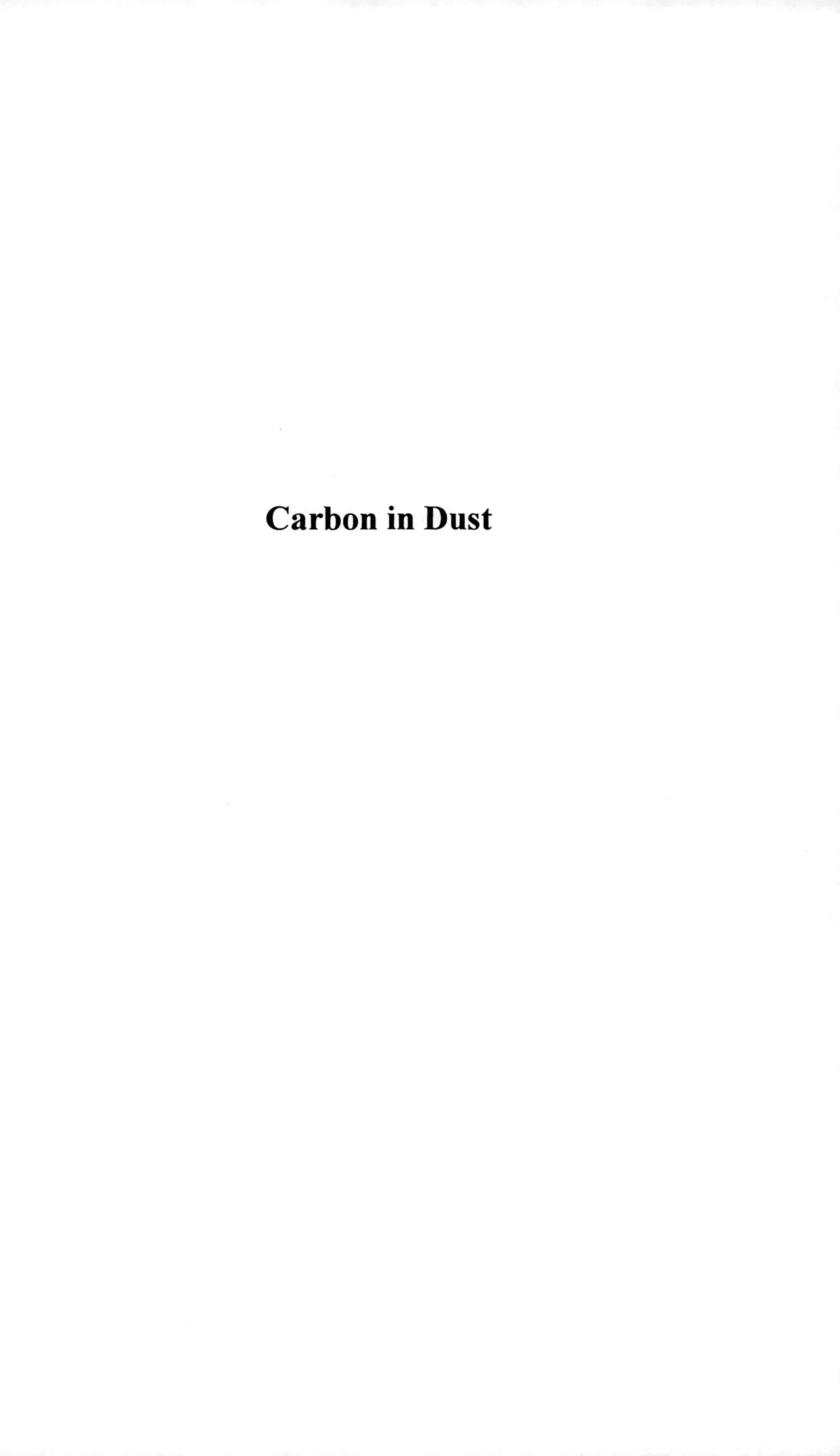

Carbon in Dust

LECTURE 5

Polycyclic Aromatic Hydrocarbons in the Interstellar Medium: A Review

F. Salama

NASA-Ames Research Center, Space Science Division, MS: 245-6, Moffett Field, CA 94035-1000, U.S.A.

Abstract. A review of the current state of knowledge regarding interstellar PAHs is presented. The interstellar distribution of PAHs is discussed. The physical and chemical properties of PAHs are reviewed according to laboratory astrophysics experiments and the astrophysical implications are discussed according to current theoretical models and observations. This survey shows the substantial progress that has been accomplished in our understanding of interstellar PAHs during the past decade.

1. INTRODUCTION

Polycyclic aromatic hydrocarbons (PAHs) have been the subject of extensive theoretical and experimental studies since the proposal, more than a decade ago, that these molecules are widespread in the interstellar medium (ISM; [1,2]). Although the "PAH proposal" is now widely accepted by the astronomical community as evidenced throughout this school, the debate is still raging regarding the form, structure, and size distribution of PAHs in various interstellar environments. An attempt is made, here, to review the current state of knowledge regarding interstellar PAHs. The interstellar distribution of PAHs is first briefly discussed on the basis of recent astronomical observations, in particular observations with the Infrared Space Observatory (ISO), which indicate that the unidentified infrared bands (UIRs), or "PAH bands" as they are often dubbed, are seen in a wide and contrasting variety of environments [3, 4]. This is followed by an overview of the theoretical modeling of the physical and chemical evolution of PAHs in the ISM. Finally, an in-depth review of the experimental information available on interstellar PAH analogs is presented and the astrophysical implications are discussed. The choice of the increased level of detail when going from observational to theoretical to laboratory studies is motivated by the fact that (i) the observational part is thoroughly discussed in other sections of these proceedings (see contributions by Boulanger and Moutou in this volume), (ii) the current generation of astrophysical models dealing with interstellar dust is limited by the lack of accurate information on the physical and chemical properties of the ISM and by the lack of quantitative laboratory data on interstellar PAH analogs, and (iii) astrophysically motivated laboratory studies of PAHs have witnessed a tremendous development during this past decade which deserves an in-depth discussion. There has been no recent extensive assessment of the laboratory work as far as the author is aware.

PAHs consist solely of fused six-membered benzenoid rings of sp^2-hybridized carbon atoms and the requisite number of hydrogen atoms attached to the periphery of the molecule (Fig. 1). They can be grouped into two large classes: pericondensed (compact) PAHs where some carbon atoms belong to three rings and catacondensed (non-compact) PAHs where no carbon atom belongs to more than two rings. Non-compact PAHs are generally expected to be less stable than compact PAHs with the same number of rings (see Sect. 4). The π electrons are delocalized over the carbon skeleton resulting in the observed high photostability of these molecules and in their survival in the harsh interstellar environments. Interstellar PAHs are thought to include neutrals and ions as well as derivatives and to follow a large size distribution ranging from small molecules ($\leq$ 25 carbon atoms) to large graphitic platelets [5, 6]. Thus, the working definition of PAHs adopted throughout this review is extensive. It includes regular PAHs (Fig. 1) as well as PAHs with an excess (or a lack) of hydrogen, PAHs where one (or more) hydrogen atom is substituted with an aliphatic side group, and PAHs with five-membered rings. PAHs are considered to form a link between the gas and the solid phase of interstellar dust and to be a key element for the coupling of stellar FUV photons with the interstellar gas (see Sect. 3).

2. ASTRONOMICAL OBSERVATIONS

Based on their spectral signature, their high photostability and the fact that they contain the cosmically abundant carbon, neutral and ionized polycyclic aromatic hydrocarbons have been invoked to explain spectral features observed in absorption and in emission in a wide range of environments in the ISM [5, 6]. These include the emission bands seen in the infrared (UIR bands) and the diffuse interstellar bands (DIBs) seen in absorption in the visible-to-near infrared range.

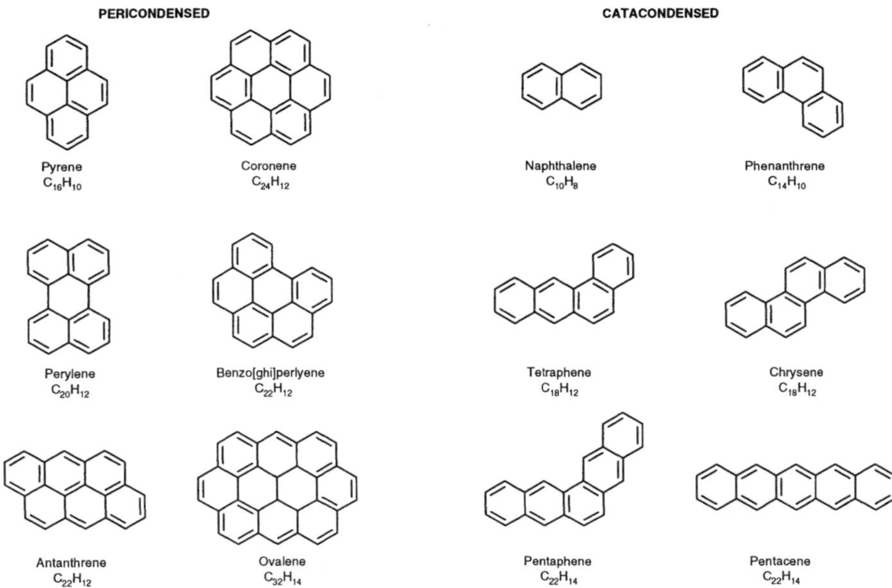

Fig. 1. — Molecular structures of some representative PAHs.

The unidentified infrared (UIR) bands are a set of infrared emission bands at 3.29, 6.2, 7.7, 8.7, 11.3, and 12.7 μm associated with a wide range of interstellar environments in our own galaxy and in other galaxies [3,5–7]. These prominent bands are often accompanied by minor, weaker, bands and underlying broad structures in the 3.1 – 3.7, 6.0 – 6.9, and 11 – 15 μm ranges (for a detailed review of the bands see [5,7]). A tentative classification of the UIR bands into two main classes (type A and type B) has been proposed to account for the pronounced variability of the bands in different environments [7]. In type A objects, which include planetary nebulae, HII regions, young stars, reflection nebulae, and the diffuse interstellar medium, the bands are strong, relatively narrow, and are associated with interstellar dust. Objects of type A are the most common. In type B objects, which include carbon-rich stars and post-AGB stars, the bands tend to be broader and appear only in circumstellar dust.

The proposed excitation mechanism of the UIRs involves the pumping of the infrared fluorescence of molecular entities by stellar UV and visible photons [5,6,8–11]. It is a one-photon mechanism leading to the transient heating of free large aromatic molecules and/or ions. The excitation mechanism implies that the emitting species must be of molecular size (*i.e.*, containing tens to a few hundred carbon atoms). The proposed identification of the UIR bands involves the molecular vibrations of polycyclic aromatic hydrocarbon (PAH) structures present either as free molecules and/or ions (for the discrete bands) or as subunits of larger carbonaceous grains (for the broad, continuum-like structures). Thus, in the case of the major features, the 3.29 μm band is associated with the aromatic C–H stretch, the 6.2 and 7.7 μm bands with C–C stretching modes, the 8.7 μm band with the aromatic C–H in-plane bend, and the 11.3 μm band with the aromatic C–H out-of-plane bend for non adjacent, peripheral hydrogen atoms. The smooth profiles of the bands indicate, however, that a *distribution* of isolated molecules and/or ions is required to fit the astronomical observations (see discussion in Sect. 4). Although no *definitive* identification of the UIR bands has yet been attained, it is now generally recognized that neutral and ionized PAHs provide the best candidates for the UIR bands (see Sect. 4).

The diffuse interstellar bands (DIBs) are absorption bands seen in the spectra of stars that are obscured by interstellar clouds. DIBs have their origin in the diffuse interstellar medium. The DIBs consist of nearly 200 bands ranging from the near ultraviolet to the near infrared with the highest density of bands found in the visible (540 – 690 nm; [12,13]). The individual bands exhibit a wide variety of strengths and widths (full width at half maximum (FWHM) values for the DIBs range from about 0.6 to 40 Å). They are too broad, however, to be associated with atomic lines. DIBs always fall at constant wavelengths, they show no profile variation and a lack of polarization among all the line-of-sights [14]. The gas phase, molecular aspect of the band carriers has been strongly supported by the observation of fine structure in some of the strong DIBs [15–17] and by the observation that part of the optical emission bands observed in the Red Rectangle (a red nebula around the star HD44179) are linked to some of the DIBs [18]. The current consensus is that the DIB carriers are most probably *large, complex, organic molecules* in the neutral and/or ionized forms (for a general overview see [19]). Large neutral PAH molecules and all ionized PAHs absorb in the visible and near IR (NIR) regions of the spectrum where most of the DIBs are found and constitute very good candidates for these bands (see discussion in Sect. 4).

3. MODELING OF INTERSTELLAR PAH PROPERTIES

Interstellar PAHs are expected to contribute to the energy balance in the interstellar medium through the absorption of stellar light. The absorbed energy is dissipated through radiative transitions (emission in the infrared) and the ejection of electrons (photoelectric effect). The photoelectrons couple with the neutral gas and lead to the heating of the interstellar gas. Models have been developed to account for the contribution of interstellar PAHs to these various processes and to account for their spectral signature in various regions of the electromagnetic spectrum, in the ultraviolet (extinction curve), in the visible and near-infrared (diffuse interstellar bands, DIBs) and in the infrared (UIR bands) ranges [5, 20–27]. The expected impact of neutral and ionized PAHs on the chemistry of interstellar gas has also been modeled [28, 29].

Models have described PAH properties as a function of their molecular size (number of carbon atoms), their structure (compact and non-compact PAHs), their ionization state (positively and negatively charged ions), and their degree of hydrogenation (PAHs with excess of hydrogen and/or partially dehydrogenated). The charge distribution of PAHs has been modeled in various interstellar environments covering a wide range of physical conditions (FUV radiation field, hydrogen density and gas temperature) under equilibrium conditions (*i.e.*, when the ionization rate is equal to the recombination rate) [5,20,21,23,24,27]. Not surprisingly, the models predict that large PAHs are mostly positively charged in high UV excitation regions such as reflection nebulae and that they are distributed between neutrals and negatively charged ions in the diffuse medium. The situation is more complicated, however, for smaller ($\leq$ 25 C-atoms) PAHs where the molecular properties (such as the electron affinity and the ionization potential) vary steeply with the structure and the size of the molecular ion making the classical approach originally developed for "grain-size" structures [30] less relevant. Thus, using the recombination rates measured in the laboratory for small PAHs [31], it is found that in the diffuse ISM small compact PAHs are distributed between neutrals and positively charged ions while non-compact PAHs can also be present in the negatively charged state [24].

The photodissociation rates of PAHs in the ISM have been extensively modeled as a function of their states of ionization and their degree of hydrogenation [25,26]. From these studies, it has been concluded that only PAHs containing more than 50 C-atoms survive in the regions of high UV field where the UIR bands are observed. In this model, the dominant channel for photodestruction is the loss of carbon through the photoejection of an acetylene (C_2H_2) molecule. Laboratory experiments on small PAHs ($\leq$ 25 C-atoms) indicate, however, that this mechanism is dominant *only* in the case of non-compact PAHs (see Sect. 4).

All astrophysical models suffer, however, from the lack of availability of precise, quantitative, information on the physical conditions prevailing in various environments in the ISM. Moreover, astrophysically relevant laboratory data on interstellar PAH analogs are scarce despite the large effort which has recently been made in this direction (see Sect. 4). Thus, the conclusions derived from these models need to be taken with precaution and must be used uniquely to *guide* further astronomical observations and laboratory experiments. The ultimate tests will come from unambiguous spectral fits between the laboratory data for relevant interstellar PAH analogs and the astronomical observations.

4. LABORATORY STUDIES

4.1. Objectives and tools

Laboratory studies are fundamental for our understanding of interstellar processes. They consist of simulating, as closely and as realistically as feasible, the conditions reigning in the ISM [32]. Among all the analytical techniques which are available for the study of interstellar matter, spectroscopy stands as the most important tool. This is because spectroscopy is the only tool which can be used to remotely probe the nature of interstellar matter. Thus it deserves the prominent position it occupies in astrophysics laboratory studies.

Since our understanding of the origin and evolution of the interstellar medium depends entirely on our ability to remotely probe the composition of interstellar matter, laboratory spectral databases of relevant materials are required for direct comparison. Schematically, the experimentalist is required to simulate the absorption of star light by a column of material lying, over interstellar distances, between the light source (star) and the observer (telescope). For example, in the case of the diffuse ISM, the intervening material is thought to be composed of free molecules and/or ions embedded in low temperature (< 100 K), very low density (25 cm^{-3}) gas cloud(s) and submitted to the average interstellar FUV radiation field of the order of 10^8 photons cm^{-2} ($G_o = 1$). Simulation of these conditions is achieved with a spectrometer system equipped with a high-vacuum, cryogenic sample chamber and an irradiation source generating energetic (10.2 eV) photons. The experimental challenge lies in simulating the interstellar PAH material itself (*i.e.*, in sample preparation) and in keeping the PAH molecules and ions isolated and unperturbed from the surrounding medium during the period of time necessary for the study of their intrinsic physical and chemical properties. This is because the interaction between the PAH (neutral or ion) and its environment induces strong perturbations of the energy levels which result in large shifts of the electronic transitions compared to the *ideal case* of a free, isolated, neutral or ionized PAH. The extent of the perturbation is directly related to the energy. Thus, this effect is particularly strong in the VUV and UV ranges.

PAHs are refractory under terrestrial conditions. Several experimental techniques are available for the study of refractory materials. They include the suspension of the material of interest in solid or liquid environments, the isolation of the material in condensed, inert-gas, matrices (Matrix Isolation Spectroscopy, MIS), the isolation of the material in the gas phase using supersonic jet expansion (JES) techniques. The degree of relevance of these techniques to the study of interstellar matter increases from the first to the last of the techniques listed. The degree of difficulty and complexity of the experiments increases also in the same direction. MIS and JES techniques are the most relevant for astrophysical simulations. In MIS experiments, PAH molecules are cold ($5 - 10$ K), fully isolated from each other and trapped in a solid cage with a minimal perturbation. Until PAHs can be routinely studied in supersonic jets, MIS remains the best tool available to simulate the conditions of the diffuse ISM. PAHs (and their ions) are minimally perturbed when isolated at low temperature (5 K) in the low-polarizability neon matrices. Neon represents the most appropriate material for a matrix because of its low polarizability. In JES experiments, PAH molecules are cold (typical translational, rotational and vibrational temperatures are of the order of 1, 10, 100 K, respectively [33]) and isolated from each other.

4.1.1. *Spectroscopy*

The discussion of the laboratory work is divided according to the wavelength range. Different types of transitions are probed within the molecular PAHs according to the wavelength (energy) of the incident photons. Vacuum ultraviolet photons ($\lambda \leq 180$ nm) provide information on the transitions which occur within the higher electronic excited states of the PAH molecules and/or ions ($\pi - \sigma^*$, $\sigma - \pi^*$, $\sigma - \sigma^*$ and Rydberg spectral transitions). Near ultraviolet, visible and near infrared photons (180 nm $< \lambda \leq$ 1200 nm) provide information on the transitions which occur within the vibronic levels of the lower electronic states of PAH molecules and/or ions (spectral transitions with a dominant $\pi - \pi^*$ character). The infrared and far infrared photons ($\lambda \geq 3.0$ μm) provide information on the vibrational modes within the fundamental (ground) state of the molecules and/or ions. The specific experimental techniques and instrumentation associated with each type of spectroscopy are described below and recapitulated in Table I.

4.1.1.1. Vacuum ultraviolet: A limited amount of information is available on the vacuum ultraviolet (VUV) spectroscopy of neutral PAHs and no information is available for their ions. Fluorescence and fluorescence excitation spectra have been measured for cold PAHs isolated in solid inert-gas matrices of argon [34] up to 75000 cm^{-1}. Absorption spectra of neutral PAHs isolated in solid organic (boron) matrices have also been reported up to 62500 cm^{-1} [35]. High temperature gas phase absorption spectra have been measured up to 143 000 cm^{-1} [36, 37]. PAH VUV absorption spectra are characterized by a single broad band peaking around 700 Å and due to the contribution of σ, π, and Rydberg transitions [36, 38]. VUV photons probe the higher excited electronic states with a dominant σ character (transitions associated with the carbon squeleton of PAHs). The transitions are intrinsically broad and are, thus, not very characteristic of a given PAH. Based on these measurements, it has been suggested, however, that PAHs may contribute to the FUV rise of the interstellar extinction curve [37].

4.1.1.2. Ultraviolet, visible, and near infrared: The situation is quite different here. A substantial amount of information is now available in this range of energy for both neutral and ionized PAHs. Absorption and fluorescence spectra have been reported for neutral and ionized PAHs isolated in inert-gas matrices of argon [34,39–41] and neon [32,42–51] and in organic (boron) matrices [35,52]. Recently, the absorption spectra of hydrogenated PAHs isolated in neon matrices have also been measured [53]. Studies of neutral PAHs have been performed in the gas phase in equilibrium at high temperature [36,69]. Limited information has also been obtained from the excitation and emission spectra of neutral PAHs measured under supersonic jet conditions [54–57]. The ionization potentials of an extensive set of PAHs have been derived from high-resolution, gas-phase, photoelectron spectroscopy (PES) measurements [58]. Photoelectron spectra also provide partial, limited, information on the gas-phase spectra of PAH ions by allowing to derive the energies of the transitions occuring from the fully occupied molecular orbitals and the partially occupied orbitals (*i.e.*, the lower energy vibronic transitions within the levels of the molecular ion). Attempts are underway to measure the spectra of PAH ions *isolated* in the gas phase (*i.e.*, in a supersonic jet expansion [59]). With the increase in sensitivity which can now be obtained with techniques such as cavity ring down laser spectroscopy [60], substantial progress is to be expected in the very near future. This is evidenced by the recent report of the first measurement of the electronic absorption spectrum of an ionized PAH, the naphthalene cation ($C_{10}H_8^+$), isolated in the gas phase using a supersonic jet coupled to a discharge and a ring down cavity [61].

The UV to NIR wavelength range is a key element for a potential identification of *specific* PAHs in the ISM because of the *unique* vibronic spectral signature of molecules and ions (as opposed to the IR transitions which are only characteristic of molecular functional side groups). The information derived from the electronic spectra of PAHs in the UV–Visible range is also essential to test the validity of the "PAH proposal" to account for the UIR bands [5, 6]. It allows a quantification of the excitation and emission mechanisms leading to the IR emission bands. Furthermore, PAHs have been proposed as potential carriers of the diffuse interstellar bands (DIBs) based on their expected abundance in the interstellar medium and their stability against UV photodissociation [62–64]. We focus, here, on the neon-matrix data because it is the most extensive set of astrophysically relevant laboratory data available (for a detailed review see [32]).

The absorption spectra of all small ($\leq$ 25 C-atoms), neutral PAHs exhibit a series of discrete transitions in the UV-NUV range (180 – 400 nm) with *no absorption in the visible-NIR range* (Fig. 2). The absorption band energies of neutral PAHs shift towards lower energies (longer wavelengths) when the molecular size increases (Figs. 2 and 4).The absorption band systems are associated with vibronic transitions between the electronic states of the molecule and can be classified, in a first approximation, into very strong (oscillator strength, f, of the order of 1), moderately strong (f of the order of 0.1) and weak (f of the order of 0.001). Note that it is important to measure the electronic spectra of PAHs (neutrals and ions) in neon matrices because neon represents the least perturbing medium relative to the gas phase. For example, in the case of naphthalene, the gas phase-to-matrix relative shift in energy is 0.25% in neon and 2.5% shift in the more polarizable argon matrix [42].

VUV irradiation of the neutral PAHs isolated in Ne matrices produces *new spectral features* in the UV-NIR range (180 – 1060 nm). All the new features are found to be associated with the PAH cation (PAH^+) formed by direct one-photon ionization of the neutral precursor (Fig. 3). The perturbation induced by the solid matrix on the electronic spectra of ions can be important compared to the case of the neutral molecules. For example, the strongest absorption of the pyrene cation $C_{16}H_{10}^+$ which falls at 439.5 nm in a Ne matrix [44] shifts to 443.5 nm in Ar [44] and 450.0 nm in an organic matrix [65]. No gas phase data is available for comparison although the shift induced by a Ne matrix is expected to be small. A gas-phase-to-Ne fractional shift of 0.3 – 0.5% is measured for $C_6F_6^+$ and its derivatives [66] and a value of 0.5% has been recently derived in the case of $C_{10}H_8^+$ [61].

In general, one can derive the absorption edge of a PAH of any size by relating the band gap energy of a cluster of fused benzenoid rings (*i.e.*, π-bonded carbon clusters) to the energy of the highest occupied molecular orbital [24] (Fig. 4).

The astrophysical implications derived for small neutral PAHs ($\leq$ 10 rings, *i.e.*, $\leq$ 25 C-atoms) are [24,32]:

- Small neutral PAHs do not absorb in the visible-NIR range and, hence, *cannot* contribute to the known DIBs (see Figs. 2 and 4).
- Neutral PAHs have strong UV absorption bands (oscillator strengths in the range 0.2 – 1.0) and could produce, if present, substructure on the UV interstellar extinction curve. In order for these species to produce substructure on the extinction curve at the 10% level would require a neutral PAH abundance in the range 1.0×10^{-7} to 2.2×10^{-7} with respect to hydrogen. A search for the stronger UV features on the interstellar extinction curve would provide a test of the PAH model and would place significant limits on the abundances of *specific* PAHs.

The astrophysical implications derived for small PAH ions ($\leq$ 10 rings, *i.e.*, $\leq$ 25 C-atoms) are [24,32]:

- Contrary to their neutral precursors, *ionized PAHs absorb in the visible and NIR and could contribute to the DIBs* (see Fig. 3).
- The strongest absorption band energy of compact PAH ions in the UV-NIR range shifts towards lower energies (longer wavelengths) when the molecular size increases.
- Large PAH ions ($\geq$ 100 rings, *i.e.*, containing more than 250 – 300 C-atoms are not expected, if present, to significantly contribute to the DIBs (see Fig. 4).
- The absorption spectrum of each PAH cation is dominated by a single band in the NUV-to-NIR region (the spectral range of interest for the DIB issue).
- In the case of non-compact PAHs (such as $C_{10}H_8^+$ and $C_{14}H_{10}^+$), the strongest absorption lies at the low energy end of the spectrum. In the case of compact PAHs (such as $C_{16}H_{10}^+$ and $C_{22}H_{12}^+$), the strongest absorption lies at the high energy end of the spectrum and its oscillator strength is of the order of 0.1.

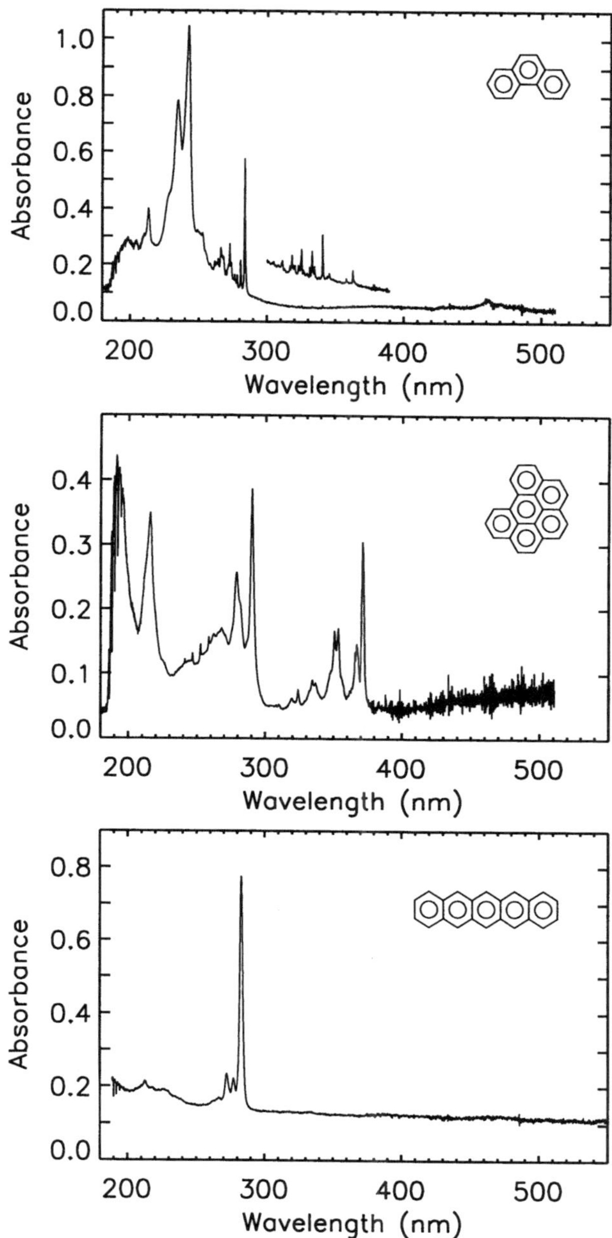

Fig. 2. — The UV-Visible absorption spectra of neutral PAHs isolated in a neon matrix at 4.2 K. (Top) phenanthrene ($C_{14}H_{10}$) [46], (middle) benzo(ghi)perylene ($C_{22}H_{12}$) [32], and (bottom) pentacene ($C_{22}H_{14}$) [47].

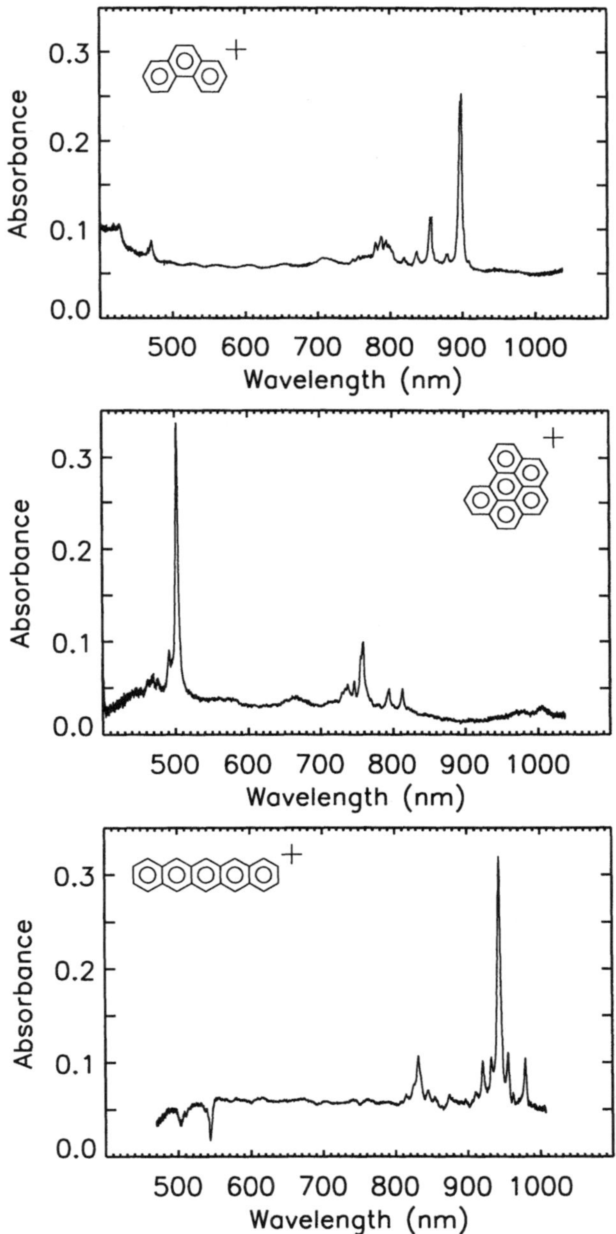

Fig. 3. — The visible-NIR absorption spectra of PAH cations isolated in a neon matrix at 4.2 K. (Top) phenanthrene ($C_{14}H_{10}^+$) [46], (middle) benzo(ghi)perylene ($C_{22}H_{12}^+$) [32], and (bottom) pentacene ($C_{22}H_{14}^+$) [47].

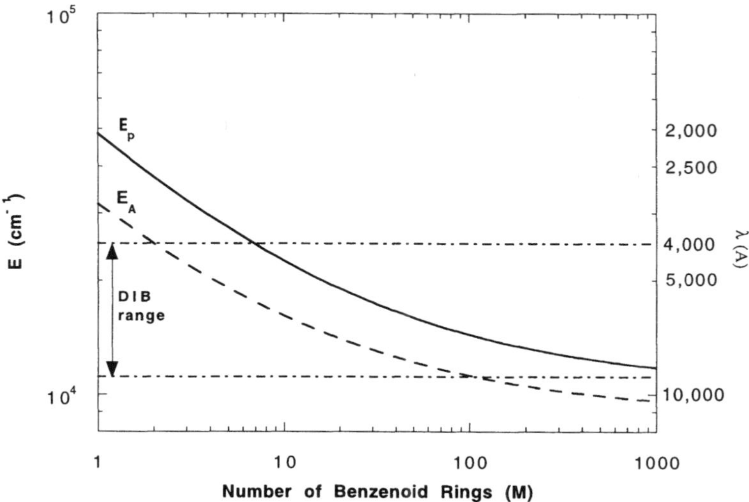

Fig. 4. — Neutral (E_P; solid line) and ionized (E_A; dashed line) compact PAH band energy as a function of the number of benzenoid rings (M) in the molecular structure. Note that E_p corresponds to the absorption edge of the neutral molecule while E_A represents the strongest absorption of the ion. The region between the two dash-dot lines encompasses the energy region where most of the known DIBs fall (4000 – 9000 Å) [24].

Comparing the stronger PAH ion transitions measured in Ne with the known DIBs, a very good agreement in wavelength positions is found between the astronomical and laboratory measurements for 8 PAHs [32, 67]. The fractional shift in energy is well within the 0.5% shift expected between Ne matrix and gas phase measurements (see above). Using the oscillator strengths measured in the laboratory, abundances of the order of 0.2 – 0.3% of the cosmic carbon are calculated for PAH ions such as naphthalene and pyrene if these ions are indeed responsible for DIB features.

4.1.1.3. Mid infrared: Mid infrared (MIR) spectroscopy of neutral and ionized PAHs encompasses the region from 3300 to 400 cm^{-1} (3.0 – 20.0 μm) where the fundamental vibrational modes of PAHs, their overtones and their combination bands are active. The IR properties of PAHs have been abundantly discussed in the literature over the past decade (see, *e.g.*, [5, 6]). Briefly, the aromatic C–H stretch mode falls near 3 μm, the aromatic C–C stretch near 6 μm, the C–H in-plane bend around 8.7 μm, and the C–H out-of plane bend in the 11 μm region. This makes the MIR range, a key region for precise comparisons with the astronomical UIR bands.

Neutral PAHs have been extensively studied in absorption in equilibrium at high temperature in the gas phase [68–72] and in the solid phase, in neon [69] and argon [40, 41, 73–82] matrices. MIS has represented the most used technique to provide the extensive set of spectra of cold, isolated, PAHs required to properly model the molecular properties expected in various IS environments. Although neon matrix data are the most astrophysically relevant, argon matrices data are a very good approximation in this case. This is because the medium-induced perturbation is weak for the vibrational energies in the fundamental (ground) state (as opposed to the case of the electronic transitions discussed above). Information is now becoming available on the emission spectra of gas-phase PAHs measured in UV laser excitation/desorption experiments [83–88]. These experimental conditions approximate the most accurately the excitation and emission mechanisms thought to occur in the ISM.

A major effort has also been devoted in the past years to the study of the MIR spectra of ionized PAHs. All the astrophysically relevant measurements have been obtained using the MIS techniques in neon [89] and argon [40, 41, 78–82, 90–93] matrices. Here too, the assumption has been made that the perturbation induced by the argon-matrix environment is small. No information is available on the IR spectra of PAH ions in the gas-phase.

The comparison between the IR spectra of neutral and ionized PAHs in the $6 - 15$ μm region indicates (Fig. 5) that (i) the spectra of the neutrals are dominated by the strong C–H out-of plane bending features in the 11 μm region with weaker features in the the aromatic C–C stretching region near 6 μm and in the C–H in-plane bending region around 8.7 μm while (ii) the IR spectra of the ions are, *at the opposite* dominated by the C–C stretch and the C–H in-plane bend. This spectral behavior had been predicted by theoretical calculations [94–96]. The comparison of the spectra in the 3 μm region (C–H stretch) is hampered by the fact that, in the matrix data, the bands associated with the ions are obscured by the bands associated with the neutral species. Theoretical calculations predict, however, that the intensity associated with this mode is moderately suppressed in the case of the ions [94–96].

The astrophysical implications derived for small neutral and ionized PAHs (≤ 20 rings, *i.e.*, ≤ 50 C-atoms) are [69, 70, 77, 88, 97, 98]:

- Both neutral and ionized PAHs produce features at the appropriate frequencies to fit the astronomical data.
- In general, neutral PAHs produce features which *do not* fit the relative strengths of the astronomical features. This conclusion has to be tempered, however, by the fact that the effects of increased molecular size and of the degree of hydrogenation of the molecules have not been measured in the laboratory. It is expected that large neutral PAHs (about 100 atoms) with a low hydrogen coverage provide a good fit to the UIR bands.
- In general, ionized PAHs produce features which fit the relative strengths of the astronomical features in the $6-15$ μm region (Fig. 6). Furthermore,

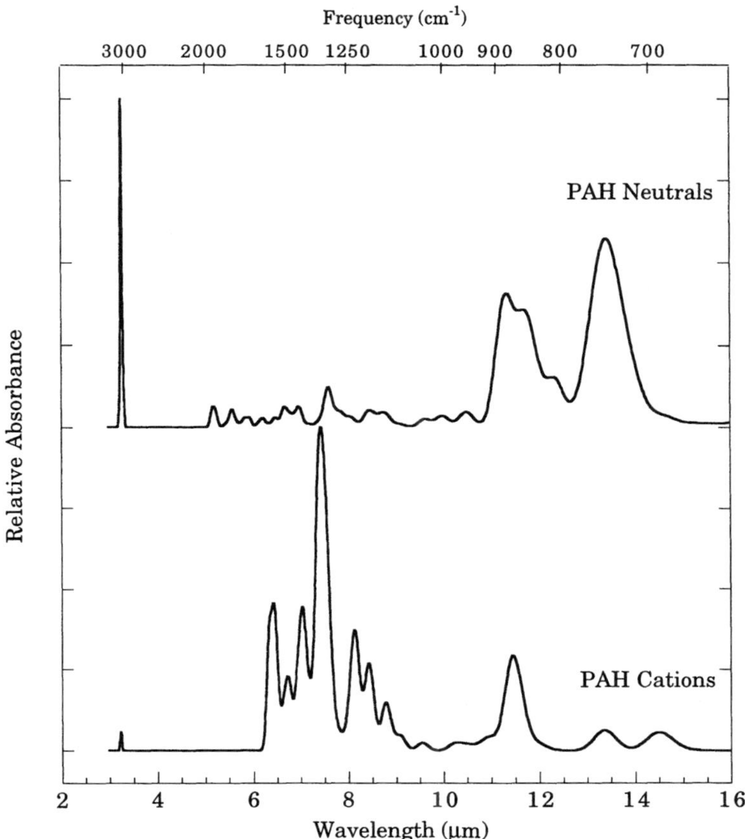

Fig. 5. — Comparison of the absorption spectrum produced by the co-addition of 6 neutral PAH spectra (top) to the spectrum produced by the same PAHs in the cationic form (bottom). The PAHs included are: anthracene ($C_{14}H_{10}$), pyrene ($C_{16}H_{10}$), tetracene ($C_{18}H_{12}$), 1,2-benzanthracene ($C_{18}H_{12}$), chrysene ($C_{18}H_{12}$), and coronene ($C_{24}H_{12}$). All PAHs are isolated in argon matrices at 9 K. This shows the extreme effect that ionization state has on relative intensities compared to its rather minor influence on vibrational frequencies (from [97]).

the size and structural (dehydrogenation) constraints noted above in the case of neutral PAHs seem to be removed for ionized PAHs. In the 3 μm region, the test of the ionized PAHs against the astronomical data awaits relevant laboratory measurements (PAH ions isolated in the gas phase).

- Composite absorption spectra calculated by combining the experimental absorption profiles measured for selected neutral and/or ionized PAHs

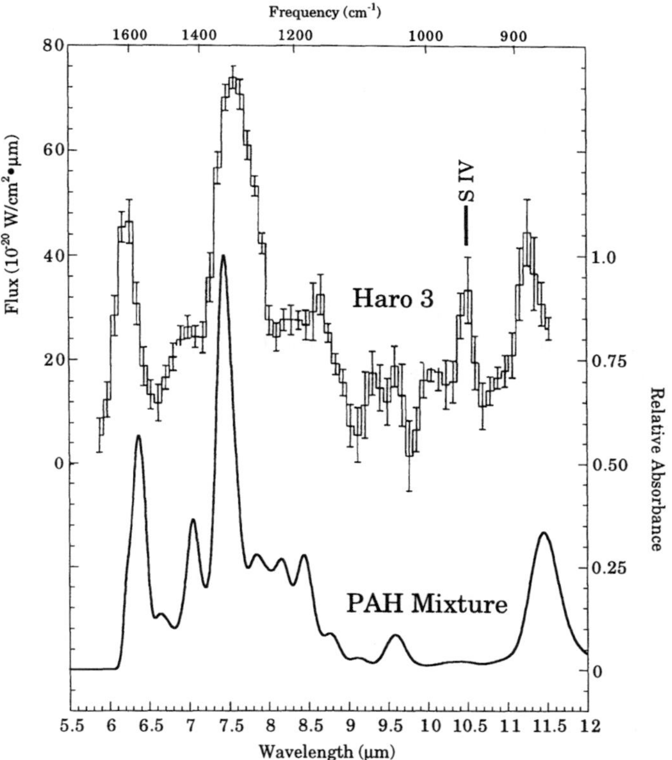

Fig. 6. — The *emission* spectrum from the compact blue dwarf galaxy Haro 3 [99] compared with the composite MIR *absorption* spectrum produced from an ionized PAH mixture. The PAH cation mixture used to produce the comparison is fully ionized and consists of: 19% dicoronylene$^+$, 14% coronene$^+$, 11% anthracene$^+$, 9% benzo(k)fluoranthene$^+$; 9% chrysene$^+$, 9% phenanthrene$^+$, 9% pyrene$^+$, 6% 9,10-dihydrobenzo(e)pyrene$^+$, 4% pentacene$^+$, 4% tetracene$^+$, 2% 1,2-benzanthracene$^+$, 2% benzo(e)pyrene$^+$, and 2% naphthalene$^+$. Parameters: 30 cm^{-1} FWHM Gaussian line profiles, Internal energy = 40 000 cm^{-1} [97].

provide a good qualitative fit to the astronomical data in the 6 – 15 μm range for objects found in a wide variety of IS environments (see Fig. 6).

4.1.1.4. Far infrared: A very limited amount of information is available on the far infrared (FIR) spectroscopy of neutral PAHs and no information is available for their ions. The FIR covers the 20 – 1000 μm range and contains the vibrations associated with the bending (out-of-plane deformation) motions of the carbon skeleton of the PAH molecules. An extensive study of the absorption spectra of neutral PAHs trapped in solid salt (CsI) pellets has been reported in the $14 - 75$ μm region [100]. Although this medium is not relevant for astrophysical studies (see above) and the individual molecules are poorly isolated, the shift of band positions, compared to the gas phase, has been estimated to be $\leq 1\%$ at this low energy [100]. The FIR absorption and emission spectra of a limited number of PAHs have also been measured in the gas phase from 2.5 to 200 μm [72]. The main conclusions of these studies are that (i) four modes, located around 16.2, 18.2, 21.2, and 23.1 μm respectively are prominent in all spectra, (ii) the FIR modes are not highly characteristic of a *specific* molecule and, (iii) the molecules can be, however, roughly classified along their molecular structure into three families (compact, non-compact, and linear) according to their FIR properties. This reflects the fact that the FIR skeletal deformations are more sensitive to molecular structure than the higher frequency, MIR, modes.

4.1.2. *Mass spectrometry*

Mass spectrometry studies have addressed the problem of interstellar PAH properties by measuring the chemical reactivity [31,101–103], photofragmentation [104–107], and formation mechanisms [108] of selected PAHs.

The dissociative recombination rates of benzene (C_6H_6), naphthalene ($C_{10}H_8$) [31] and anthracene ($C_{14}H_{10}$) [101] have been measured at room temperature with a flowing afterglow system combined to a mass spectrometer. Studies of the electron attachment to a limited number of small PAHs [109, 110] have shown that the associated rates are low thus pointing to the dominant role of PAH cations in interstellar cloud chemistry. These measurements are essential for the modeling of the ion chemistry of PAHs in interstellar clouds and, in particular, for a correct estimate of the rate for PAH ion-electron recombination. No information is available for larger PAHs and for their derivatives (branched PAHs). Recent studies [102, 103] have studied the chemical reactivity of small PAH cations in the gas phase in a tandem-flowing afterglow- selected ion flow tube (FA-SIFT). The ion-atom reaction rates of benzene ($C_6H_6^+$), naphthalene ($C_{10}H_8^+$), pyrene ($C_{16}H_{10}^+$) and some of their hydrogenated and dehydrogenated derivatives with neutral reactants (H_2, H, O, N) known to be abundant in the diffuse interstellar medium have been measured. From the product distribution, it was found that these small PAH cations react rapidly with hydrogen leading to the formation of protonated PAHs. These results indicate that protonated PAH cations may contribute substantially to the diffuse ISM PAH distribution and, hence, to the carriers of the diffuse interstellar bands (DIBs). Preliminary laboratory experiments indicate that small protonated PAHs absorb in the range of the DIBs [53]. The effects of important competing physical processes (electron recombination, photoionization and photodissociation) which counterbalance the protonation mechanisms are, however, unknown and await dedicated laboratory experiments.

The fragmentation mechanisms of small PAHs (anthracene ($C_{14}H_{10}$) and pyrene ($C_{16}H_{10}$)) by multiphoton absorption have been studied under high isolation conditions using an ion trap [104,105]. These experiments provide a collisionless environment appropriate for the study of the nature of interstellar PAH ions. Although this type of experiment does not simulate interstellar irradiation conditions (one-photon processes), it shows, however, that the dominant fragmentation channel for the compact PAH (pyrene) is dehydrogenation while the dominant fragmentation channel for the non- compact PAH (anthracene) occurs through the loss of C_2H_2. The same experiments also indicate that dehydrogenated PAHs are stable. This would favor the dominance of compact PAHs *versus* non-compact species in the ISM. The same general conclusions have been reached in a subsequent study of the dehydrogenation of a series of small PAHs (up to 24 C-atom molecules) using Fourier transform ion cyclotron resonance mass spectrometry (FT-ICRMS) [106,107]. Production of large (up to 200 C-atoms), highly dehydrogenated, PAHs by laser ablation of pyrolysed coronene ($C_{24}H_{12}$) has also been reported using the FT-ICRMS technique [108]. These large dehydrogenated oligomers, which may well sample the larger sized interstellar analogs of PAHs, have not yet been spectroscopically characterized.

5. CONCLUSION

A survey of the current state of knowledge regarding the physical and chemical properties of PAHs and their astrophysical implications has been attempted. This survey shows that substantial progress has been accomplished in this domain during the past decade. Thanks to the combined efforts which have been made on the observational, laboratory and theoretical fronts, the PAH model has been extensively tested and strengthened over the past few years. Thus, it is clear today that a distribution of neutral and/or ionized PAHs (or PAH-like molecular entities) can provide a solid interpretation for the UIR bands in terms of both the spectral fit and the excitation mechanism. It is also true that the PAH model does not provide an unambiguous identification for these bands. This cannot be expected, however, from the infrared spectra alone which are only representative of *structural* chemical groups and a firm identification must await the detection of *specific*, individual, interstellar PAH molecules and/or ions through their signature in other spectral regions. Preliminary studies of neutral PAHs in the VUV have shown that this wavelength range is not characteristic of specific PAH molecules (see Sect. 4). The UV and visible ranges, however, may hold the key for the identification of specific interstellar PAH molecules and/or ions because the electronic spectra are characteristic of individual species (Sect. 4). Thus, the study of the interstellar extinction curve (and, in particular, of the DIBs) may hold the key to the identification of specific interstellar PAHs. Until then, however, one will be left with an open debate on the nature (charge state) and the structure (size, degree of hydrogenation) of the carbonaceous materials thought to be responsible for many of the ubiquitous bands seen in the interstellar medium.

Table I. — Laboratory techniques used to study the spectroscopy of interstellar PAH analogs.

	Spectral domain			
PAH Sample[a,b]	VUV	UV-Visible-NIR	MIR	FIR
Neutral	S, MIS, G	S, MIS, G, G^*	S, MIS, G, G^*	S, G
Ion (+)	—	MIS, G^*	MIS	—
Ion (−)	—	MIS	MIS	—
Hydrogenated	—	MIS	MIS	—
Dehydrogenated	—	—	—	—

[a]Laboratory PAH samples are all ≤ 50 C-atoms.
[b]S: Solid; MIS: (Ne, Ar) Matrix-Isolated Sample; G: Gas; G^*: Jet.

In summary, as a result of the extensive laboratory effort dedicated to the study of interstellar PAH analogs (*i.e.*, to the study of PAH samples in laboratory environments which are astrophysically *realistic*), the spectral properties of neutral and ionized PAHs containing from 10 carbon atoms up to about 50 carbon atoms are now well documented over a broad wavelength range (see Tab. I). Among all the analytical techniques which have been used to study and characterize neutral and ionized PAHs, the techniques of Matrix Isolation Spectroscopy (MIS) have largely dominated (Tab. I). This is because MIS provides a unique tool to routinely probe the properties of PAHs *isolated* at low temperature (see Sect. 4). Moreover, in certain cases, MIS studies have paved the way for future experiments (larger, neutral and ionized, PAHs; PAH ions in supersonic jet expansions, *etc.*).

Among the many directions identified for future laboratory studies of interstellar PAH analogs, the four routes listed below appear to be of major importance:

- Gas phase studies of the spectroscopy of an extensive set of PAHs (neutral and ionized) isolated in supersonic free-jet expansions. Based on the information available from the MIS study of PAHs containing ≤ 25 carbon atoms, it is now possible to study these same PAHs in jets and to quantitatively measure the small perturbation induced by the solid matrix. The feasibility of this type of experiment, which was until recently in the wishful thinking domain, has recently been demonstrated by the measurement of the spectrum of $C_{10}H_8^+$ in a slit jet discharge by cavity ring down [61] and by the subsequent measurement of the spectrum of $C_{14}H_{10}^{+}$ in a supersonic jet expansion [111]. These measurements are essential because they provide data which can be used for a direct, unambiguous, comparison with astronomical spectra and to generate more accurate astrophysical models.

- Generate, isolate and spectrally characterize large, neutral and ionized, PAHs (*i.e.*, PAHs $\geq$ 50 carbon atoms and containing up to a few 100 carbon atoms). This type of study is expected to be quite challenging because larger PAHs cannot be easily generated (synthesized) in the laboratory and it will probably require the combination of various techniques such as mass spectrometry, MIS, supersonic jet expansions and cavity ring down spectroscopy. In these experiments, large PAH molecules (and radicals) can be generated in a resistively heated source or through laser ablation, mass selected in a mass spectrometer and subsequently trapped in an inert-gas matrix or expanded in a supersonic jet coupled to a cavity ring down to measure their spectra. In an alternative scenario, the large PAH molecules and/or ions can be first expanded in a supersonic jet and subsequently deposited in an inert gas matrix to measure their spectra. The feasibility of this type of studies for large particles has recently been demonstrated in the case of carbon nanoparticles [112]. These measurements are needed to quantitatively measure the impact of size variation on the molecular properties of PAHs and, hence, to achieve a more accurate description of the size distribution of interstellar PAHs in astrophysical models. This line of research is also of fundamental interest because it may provide information on intermediate size particles and help close the gap between the domains of molecular and solid state physics.
- The study of the VUV and FIR spectroscopy of neutral and ionized PAHs (preferentially in astrophysically relevant media (gas phase or neon matrices)). No information is available at all on PAH ions in these wavelength ranges (see Tab. I). Although the information derived from the VUV and FIR spectra is not expected to be characteristic of specific, individual, molecular species (see Sect. 4), these studies are important for the purpose of deriving accurate, quantitative, energy budgets for interstellar PAHs.
- The multiwavelength study of PAH derivatives (*i.e.*, hydrogenated and dehydrogenated PAHs). Although some limited, preliminary, information is becoming available on the spectroscopy [53] and chemical reactivity [102, 103] of hydrogenated PAHs, no information exists regarding the spectroscopy of dehydrogenated PAHs. In this latter case, a combination of mass spectrometry (see Sect. 4) with MIS and/or jet spectroscopy should provide an appropriate tool for the study of these species in the laboratory. Dehydrogenated PAHs have recently been generated and mass detected using Fourier transform ion cyclotron resonance mass spectrometry [106–108].

ACKNOWLEDGMENTS

The author wishes to thank Lou Allamandola and Doug Hudgins for numerous discussions regarding the infrared properties of PAHs and for providing Figures 5 and 6 prior to publication. Part of the research described in this article is supported by NASA, Office of Space Science (Astrophysics program).

REFERENCES

[1] Allamandola L.J., Tielens A.G.G.M. and Barker J.R., *ApJ* **290** (1985) L25.
[2] Léger A. and Puget J.L., *A&A* **137** (1984) L5.
[3] First ISO Results, *A&A* **315** (1996) L27.
[4] Onaka T. and the IRTS team, Diffuse Infrared Radiation and the IRTS, edited by H. Okuda, T. Matsumoto and T.L. Roellig (ASP, San Francisco, 1997) p. 170.
[5] Allamandola L.J., Tielens A.G.G.M. and Barker J.R., *ApJS* **71** (1989) 733.
[6] Puget J.L. and Léger A., *ARA&A* **27** (1989) 161.
[7] Geballe T., From Stardust to Planetesimals, ASP Conf. 122, edited by Y.J. Pendleton and A.G.G.M. Tielens (ASP, San Francisco, 1997) p. 303.
[8] Buss R.H. Jr., Tielens A.G.G.M. and Snow T.P., *ApJ* **372** (1993) 281.
[9] Sellgren K., Werner M.W. and Allamandola L.J., *ApJS* **102** (1996) 369.
[10] Justtanont K., Barlow M.J., Skinner C.J., *et al.*, *A&A* **309** (1996) 612.
[11] Uchida K.I., Sellgren K. and Werner M., *ApJ* **493** (1998) L109.
[12] Herbig G.H., *ARA&A* **33** (1995) 19.
[13] Jenniskens P. and Désert X., *A&AS* **106** (1994) 39.
[14] Martin P.G., The Diffuse Interstellar Bands, edited by A.G.G.M. Tielens and T.P. Snow (Kluwer, Dordrecht, 1995) p. 263.
[15] Jenniskens P., Porceddu I., Benvenuti P. and Désert X., *A&A* **313** (1996) 649.
[16] Ehrenfreund P. and Foing B.H., *A&A* **307** (1996) L25.
[17] Krelowski J. and Schmidt M., *ApJ* **477** (1997) 209.
[18] Sarre P.J., Miles J.R. and Scarrott S.M., *Science* **269** (1995) 674.
[19] Tielens A.G.G.M. and Snow T.P., The Diffuse Interstellar Bands, edited by A.G.G.M. Tielens and T.P. Snow (Kluwer, Dordrecht, 1995) p. 1.
[20] Omont A., *A&A* **164** (1986) 159.
[21] d'Hendecourt L.B. and Léger A., *A&A* **180** (1987) L9.
[22] Schutte W., Tielens A.G.G.M. and Allamandola L.J., *ApJ* **415** (1993) 397.
[23] Bakes E.L. and Tielens A.G.G.M., *ApJ* **427** (1994) 822.
[24] Salama F., Bakes E., Allamandola L.J. and Tielens A.G.G.M., *ApJ* **458** (1996) 621.
[25] Allain T., Leach S. and Sedlmayr E., *A&A* **305** (1996) 602.

[26] Allain T., Leach S. and Sedlmayr E., *A&A* **305** (1996) 616.
[27] Dartois E. and d'Hendecourt L., *A&A* **323** (1997) 534.
[28] Lepp S., Dalgarno A. and van Dishoeck E.F., *ApJ* **329** (1988) 418.
[29] Millar T.J., *MNRAS* **259** (1992) 35.
[30] Draine B.T. and Sutin B., *ApJ* **320** (1987) 803.
[31] Abouelaziz H., Gomet J.C., Pasquerault D. and Rowe B.R., *J. Chem. Phys.* **99** (1993) 237.
[32] Salama F., Low Temperature Molecular Spectroscopy, NATO/ASI Series, Series C: Mathematical and Physical Sciences, edited by R. Fausto (Kluwer, Dordrecht, 1996) p. 169.
[33] Jost R., Low Temperature Molecular Spectroscopy, NATO/ASI Series, Series C: Mathematical and Physical Sciences, edited by R. Fausto (Kluwer, Dordrecht, 1996) p. 249.
[34] Gudipati M.S., Daverkausen J. and Hohlneicher G., *Chem. Phys.* **173** (1993) 143.
[35] Robinson M.S., Beegle L.W. and Wdowiak T.J., *ApJ* **474** (1997) 474.
[36] Joblin C., PhD thesis (Université Paris 7, Paris, 1992) p. 139.
[37] Joblin C., Léger A. and Martin P., *ApJ* **393** (1992) L79.
[38] Leach S., *Planet. Space Sci.* **43** (1995) 1153.
[39] Andrews L., Kelsall B.J. and Blankenship T.A., *J. Phys. Chem.* **86** (1982) 2916.
[40] Szczepanski J., Roser D., Personette W., *et al.*, *J. Phys. Chem.* **96** (1992) 7876.
[41] Szczepanski J., Drawdy J., Wehlburg C. and Vala M., *Chem. Phys. Lett.* **245** (1995) 539.
[42] Salama F. and Allamandola L.J., *J. Chem. Phys.* **94** (1991) 6964.
[43] Salama F. and Allamandola L.J., *ApJ* **395** (1992) 301.
[44] Salama F. and Allamandola L.J., *Nature* **358** (1992) 42.
[45] Salama F. and Allamandola L.J., *J. Chem. Soc. Faraday Trans.* **89** (1993) 2277.
[46] Salama F., Joblin C. and Allamandola L.J., *J. Chem. Phys.* **101** (1994) 10252.
[47] Salama F., *Orig. Life Evol. Bio.* **28** (1998) 349.
[48] Ehrenfreund P., d'Hendecourt L., Verstraete L., *et al.*, *A&A* **259** (1992) 257.
[49] Ehrenfreund P., Foing B.H., d'Hendecourt L., Jenniskens P. and Désert X., *A&A* **299** (1995) 213.
[50] Léger A., d'Hendecourt L.B. and Défourneau D., *A&A* **293** (1995) L53.
[51] Joblin C., Salama F. and Allamandola L., *J. Chem. Phys.* **102** (1995) 9743.
[52] Shida T., Electronic Absorption Spectra of Radical Ions, edited by T. Shida (Elsevier, Amsterdam, 1988) p. 1.
[53] Halasinski T.M., Salama F. and Allamandola L.J., *J. Chem. Phys.* (1998) submitted.
[54] Yu H., Joslin E. and Phillips D., *Chem. Phys. Lett.* **203** (1993) 515.

[55] Schwartz S.A. and Topp M.R., *Chem. Phys.* **86** (1984) 245.
[56] Moreels G., Clairemidi J., Hermine P., Bréchignac P. and Rousselot P., *A&A* **282** (1994) 643.
[57] Babbitt R.J., Ho C.-J. and Topp M.R., *J. Phys. Chem.* **92** (1988) 2422.
[58] Schmidt W., *J. Chem. Phys.* **66** (1977) 828.
[59] Moutou C., Verstraete L., Bréchignac P., Piccirillo S. and Léger A., *A&A* **319** (1997) 331.
[60] O'Keefe A., Scherer J.J., Cooksy A.L., *et al.*, *Chem. Phys. Lett.* **172** (1990) 214.
[61] Romanini D., Biennier L., Salama F., *et al.*, *Chem. Phys. Lett.* (1999) in press.
[62] Van der Zwet G.P. and Allamandola L.J., *A&A* **146** (1985) 76.
[63] Léger A. and d'Hendecourt L.B., *A&A Lett.* **146** (1985) 81.
[64] Crawford M.K., Tielens A.G.G.M. and Allamandola L.J., *ApJ* **293** (1985) L45.
[65] Shida T. and Iwata S., *J. Am. Chem. Soc.* **95** (1973) 3473.
[66] Bondybey V.E. and Miller T.A., Molecular Ions: Spectroscopy, Structure and Chemistry, edited by T.A. Miller and V.E. Bondybey (North Holland Publishing Company, Dordrecht, 1983) p. 138.
[67] Salama F., Joblin C. and Allamandola L.J., *Planet. Space Sci.* **43** (1995) 1165.
[68] Kurtz J., *A&A* **255** (1992) L1.
[69] Joblin C., d'Hendecourt L., Léger A. and Défourneau D., *A&A* **281** (1994) 923.
[70] Joblin C., Boissel P., Léger A., d'Hendecourt L. and Défourneau D., *A&A* **299** (1995) 835.
[71] Robinson M.S., Beegle L.W. and Wdowiak T.J., *Planet. Space Sci.* **43** (1995) 1293.
[72] Zhang K., Guo B., Colarusso P. and Bernath P.F., *Science* **274** (1996) 582.
[73] Bauschliker C.W., Langhoff S.R., Sandford S.A. and Hudgins D.M., *J. Phys. Chem.* **101** (1997) 2414.
[74] Hudgins D.M. and Sandford S.A., *J. Phys. Chem.* **102** (1998) 329.
[75] Hudgins D.M. and Sandford S.A., *J. Phys. Chem.* **102** (1998) 353.
[76] Hudgins D.M. and Sandford S.A., *J. Phys. Chem.* **102** (1998) 329.
[77] Bernstein M.P., Sandford S.A. and Allamandola L.J., *ApJ* **472** (1996) L127.
[78] Szczepanski J. and Vala M., *ApJ* **414** (1993) 646.
[79] Szczepanski J., Chapo C. and Vala M., *Chem. Phys. Lett.* **205** (1993) 434.
[80] Szczepanski J., Vala M., Talbi D., Parisel O. and Ellinger Y.J., *J. Chem. Phys.* **98** (1993) 4494.
[81] Szczepanski J., Wehlburg C. and Vala M., *Chem. Phys. Lett.* **232** (1995) 221.
[82] Vala M., Szczepanski J., Pauzat F., Parisel O., Talbi D. and Ellinger Y.J., *J. Phys. Chem.* **98** (1994) 9187.

[83] Shan J., Suto M. and Lee L.C., *ApJ* **383** (1991) 459.
[84] Cherchneff I. and Barker J.R., *ApJ* **341** (1989) L21.
[85] Brenner J.D. and Barker J.R., *ApJ* **388** (1992) L39.
[86] Schlemmer S., Cook D.J., Harrison J.A., *et al.*, *Science* **265** (1994) 1686.
[87] Cook D.J., Schlemmer S., Balucani N., *et al.*, *Nature* **380** (1996) 227.
[88] Cook D.J. and Saykally R.J., *ApJ* **493** (1998) 793.
[89] d'Hendecourt L.B. and Léger A., The First Symposium on the Infrared Cirrus and Diffuse Interstellar Clouds, ASP Conf. 58, edited by R.M. Cutri and W.B. Latter (ASP, San Francisco, 1994) p. 303.
[90] Hudgins D.M., Sandford S.A. and Allamandola L.J., *J. Phys. Chem.* **98** (1994) 4243.
[91] Hudgins D.M. and Allamandola L.J., *J. Phys. Chem.* **99** (1995) 3033.
[92] Hudgins D.M. and Allamandola L.J., *J. Phys. Chem.* **99** (1995) 8978.
[93] Hudgins D.M. and Allamandola L.J., *J. Phys. Chem.* **101** (1997) 3472.
[94] Pauzat F., Talbi D., Miller M.D., deFrees D.J. and Ellinger Y., *J. Phys. Chem.* **96** (1992) 7892.
[95] deFrees D.J., Miller M.D., Talbi D., Pauzat F. and Ellinger Y., *ApJ* **408** (1993) 530.
[96] Langhoff S.R., *J. Chem. Phys.* **100** (1996) 2819.
[97] Allamandola L.J., Hudgins D.M. and Sandford S.A., *ApJ* (1998) in press.
[98] Hudgins D.M., Allamandola L.J. and Sandford S.A., *Adv. Space Res.* **19** (1997) 999.
[99] Metcalfe L., Steel S.J., Barr P., *et al.*, *A&A* **315** (1996) L105.
[100] Moutou C., Léger A. and d'Hendecourt L.B., *A&A* **310** (1996) 297.
[101] Canosa A., Laube S., Rebrion C., *et al.*, *Chem. Phys. Lett.* **245** (1995) 407.
[102] Le Page V., Keheyan Y., Bierbaum V. and Snow T.P., *J. Am. Chem. Soc.* **119** (1997) 8373.
[103] Snow T.P., Le Page V., Keheyan Y. and Bierbaum V., *Nature* **391** (1998) 259.
[104] Boissel P., Lefèvre G. and Thiébot Ph., Physical Chemistry of Molecules and Grains in Space, edited by I. Nenner (American Institute of Physics, New York, 1994) p. 667.
[105] Boissel P., de Parseval P., Marty P. and Lefèvre G., *J. Chem. Phys.* **106** (1997) 4973.
[106] Ekern S.P., Marshall A.G., Szczepanski J. and Vala M., *ApJ* **488** (1997) L39.
[107] Ekern S.P., Marshall A.G., Szczepanski J. and Vala M., *J. Phys. Chem.* **102** (1998) 3498.
[108] Joblin C., Masselon C., Boissel P., *et al.*, *Rapid. Commun. Mass Spectrom.* **11** (1997) 1619.
[109] Tobita S., Meinke M., Illenberg E., *et al.*, *Chem. Phys.* **161** (1992) 501.
[110] Canosa A., Parent D.C., Pasquerault D., *et al.*, *Chem. Phys. Lett.* **228** (1994) 26.
[111] Bréchignac Ph. and Pino Th., *A&A* (1999) in press.
[112] Schnaiter M., Mutschke H., Dorschner J., Henning Th. and Salama F., *ApJ* **498** (1998) 486.

LECTURE 6

PAHs in Reflection Nebulae, Fullerenes in the ISM

C. Moutou[1], K. Sellgren[2], A. Léger[3], L. Verstraete[3] and P. Le Coupanec[4]

[1] *Observatoire de Haute Provence, CNRS, 04870 Saint-Michel-l'Observatoire, France*
[2] *Astronomy Department, Ohio State University, Columbus, OH 43210, U.S.A.*
[3] *Institut d'Astrophysique Spatiale, CNRS, 91400 Orsay, France*
[4] *DESPA, Observatoire de Paris, 92195 Meudon, France*

Abstract. The problem of the identification of the mid-IR emission band carriers is reconsidered in the light of the recent data obtained by the Infrared Space Observatory (ISO). These data further constrain the interpretation and provide new insights into our knowledge of the nature of large carbon-bearing molecules and their role in interstellar environments. Some of the recent advances in this field are reported, including: the behaviour of the IR emission bands in reflection nebulae, the detection of new aromatic features and their possible assignment, and the plausible presence of fullerenes (in particular C_{60} and its cation) in interstellar clouds, an identification which is still a matter of debate. The discussion will be mostly based on the high resolution spectrum of the bright reflection nebula NGC 7023. We will also briefly discuss the behaviour of the diffuse interstellar bands in reflection nebulae, since these bands are also related to the presence of large molecules in the interstellar medium.

1. INTRODUCTION

The identification of the carriers of the mid-IR emission bands at 3.3, 6.2, 7.7, 8.6, and 11.3 μm has been an active subject for more than 10 years, especially because the amount of energy emitted by these features can be as high as 30% of the total energy absorbed and re-emited by dust. The PAH model attributes the mid-IR emission bands to polycyclic aromatic hydrocarbons (PAHs) transiently heated by the absorption of single UV photons, this mechanism being implied by the interpretation of the observations of Sellgren [1]. However, the identification of individual interstellar PAHs has still to be achieved. A review on the present status of the PAH model is given in this book by Salama. In this paper, we present some ISO observations of the mid-IR emission bands and interpret them within the framework of the PAH model, using comparisons with laboratory data obtained on PAH compounds.

2. PAHs IN REFLECTION NEBULAE

Reflection nebulae (RN) are particularly interesting objects to study interstellar matter for the following reasons: (i) the matter can be viewed as "typical" interstellar material, (ii) the geometry of the nebulae are simple, because the exciting source is usually a single star. This type of object is thus well suited for assessing the nature of the variations in the observed mid-IR spectra (features, continuum, line ratios, ...) as a function of the incident flux.

2.1. The reflection nebula NGC 7023

2.1.1. *Observations*

The bright reflection nebula NGC 7023 was observed at a position 27″ W and 34″ N of the illuminating star HD200775 ($T_* = 17000$ K) with the Short Wavelength Spectrometer (SWS) on board ISO. The beam size of the spectrometer which is typically $14'' \times 20''$ includes the brightest part of the photodissociation region (PDR). The full ISO-SWS spectrum of this object is described by Sellgren *et al.* [2] (Fig. 1). The achieved spectral resolution is typically 1800 and has been rebinned at a resolution 500 in Figure 1. Two independent scans were acquired separately with increasing and decreasing wavelengths respectively. The comparison of both scans helps in determining real features from spurious noise features.

2.1.2. *General view of the spectrum*

Several pure rotational H_2 lines are observed in the spectrum of NGC 7023. An excitation temperature of 400 ± 90 K can be derived for the warm molecular gas. Models with gas densities of $\sim 10^6$ cm^{-3} and an ultraviolet field 10^4 times the interstellar value provide the best match to the H_2 observations.

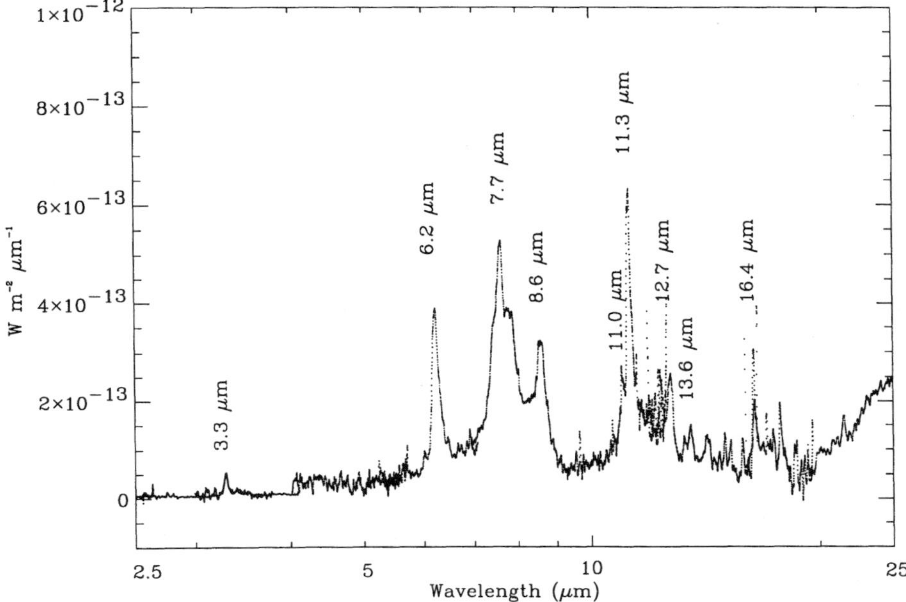

Fig. 1. — SWS spectrum of the reflection nebula NGC 7023 in the wavelength range 2.5 – 25 μm.

In the 2.5 – 45 μm wavelength range, 60% of the energy is emitted in the bands and 40% in the continuum. The continuum is detected over the entire mid-IR range (2.5 – 20 μm) and strongly increases longwards of 20 μm due to the increasing contribution of cool large grains in thermal equilibrium with the radiation field. Désert *et al.* [3] have modelled the infrared emission spectrum of the diffuse interstellar medium (ISM) including a continuous size distribution of grains composed of PAHs, very small grains, and big grains. The comparison of the spectrum of the diffuse ISM with that of NGC 7023 shows a significant dip in the continuum emission of NGC 7023 between 10 and 20 μm. This suggests that the population of "very small grains" is depleted in this object.

Several remarks can be made concerning the emission bands; these are possible because of the high resolution, good quality and the extensive spectral coverage of the obtained NGC 7023 spectrum:

- No substructures are observed in the profiles of the main mid-IR bands at 3.3, 6.2, 8.6, 11.3 and 12.7 μm. The smoothness of the band profiles may be induced by temperature effects [4] and/or by overlapping transitions from a large number of molecular species.

- A significant asymmetry in the 6.2 and 11.3 μm bands is observed which might be explained by anharmonicity effects.
- The structure of the 7.7 μm feature seems very sensitive to the radiation field, particularly the ratio of the two main components at 7.6 and 7.8 μm (*cf.* Boulanger this volume). A third component at 7.45 μm is detected for the first time in the spectrum of NGC 7023.

2.1.3. Beyond 12 μm

Due to atmospheric absorption, the spectral range above 12 μm was poorly known before ISO, making difficult the identification of additionnal spectral features related to the carriers of the Unidentified Infrared Emission Bands (UIBs) between 3 and 12 μm. The only observational investigations were at low spectral resolution and on bright and compact sources [5–7]. ISO has performed the spectroscopic exploration of this domain on a large number of objects.

The lack of astronomical data in the spectral range above 12 μm was accompanied by a lack of spectral data from laboratory measurements. Recently, a more extensive study was performed in the laboratory in order to prepare for the interpretation of ISO data. Moutou *et al.* [8] showed that, for a sample of 40 aromatic species, bands at 16.2, 18.2, 21.3 and 23.2 μm are present in the spectra of many molecules. This relative concentration of bands means that they are associated with modes which do not depend too much on the molecular shape. Thus, a search for such bands in astronomical spectra appeared warranted.

The spectrum displayed in Figure 1 contains new features, located at 13.3, 13.6, 15.8 and 16.4 μm. The bands at 13.3 and 13.6 μm could be due to the C-H vibrations of aromatic cycles containing at least four neighboring hydrogens, as observed in the spectra of PAHs [8]. We do not discuss further these bands but focus in the following on the features at 15.8 and 16.4 μm (*cf.* Fig. 2). These bands are present in the two independent scans performed by the SWS spectrometer at a level higher than 10 times the standard deviation over this spectral range. Their profiles are slightly asymmetrical. Their widths (approximately 0.15 μm) are comparable to that of the 11.3 μm feature, which suggests a similar size and excitation mechanism for the carriers of these bands.

In a sample of 40 neutral PAHs Moutou *et al.* [8] showed that many vibrational modes are observed in the range 600 – 640 cm^{-1} (15.6 – 16.6 μm). Twenty of the studied molecules have modes in this frequency range, with varying absorption cross-sections and central wavelengths. These modes seem to be associated with a specific vibrational motion of the skeleton of polycylic aromatic molecules.

It is interesting to note that the band near 16 μm is absent in some well-known compact molecules, namely coronene ($C_{24}H_{12}$), pyrene ($C_{16}H_{10}$), peropyrene ($C_{26}H_{14}$), and perylene ($C_{20}H_{12}$). It is strong in ovalene ($C_{32}H_{14}$), fluoranthene ($C_{16}H_{10}$), decacyclene ($C_{36}H_{18}$), and circumbiphenyl ($C_{38}H_{16}$)

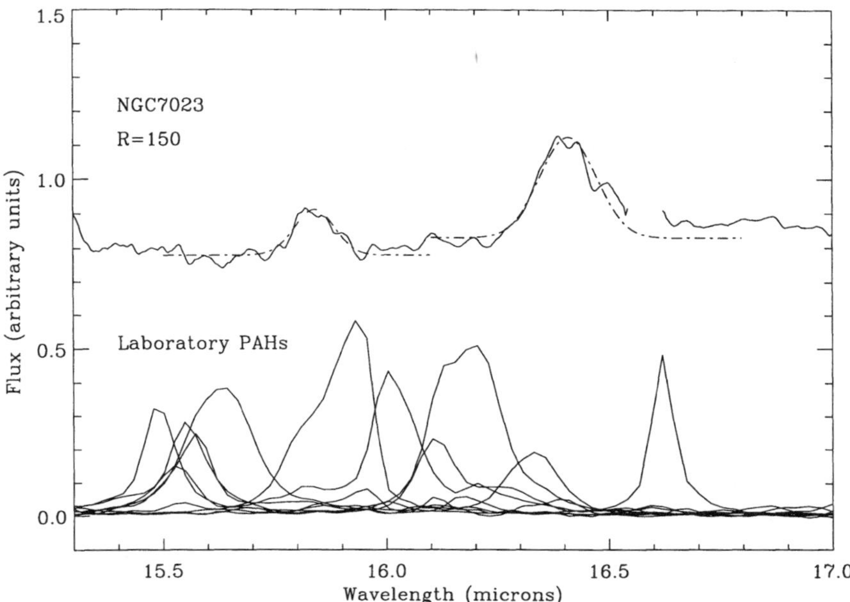

Fig. 2. — Comparison of the laboratory absorption spectra of 9 PAH molecules, isolated in CsI pellets, with the observed spectrum of NGC 7023 which has been rebinned to a resolution of 150. The lack of data at 16.6 μm in the NGC 7023 spectrum corresponds to an instrumental artefact. Two Gaussian curves have been superimposed on the observed spectrum.

and is present (although weak) in hexabenzocoronene ($C_{42}H_{18}$) and many other more "exotic" species. This suggests that the band near 16 μm (15.8 – 16.4 μm) is associated with compact molecules having more than 30 carbon atoms, and also with molecules containing pentagonal rings. More work has to be done before any firmer identifications can be made.

The absorption spectra of some molecules studied in the laboratory are given in Figure 2. The spectra are taken from Moutou *et al.* [8]. In most of these cases the 15.8 – 16.4 μm band is the strongest mode observed in the low-frequency range. Finally, we emphasize that the laboratory spectra which have been measured in absorption and in the solid phase, are compared to the emission spectra of interstellar species in gas phase at high temperature. Differences in the band profiles and positions are therefore not unexpected [4].

Additional modes between 700 and 500 cm^{-1} (14.3 – 20 μm) are observed in some molecules. These features have no counterpart in the spectrum of NGC 7023. However, the achieved signal-to-noise ratio is quite low in this object up to 20 μm. It would be interesting to search for such features in other astronomical sources for which the mid-IR flux is higher. Since these bands are molecule-specific, their detection may enable the identification of individual PAHs.

2.2. PAHs in cool reflection nebulae

Uchida *et al.* [9] have recently reported the first IR spectrum of a reflection nebula illuminated by a cool star. The spectrum measured with ISOCAM using the circular variable filter (CVF) is displayed in Figure 3. VdB133 surrounds a binary star whose two components are at effective temperatures of 6500 and 12500 K. Uchida *et al.* have demonstrated that only 21% of the stellar flux is emitted in the UV ($\lambda < 400$ nm), and that the radiation is predominantly visible light. The measured IR spectrum is surprisingly similar to the spectrum observed in the diffuse medium [10] or in the reflection nebula NGC 7023 (this paper and [11]), where UV photons dominate. The bandwidths and band ratios in the two spectra are comparable in first approximation. The similarities in the IR spectra could imply that the band carriers are able to absorb efficiently over a broad spectral range including the visible. The consequences of this for the excitation mechanism and the nature of the emitters have yet to be investigated. This constraint is not incompatible with the PAH hypothesis, and favours PAH ions and radicals, for which the first electronic absorption band lies in the visible [12, 13]. Another possibility is that the average size of emitting PAHs is larger around cooler stars, so that their first electronic absorption band is displaced to longer wavelengths.

2.3. Diffuse bands in reflection nebulae

The diffuse interstellar bands (DIBs) are absorption features in the visible and near-IR range which are most likely due to large molecules present in the interstellar medium (see Herbig [14] for a review). Their intensity is roughly proportional to the column density of interstellar matter along the line of sight. In particular, the species responsible for the IR emission bands may also carry some of the DIBs (*cf.* contribution by Salama this volume). PAH derivatives (ions or radicals) are good candidates for carrying some of the DIBs [12, 13]. Rouan *et al.* [15] have proposed that an efficient "rocket effect" (due to H atom or H_2 molecule ejection) can lead to the suprathermal rotation of PAH molecules and therefore to a broadening of the DIBs in the Red Rectangle. In a typical reflection nebula, for a 40 carbon atom molecule, the model predicts that the rotational temperature is around 100 K and thus a $\approx 40\%$ broadening is expected with respect to the diffuse medium. To test this prediction, we observed the narrow DIB at 5797 Å towards a sample of reflection nebulae, and compared its profile with the "canonical" band of the standard diffuse medium [16]. Some results are shown in Figure 4. The profile has been corrected for the Doppler component of the clouds present on the line of sight. It has been obtained by deconvolution of the velocity profile observed in the interstellar

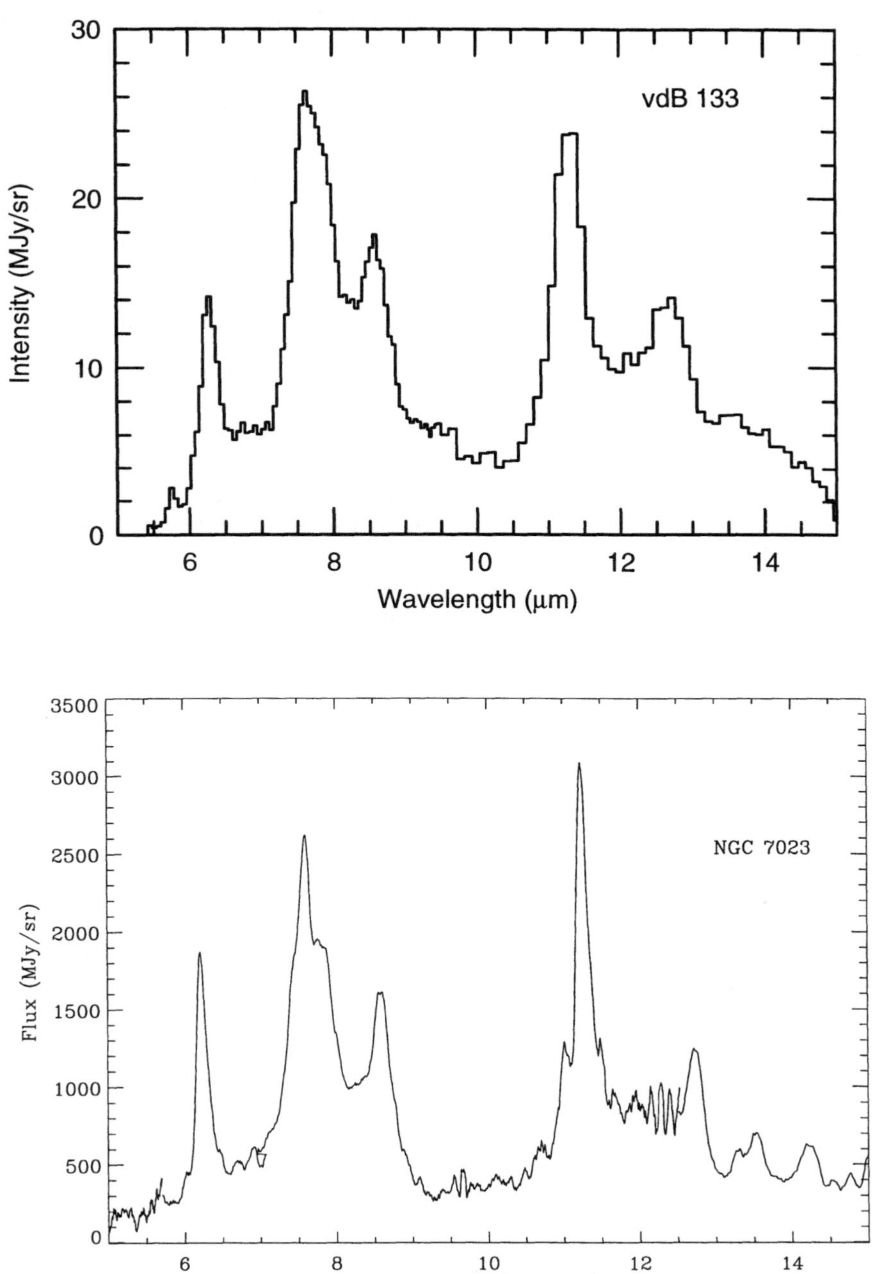

Fig. 3. — The ISOCAM spectrum of VdB 133, a reflection nebula illuminated mostly by a cool star (6500 K). For a direct comparison, the spectrum of NGC 7023 (illuminated by a 17000 K star) is also shown at a similar resolution. VdB 133 spectrum courtesy of Uchida *et al.* [9].

sodium lines. The residual broadening observed for the stars HD 34078 and HD 216200 is therefore real and may correspond to the rotational broadening predicted by the model of Rouan *et al.* [15]. A more detailed analysis of these observations is given by Le Coupanec *et al.* [16].

Other observations of DIBs towards RN [17, 18] show in some cases a weakening of the band strength compared to the reddening (which characterizes the column density of interstellar matter), especially when the incident flux is strong, probably leading to the destruction of the carriers. For instance, very few DIBs are detected in NGC 7023, which is illuminated by a Be star. For the case of a PAH molecule carrier, this could imply that the molecule is either photodissociated due to the ejection of a small fragment, such as H, H_2, C_2H_2 *etc.*, or is photoionised.

The comparison between the IR emission band spectrum and the diffuse band spectrum is certainly of interest in constraining the physical nature of the interstellar species. However, we should not expect a strong correlation between the DIB and the IR band intensities, even if the bands imply similar carriers. DIBs are electronic transitions, and they are much more specific to individual molecules than the vibrational modes given by the IR emission spectrum. It is, however, interesting to look for correlations such as, for example, a correlation between the 12.7/11.3 μm band ratio (which is assumed to quantify the hydrogen coverage of PAHs) with the DIB widths. This comparison could give evidence for the hydrogen ejection mechanism in strong UV fields.

3. FULLERENES IN THE ISM

3.1. Previous searches

Kroto [19] proposed that the fullerene species (C_{60} and related species) play an important role in interstellar physics and chemistry. This argument is based on the stability of the "buckyball" structure of fullerene compounds, as observed in the laboratory and confirmed by ab-initio calculations. The formation mechanisms invoked to produce such spherical carbonaceous structures are usually based on reactions in circumstellar shells, involving precursors such as acetylene and corannulene [20–22]. The photoerosion of hydrogenated amorphous carbon grains could also lead to the formation of fullerenes, together with PAHs [23]. Fullerene or fullerane species have been searched for at various wavelengths in interstellar spectra. They have also been proposed to account for many unidentified interstellar features: the extinction bump at 2175 Å, some of the diffuse interstellar bands, the infrared emission features, and the extended red emission [24–27].

The C_{60} absorption spectrum shows an absorption band in the ultraviolet (UV) wavelength range, at 3860 Å [28]. Snow and Seab [29] and Somerville

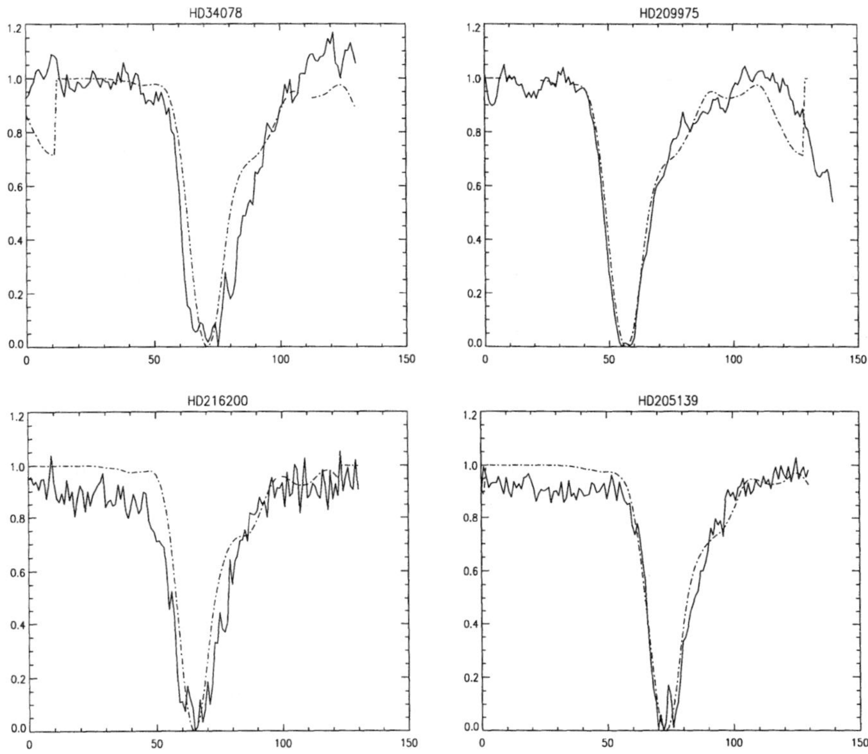

Fig. 4. — The profile of the 5797 Å DIB along a few lines of sight. At right, interstellar "standards" with negligible broadening; at left, reflection nebulae where the band is broadened by 40%, as predicted by the "rocket effect" model [16] for these objects. The x-axis is in pixel units (one pixel = 0.03 Å), and the intensity is normalised. The Doppler profile was deconvolved from the observation of the sodium lines along the same line of sight. The dotted lines represent the standard interstellar star HD 30614 for comparison.

and Bellis [30] have searched for this feature without any clear evidence of its presence. They derived an upper limit for the abundance of C_{60} of the order of 0.01% of the cosmic carbon abundance in the diffuse medium, and of approximately 0.7% of the carbon abundance in Mira stars.

More recently, Clayton *et al.* [31] have searched for the IR emission features of C_{60} at 8.6 μm in the spectra of carbon stars. No feature was seen in the spectrum of RCB stars (carbon stars in which C_{60} is expected to be very abundant because of the low H abundance) at a level of 2% of the continuum. Only the spectrum of the evolved star IRC+10216 possibly shows a 8.6 μm band that may be assigned to C_{60}.

The search for fullerene and fullerane species in the interstellar spectra is today limited by the lack of laboratory data. Only a few species have been studied on specific wavelength ranges. Progresses in this field are expected in the coming years.

The ionization state of C_{60} in the interstellar medium has been modeled [32–34]. The authors showed that C_{60} would be mostly neutral and anionic in the diffuse medium, and cationic in more irradiated clouds such as reflection nebulae associated to B stars.

Laboratory work by Petrie *et al.* [35] has shown that C_{60}^+ is the most stable species among 60 atom fullerenes, because of its "exceptional unreactivity" with interstellar gas.

The electronic absorption spectrum of C_{60}^+ shows two bands at 9583 and 9645 Å (10435 and 10368 cm^{-1}) in a neon matrix [36]. The corresponding absorption bands have been searched for in the interstellar medium, taking into account the wavelength shift induced by the matrix polarisability. The two interstellar bands detected at 9577 and 9632 Å (10442 and 10382 cm^{-1}) by Foing and Ehrenfreund [37,38] have been tentatively assigned to C_{60}^+. The most reliable estimates of the oscillator strength ($f = 0.003 - 0.006$ [36]) lead to a total abundance of 0.6 to 1.2% of the cosmic carbon in this species [39].

More recent observations of the 9577 and 9632 Å DIBs [40] lead to a lower value of the measured intensity, due to a different estimate of the telluric contribution, and thus lead to a lower abundance range for C_{60}^+ of 0.25 to 0.52% of the cosmic carbon.

3.2. IR signature in NGC 7023

C_{60}^+ species would re-emit any energy absorbed from a neighboring star through its IR modes. Although four modes are predicted theoretically, only two of them have been measured in the laboratory at 7.11 and 7.51 μm [36]. Assuming that the two other modes are identical to those of C_{60} at 17.3 and 18.9 μm, Moutou *et al.* [39] found that approximately 35% of the emitted power emerges at 7.11 μm, and amother 25% is emitted at 7.51 μm. Less than 40% of the total energy is emitted in the low-frequency modes.

A higher signal-to-noise ratio (up to 20) was obtained in the reduced wavelength range 7.0 – 8.7 μm of the spectrum of NGC 7023. The spectrum has been rebinned to a resolution $R = \lambda/\Delta\lambda = 1150$ (Fig. 5).

We expect any features of gas phase C_{60}^+ at high temperature to be broadened and shifted to longer wavelength by up to 20 cm^{-1} (up to 0.1 μm) compared to the neon matrix measurement [4]. The IR bands of C_{60}^+ are not detected in NGC 7023. An upper limit for the abundance $N_{C_{60}^+}/N_H$ was derived to be 8.3×10^{-7}. If we use the total carbon abundance measured by Cardelli *et al.* [41] in the local interstellar medium (2.6×10^{-4} relative to hydrogen), we find an upper limit of 0.32% of carbon contained in C_{60}^+. This value is slightly lower than the abundance deduced from the near-IR observations [38]. This discrepancy can be due either to an overestimate of the oscillator strength of

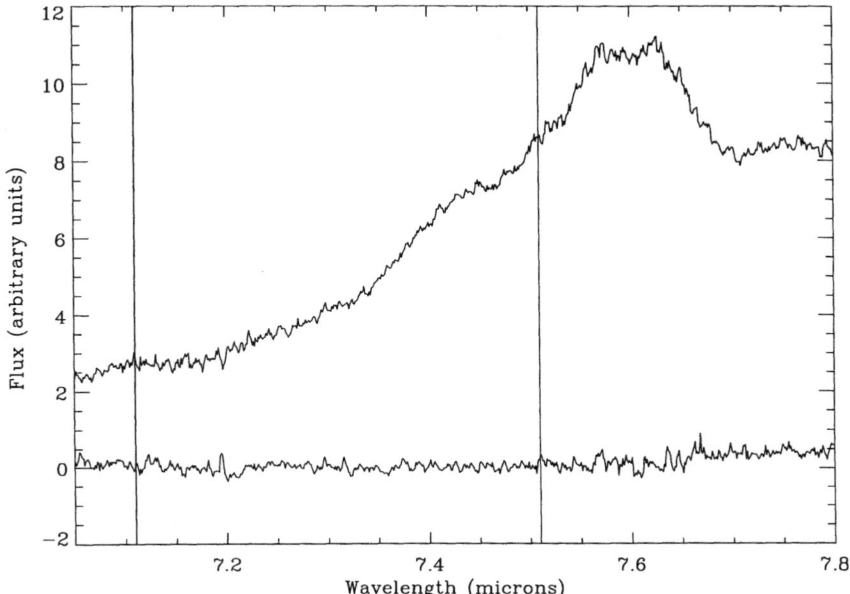

Fig. 5. — Spectrum of the reflection nebula NGC 7023 in the wavelength range 7.0 – 7.8 μm. The expected locations for the C_{60}^+ emission bands [36] are marked by vertical lines. The spectrum is the mean value of up and down scans, rebinned at $R = \lambda/\Delta\lambda = 1200$. The half difference between the up and down scans is also plotted to provide an estimate of the noise (lower curve).

the 9583 and 9645 Å bands, or to the destruction of C_{60}^+ in the environment of NGC 7023 (for example by a double ionisation process). The measurement of the 9577 and 9632 Å bands towards the star HD 200775, which illuminates the reflection nebula NGC 7023, would certainly help to answer this question. As mentioned above (Sect. 2.3), the DIBs are not strong relative to the reddening towards NGC 7023, and we may expect that the 9577 and 9632 Å DIBs would suffer the same weakeness. This would be consistent with our IR observations.

4. CONCLUSIONS

The conclusions obtained from the analysis of PAH and fullerene signatures in reflection nebulae are summarised below:

- The energy distribution in NGC 7023 can be compared to the average spectrum of the interstellar medium. The continuum is detected throughout the IR range and appears slightly weaker between 6 and 20 μm compared to the spectrum of the diffuse medium modeled by Désert *et al.* [3]. This difference could be explained by a relative depletion of very small grains, which are the intermediate population between molecules and large grains.

- The emission band profiles are characterised by their smoothness, the rarity of prominent substructures, their asymmetry (especially for the 6.2 and 11.3 μm bands). These characteristics could be explained by temperature effects. A new band is reported at 7.45 μm, in the blue wing of the 7.7 μm band.
- New bands at 15.8 and 16.4 μm are observed in the spectrum of NGC 7023. Their characteristics are close to those of the aromatic features. These bands could be interpreted as low-frequency modes of the carbonaceous skeleton of interstellar PAHs. From a comparison with a laboratory data base of aromatic low-frequency modes, we found that the most likely carriers are large and compact PAHs, perhaps containing pentagonal rings.
- The observation of the 5797 Å DIB in reflection nebulae shows evidence of a broadening and a weakening of this absorption band. The weakening may be an indication that the species are at a higher ionisation levels when UV-illuminated. The broadening might be evidence for the rocket effect, which predicts a high rotational temperature for PAHs in UV-irradiated regions.
- A cool star illuminating a reflection nebula can produce a similar IR spectrum than that measured in the diffuse medium or in nebulae surrounding hotter stars. This observation, provided by ISO, leads to the conclusion that the IR band carriers absorb efficiently in the visible range, and not exclusively in the UV. This is the case for PAH cations or radicals.
- An upper limit of the C_{60}^+ abundance of 0.32% of the cosmic carbon was derived in NGC 7023. This result is not compatible with the attribution of the diffuse medium 9577 and 9632 Å bands to C_{60}^+ by Foing and Ehrenfreund [37]. Further investigations of the near-IR oscillator strengths would be useful, especially in the gas phase. It is also possible that the C_{60}^+ abundance differs between the diffuse medium (where the DIBs are observed) and the reflection nebulae (where the vibrational modes of the cation have been searched for), due for example to a variation of the ionisation state.

ACKNOWLEDGEMENTS

We are grateful to ESA member states for funding the ISO mission and instruments. Thanks to K. Uchida for providing the ISOCAM spectrum of VdB133. We gratefully acknowledge NATO Collaborative Research Grant 951347.

REFERENCES

[1] Sellgren K., *ApJ* **277** (1984) 623.
[2] Sellgren K., Moutou C., Léger A., *et al.* (1998) in preparation.
[3] Désert F.X., Boulanger F. and Puget J.L., *A&A* **237** (1990) 215.
[4] Joblin C., Boissel P., Léger A., d'Hendecourt L. and Défourneau D., *A&A* **299** (1995) 835.
[5] McCarthy J.F., Forrest W.J. and Houck J.R., *ApJ* **224** (1978) 109.
[6] Justtanont K., Barlow M.J., Skinner C.J., *et al.*, *A&A* **309** (1996) 612.
[7] Cox P., *A&A* **236** (1990) L29.
[8] Moutou C., Léger A. and d'Hendecourt L., *A&A* **310** (1996) 297.
[9] Uchida K., Sellgren K. and Werner M.W., *ApJ* **493** (1998) L109.
[10] Boulanger F., Reach W.T., Abergel A., *et al.*, *A&A* **315** (1996) L325.
[11] Cesarsky D., Lequeux J., Abergel A., *et al.*, *A&A* **315** (1996) L305.
[12] Léger A. and d'Hendecourt L., *A&A* **146** (1985) 81.
[13] Allamandola L.J., Tielens A.G.G.M. and Barker J.R., *ApJ* **290** (1985) L25.
[14] Herbig G.H., *ARA&A* **33** (1995) 19.
[15] Rouan D., Léger A. and Le Coupanec P., *A&A* **324** (1997) 661.
[16] Le Coupanec P., Rouan D., Léger A. and Moutou C., *A&A* **338** (1998) 217.
[17] Josafatsson K. and Snow T.P., *ApJ* **319** (1987) 436.
[18] Jenniskens P., Ehrenfreund P. and Foing B., *A&A* **281** (1994) 517.
[19] Kroto H.W., Heath J.R., O'Brien S.C., Curl R.F. and Smalley R.E., *Nature* **318** (1985) 162.
[20] Goeres A. and Seldmayr E., *A&A* **265** (1992) 216.
[21] Kroto H.W and Jura M., *A&A* **263** (1992) 275.
[22] Bettens R.P.A. and Herbst E., *ApJ* **478** (1997) 585.
[23] Scott A., Duley W.W. and Pinho G.P., *ApJ* **489** (1997) L193.
[24] Webster A., *MNRAS* **263** (1993) L55.
[25] Webster A., *MNRAS* **264** (1993) 121.
[26] Webster A., *MNRAS* **264** (1993) L1.
[27] Webster A., *MNRAS* **282** (1996) 1372.
[28] Heath J.R., Curl R.F. and Smalley R.E., *J. Chem. Phys.* **87** (1987) 4236.
[29] Snow T.P. and Seab C.G., *A&A* **213** (1989) 291.
[30] Somerville W.B. and Bellis J.G., *MNRAS* **240** (1989) 41.
[31] Clayton G.C., Kelly D.M., Lacy J.H., *et al.*, *AJ* **109** (1995) 2096.
[32] Bakes E.L.O. and Tielens A.G.G.M., The Diffuse Interstellar Bands, edited by A.G.G.M. Tielens and T.P. Snow (Kluwer, Dordrecht, 1995) p. 315.
[33] Salama F., Bakes E.L.O, Allamandola L.J. and Tielens A.G.G.M., *ApJ* **458** (1996) 621.
[34] Dartois E. and d'Hendecourt L., *A&A* **323** (1997) 534.
[35] Petrie S., Javahery G. and Bohme D.K., *A&A* **271** (1993) 662.
[36] Fulara J., Jakobi M. and Maier J.P., *Chem. Phys. Lett.* **211** (1993) 227.

[37] Foing B. and Ehrenfreund P., *Nature* **369** (1994) 296.
[38] Foing B. and Ehrenfreund P., *A&A* **317** (1997) L59.
[39] Moutou C., Léger A., d'Hendecourt L. and Maier J., *A&A* **311** (1996) 968.
[40] Jenniskens P., Mulas G., Porceddu I. and Benvenuti P., *A&A* **327** (1997) 337.
[41] Cardelli J.A., Meyer D.M., Jura M. and Savage B.D., *ApJ* **467** (1996) 334.

LECTURE 7

Present Situation of the Coal Model for Interstellar Carbon Dust

O. Guillois, G. Ledoux, I. Nenner, R. Papoular and C. Reynaud

Service des Photons, Atomes et Molécules,
CEA-Saclay, Bât. 522, 91191 Gif-sur-Yvette Cedex, France

Abstract. Examining the available ISO spectra, especially those recorded at high resolution with SWS, it is clear that the width and the shape of the UIR bands are consistent with solid-state emission and are very similar to those observed in a number of solid-state hydrogenated carbonaceous materials, including natural coals. Here we summarise the spectroscopic properties of coals, from the UV to far-IR, and show that this family of materials offers satisfactory spectroscopic agreement with the astronomical observations. We describe how the evolution of carbon dust in different environments within the interstellar medium (ISM) can be modelled by the time and temperature evolution of coal. The problem of the excitation mechanism for the infrared emission of small solid grains in the different environments of the ISM is discussed.

1. INTRODUCTION

The Unidentified Infrared (UIR) bands observed in emission at 3.3, 6.2, 7.7, 8.6 and 11.3 μm are attributed to vibrational bands due to CH and CC aromatic bonds [1] but the exact nature of their carriers is still a matter of debate.

Thanks to the ISO mission, there is at present a large amount of observational data on the UIR bands [2]. They are smooth and intrinsically wide, even when observed with the high spectral resolution mode of the short wavelength spectrometer (SWS). In general, the profiles and the central wavelengths are uniform from source to source. They are nearly insensitive to the local radiation field intensity over a range of several orders of magnitude (see contribution by Boulanger this volume) and can be observed in regions where exciting radiation wavelength are in the UV or in the visible range [3]. To date a relatively small dispersion of widths, peak wavelengths and relative intensities of the features have been observed in PhotoDissociation Regions (PDRs), but these spectral variations are limited in scope. Thus the emitting material has a fairly high degree of uniformity regardless of physical conditions of the sources.

There are still no clearly identified observations of specific gas-phase polycyclic aromatic hydrocarbons (PAHs) at IR, visible and UV wavelengths, which casts some doubts on the interpretation of the UIR bands in term of a collection of isolated PAHs molecules as proposed [4,5] and reviewed in the present volume by Salama. A spectral fit to the data has not been obtained up to now [6]. By contrast, solid carbonaceous materials fit the data quite well and grains made of hydrogenated amorphous carbon (HAC) [1], quenched carbonaceous compounds (QCC) [7] or naturals coals [8] have been proposed as model to fulfil this criteria [9].

In this paper, we summarise the interpretation of astronomical observations in the framework of solid-state models and present a review of the optical properties of coals. We show how coals can reproduce the observations. We first describe the structural nature of the coal family and its diversity. Then we show that the coal model of carbon cosmic dust is able to meet the diversity of the astronomical observations in the UV (modelling of the interstellar extinction curve [10]) as well as in the IR range. We discuss a plausible IR emission mechanism. We show that radiative equilibrium heating is able to produce agreement not only with observed shapes and contrasts, but also with absolute intensities in the case of strongly illuminated dust, such as in circumstellar environments [11]. Finally we discuss the difficulties encountered in simulating emission spectra in the ISM where stellar photons are scarce and the continuum is very low. We emphasise the need for further experimental work.

2. STRUCTURAL PROPERTIES AND EVOLUTION OF COALS

Coal properties have been extensively studied in the past [12, 13] and compared recently to the other carbon solids as a model for cosmic carbon dust by Papoular *et al.* [9] where details can be found.

We are interested here only in the residual part of the material, after the minerals and free molecules pre-existing in the pores have been removed by suitable chemical and physical procedures. These materials are natural solids made up essentially of C, H and O and are classified according to their hydrogen and oxygen content. With time and temperature, an evolution arises which is marked by a progressive release of heteroatoms (reduction of H/C and O/C). This favours the aromatisation of the micro-structure (an increase in the aromatic/aliphatic carbon content) and leads to the progressive formation

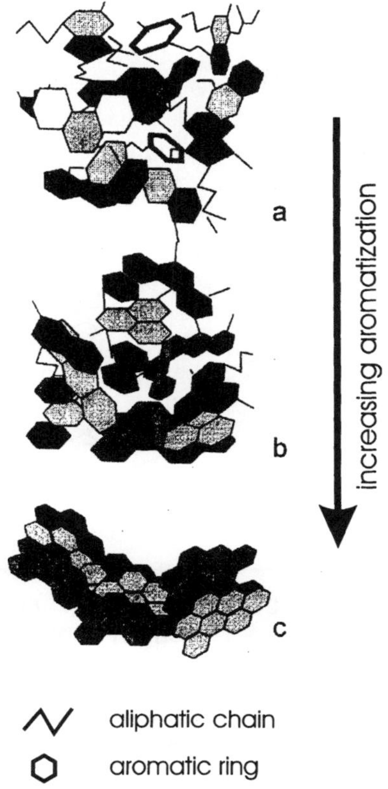

Fig. 1. — An idealised representation of the transformation of the coal structure during its evolution (adapted from [14]).

of planar clusters of benzene-type rings (Fig. 1), followed by progressive stacking of these aromatic sheets to form randomly oriented basic structural units (BSU). In the latter stages of evolution medium range order develops through the rearrangement (re-orientation) of adjacent BSUs.

Due to this structural evolution, coals form a family of candidate materials, with a continuous range of optical properties, from which it is possible to choose those which best reproduce a given astronomical observation. Moreover, the resulting spectral diversity is not erratic but regular and monotonous: the positions of the bands (from the UV to the IR) vary only slightly as the coal rank changes, whereas the relative intensities change considerably and regularly. Following the international classification of coals, the main steps of evolution of this material have been defined as bituminous, semi-anthracite, anthracite and semi-graphite (*i.e.*, polycrystalline graphite). We refer to this classification in the rest of the paper.

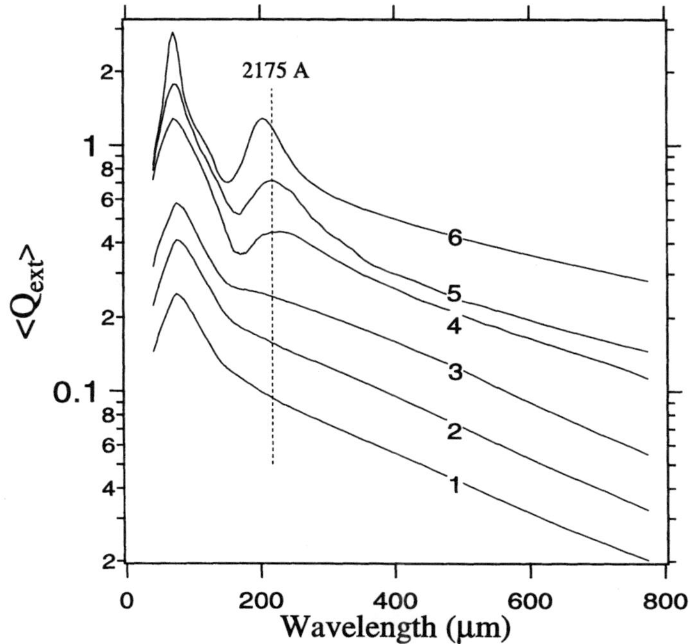

Fig. 2. — From top to bottom, optical extinction efficiencies of Highly Ordered Pyrolytic Graphite (HOPG) (6), polycrystalline graphite (5), and coals of decreasing stage of evolution: anthracite (4), semi-anthracite (3) and two bituminous (2,1).

3. COALS AND THE UV EXTINCTION CURVE

Based on UV-visible reflectometry of various graphites and more or less evolved coals, the complex index of refraction of the corresponding samples have been obtained [10]. Then, average optical extinction efficiencies, $Q_{\rm ext}(\lambda)$, have been deduced (Fig. 2) for a grain population with a classical size distribution [15]. The dust radius a runs from 2 nm to 0.25 μm with a power-law size distribution given by $a^{-3.5}$ [15]. The UV/visible extinction coefficient of coals evolves accordingly with the increase of the aromatic character. The bump at 2175 Å due to $\pi - \pi^*$ transition of sp^2 carbon in the graphitic network becomes more and more prominent. The maxima at 800 Å (0.8 μm) corresponds to the $\sigma - \sigma^*$ resonance due to both sp^2 and sp^3 carbon atoms. This resonance is present in all of the carbon compounds and is likely to be responsible for the rise on the short wavelength side of the bump.

With a mixing of different coals, we have been able to reproduce the mean galactic interstellar extinction curve (Fig. 3). Details of the calculation can be found in [10]. A mixture of coals at different stages of evolution is a valid approach to the carbon component seen in extinction, since both aliphatic

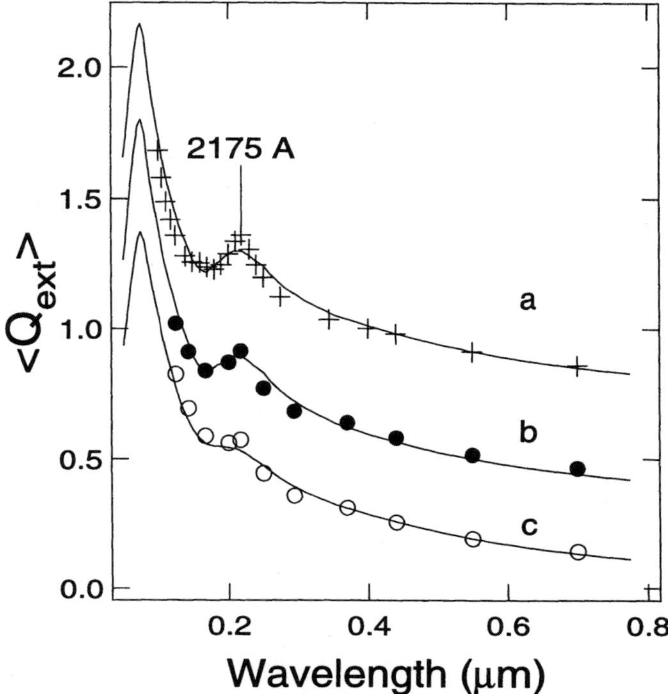

Fig. 3. — a) Fit of the average galactic interstellar extinction curve [16] obtained by mixing equal amounts of polygraphite and young coal. b) and c) Fit of the extinction curve for different lines of sight in the Large Magellanic Cloud [10].

and aromatic carbon dust are present in the ISM. Changing the ratio between the more- and less-evolved coals allows a reproduction of the diversity of the interstellar extinction curves for different lines of sight, as in the case of the Large Magellanic Cloud (Fig. 3). The data and model agree to better than 10%. A problem with the fit is the excessive width of the 2175 Å hump. Nevertheless, the quality of the fits in Figure 3 is good enough to convey the global message: that highly aromatised coal can provide the bump in the interstellar extinction curve, weakly aromatised coal can provide the underlying UV-visible continuum and both contribute to the far UV rise. No spurious features are introduced anywhere in the spectrum. No fine tuning of grain size or shape is required [10].

4. INFRARED SPECTROSCOPY OF COALS AND OBSERVATIONAL DIVERSITY

Tokunaga [17] has suggested a simple classification scheme for UIR bands based on the $3-14$ μm spectral region in an effort to summarise the present status of observations. Here, we will focus only on the A and B types sources described in his paper.

Type A are characterised by strong emission bands at 3.3, 6.2, 7.7 – 7.8, 8.6 and 11.3 μm features. Typical sources which show Type A spectra are planetary nebulae, HII regions, reflection nebulae and diffuse interstellar emission. Minor emission bands at 3.4, 5.2, 5.6 and 12.7 μm are also exhibited by Type A sources. Type B sources are characterised by a broad 3.4 μm emission feature as well as 6.2, 6.9, 8 and strong 11.5 μm features. At longer wavelengths, very strong bands at 21 and 30 μm are observed but there is no clear relationship with the bands at shorter wavelengths and they could arise from different class of materials. Another characteristic of type B spectra is their high 3.4/3.3 μm ratio when compared with those demonstrated by type A sources. Generally, sources with a type B spectrum are stars which are thought to be in transition towards the planetary nebulae stage and are designated as proto-planetary nebulae (PPN) [18].

Figure 4A presents different sets of observations in the near-IR. When compared, the IR features of all the presented objects demonstrate similarities and differences that, therefore, contribute to the large diversity of observed IR spectra. The two sources IRAS 05341+0852 and IRAS 04296+3429 are two proto-planetary nebulae. Both these objects exhibit type B spectra, characterised in this wavelength region by a high 3.4/3.3 μm band ratio. By contrast, the reflection nebula NGC 2023 demonstrates a dominant 3.3 μm aromatic feature, typical of class A sources. For comparison, we show at the top of Figure 4A the broad 3.4 μm band observed in absorption towards the Galactic Centre [19].

In view of the description of coal evolution summarised in Section 2, it is understandable that low ranking coals such as bituminous coals (characterised by high H/C and disorder) exhibit strong and mainly aliphatic (sp^3) IR vibrational features, while higher ranking coals exhibit mainly aromatic (sp^2) IR features. Figure 4B illustrates this point by showing the evolution of the aliphatic 3.4 – 3.5 μm blend and aromatic 3.3 μm bands with coal rank. Hence, in the near-IR wavelength range, bituminous coals exhibit only a strong 3.4 μm aliphatic band, very similar to that observed in absorption in direction of the galactic centre. By contrast, semi-anthracite coals ($[H/C]_{at} = 0.5$, $[O/C]_{at} = 0.03$) exhibit both aromatic and aliphatic stretching bands, with relative intensities similar to those observed in PPN spectra (Type B). Finally, more organised coals (curves d and e) such as anthracites ($[H/C]_{at} = 0.2$, $[O/C]_{at} = 0.02$) are needed to reproduce the type A spectra with a strong and dominant 3.3 μm aromatic band.

Spectral evolution is also observed in the mid-IR part of the spectrum. For instance, the progressive aromatisation of coal leads to the reduction and disappearance of the 6.9 μm aliphatic band. The 10 – 15 μm region is characterised by the presence of three strong bands at 11.45, 12.2 and 13.3 μm (respectively, the solo, duo/trio and quarto features) which arise from out-of-plane bending vibrational modes of aromatic C–H bond: their peak position depending on the number of neighbouring C–H groups at the periphery of a given aromatic ring (respectively one, two, three or four adjacent aromatic C–H

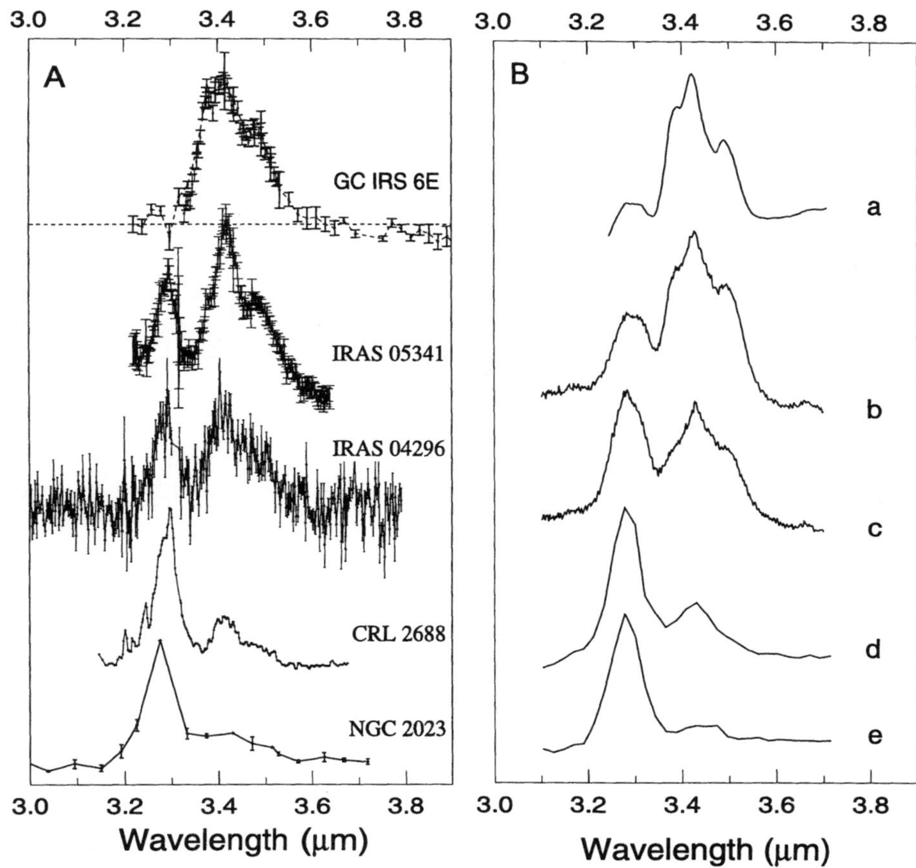

Fig. 4. — Comparison in the near-IR wavelength range between a set of observational data and a set of coal IR spectra. On the left, from top to bottom, extinction spectrum toward the Galactic Centre IR source GCIRS6E [19], emission spectra of three post-AGB nebulae: IRAS 05341+0852 [20], IRAS 04296+3429 and CRL2688 [21], and the emission spectrum of the reflection nebula NGC 2023 [22] respectively. On the right there are spectra of coals of increasing stage of evolution from top to bottom respectively.

bonds). This spectral region is therefore particularly sensitive to the hydrogen content of the material. These three bending modes are conspicuous in the spectra of semi-anthracite coals. Because of the progressive release of hydrogen atoms which takes place during the latest stages of coal evolution, the relative intensities of these three bands are dramatically modified when coal evolves further towards anthracite. Both the quarto and duo/trio bands are progressively and considerably reduced, while the now dominant solo band due to isolated

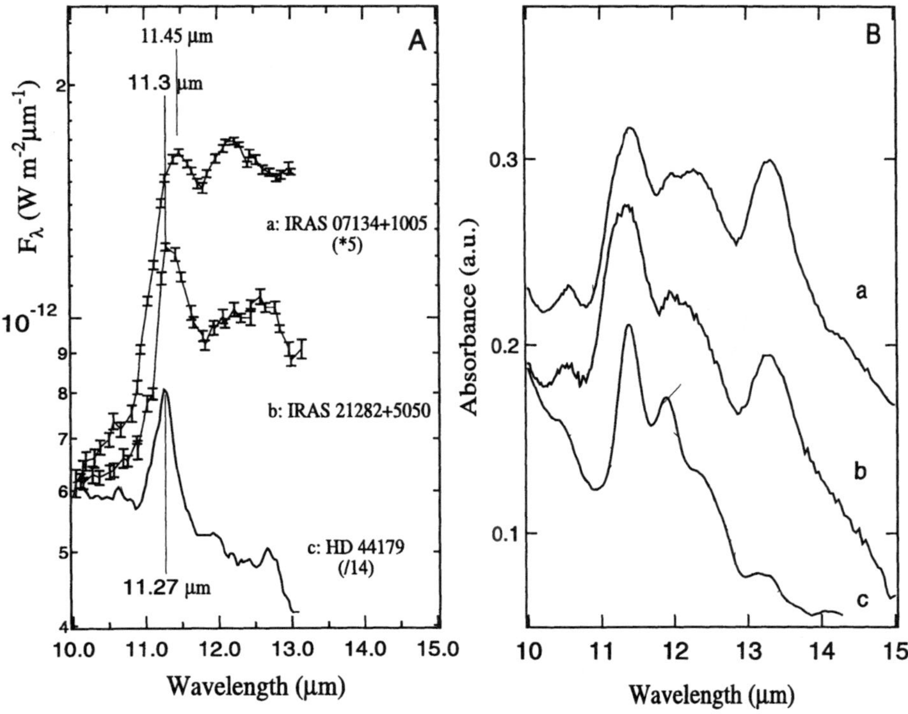

Fig. 5. — Comparison in the 11 – 14 μm wavelength range for a set of observational data [23, 24] and a set of coals in absorbance: a) semi-anthracite, b) heat-treated semi-anthracite and c) anthracite.

CH groups shifts from 11.45 to 11.3 μm in anthracite. This is illustrated by Figure 5B in which both spectra of semi-anthracite and anthracite coals have been reproduced between 10 and 15 μm, together with that of a semi-anthracite heat treated at 600 °C. The progressive reduction of the highly hydrogenated bands is clearly apparent, especially that of the 13.3 μm band which displays a large range of intensities. It is also, worthwhile to note that the anthracite coal considered here demonstrates in this wavelength range two residual 13.3 and 14.2 μm features similar to those observed with SWS instrument by Waters *et al.* [25] in the Red Rectangle.

Thus, from the spectral evolution of coal with increasing organisation, it seems natural to invoke semi-anthracite coals as candidate carrier of observed Type B spectra, and detailed spectroscopic comparisons can be obtained [11]. Moreover, because of the spectral changes which take place from the near-IR to the mid-IR when semi-anthracites evolve further (due to increasing aromatisation and decreasing H/C), the IR spectra of more evolved coals are simpler, and every band in the spectra of anthracites coincides with one of the UIR bands

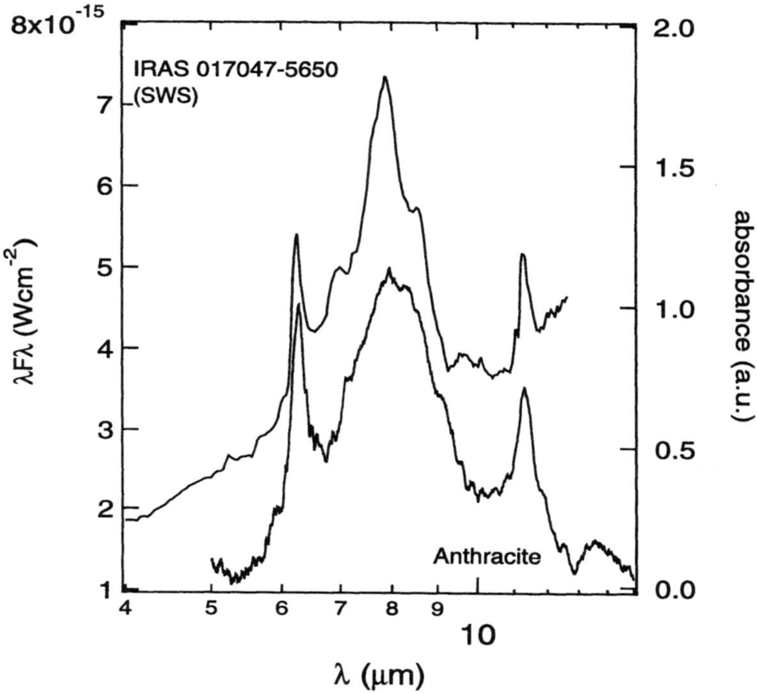

Fig. 6. — Mid-IR spectrum of an anthracite coal – continuum subtracted – compared to the corresponding part of the ISO emission spectrum of IRAS 017047–5650 (courtesy of L. d'Hendecourt) measured with SWS.

characteristic of type A. This is shown in Figures 6 and 7, which present comparisons of the IR spectroscopy of anthracite coals with the observed emission spectra of a planetary nebula (PN) and the Antennae galaxy respectively.

It is remarkable that the same spectral changes that occur when semi-anthracite evolves toward anthracite seem to be observed, from the near to the mid-IR, when PPN dust evolves into PN dust. Thus the differences between the circumstellar bands of PPN spectra, as compared to PN spectra, could be understood in terms of an evolution of carbonaceous grains similar to the natural evolution of coal on earth. It is possible to simulate the natural evolution of coals by heat-treatment or UV-visible light irradiation [27], and so, dust processing by increasing interstellar UV irradiation could correspond to temperature effects responsible for the coal evolution.

Hence, hydrogenated carbonaceous materials like coals correctly mimic the UIR emission and absorption bands from various directions in space, provided their hydrogen content and aromaticity are adequately chosen. This ability to reproduce many of the observed characteristics of the UIR bands (peak

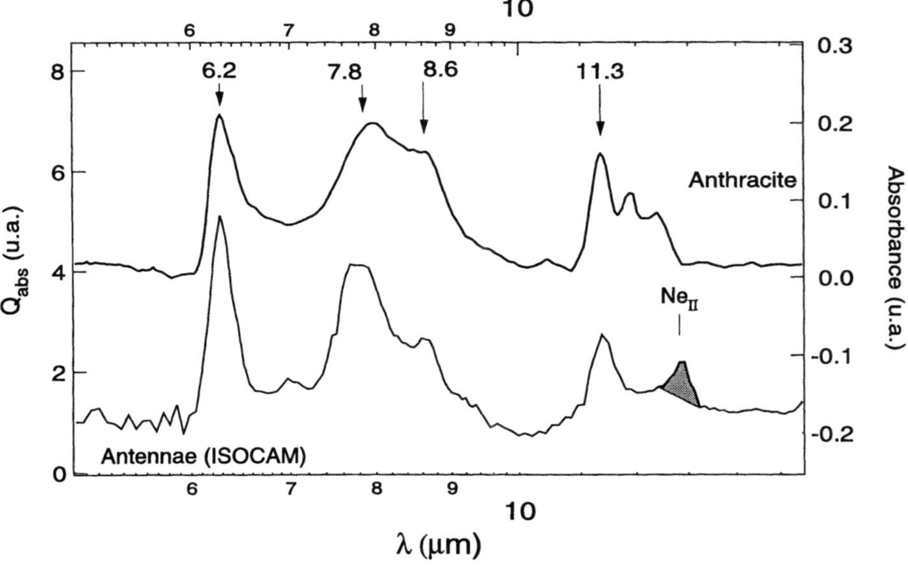

Fig. 7. — Comparison between the mid-IR absorbance spectrum of an anthracite coal and the ISOCAM spectrum of the Antennae [26].

position, widths and limited number) is due to the three dimensional micro structure of these materials, and this sheds light on the solid state nature of the observed IS features (as opposed to molecular signatures).

5. IR EMISSION MECHANISM

The second issue to be addressed in the course of the interpretation of the UIR bands is the emission/excitation mechanism. Dust has to be excited to be able to emit the infrared bands. There are several strong observational constraints. First, similar spectra are observed whatever the wavelength of the exciting radiation: visible or UV radiation field are both able to excite the IR emission bands [3]. This constraint is easily overcome for solid particles, since the absorptivity in the visible ($\langle\beta\rangle > 2\times10^4$ cm^{-1}) is essentially the same than the absorptivity in the UV (about 2×10^5 cm^{-1}) (Fig. 8). It is not the case for molecular entities which generally absorb poorly in the visible and are therefore more efficiently heated by UV photons. Secondly, the emission IR spectrum is similar in regions with very different radiation field intensities (over 5 decades in intensity), see contribution by Boulanger this volume. This constraint is now discussed in more detail.

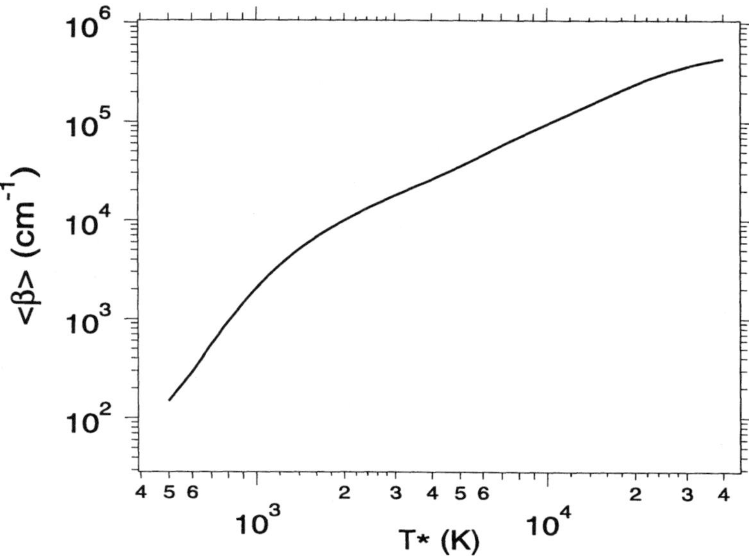

Fig. 8. — Mean absorptivity of coal as a function of exciting star temperature.

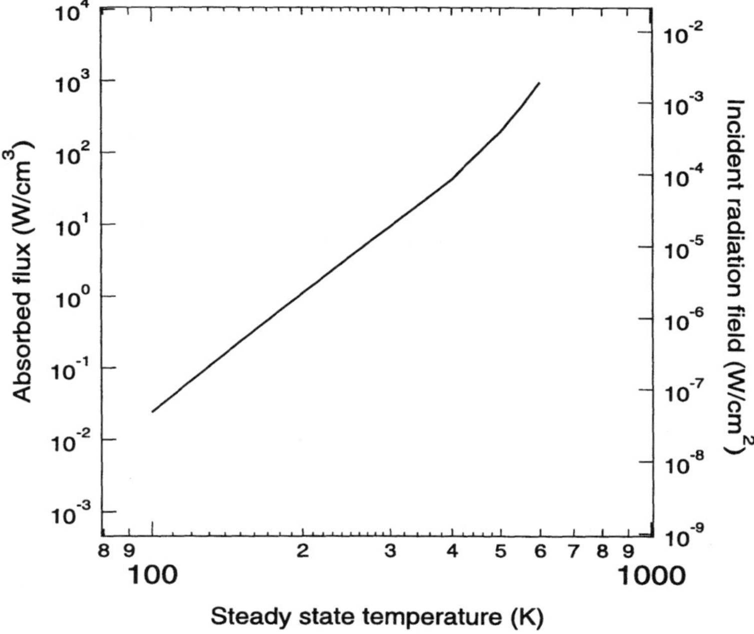

Fig. 9. — Relation between the steady state temperature $T_{\rm d}$ of an optically thin coal grain ($a < 100$ nm) and the ambient radiation field intensity I (right-hand ordinate). For the estimated intensities in excited nebulae, $10^{-6} < I < 10^{-4}$, $T_{\rm d}$ ranges between 200 and 400 K.

As a first step, we have investigated the radiative equilibrium heating hypothesis. Thanks to the measurement in the laboratory of the coal absorptivity on an absolute scale from UV to far-IR, we were able to determine the temperature reached by a grain as a function of the ambient radiation field (Fig. 9) [27]. This determination is made in the limit of the Rayleigh approximation. Because coals are semi-conductors absorbing strongly in the UV/visible range and emit weakly in the IR range, the temperature reached by coal grains under typical circumstellar radiation fields is in the 200 – 400 K range. At such temperatures, coal grains evolve very slowly and are not destroyed.

This excitation mechanism works with success in the case of circumstellar environments (PN and PPN), characterised by high radiation field intensities, and by low contrasts of the features (*i.e.*, strong IR continuum). This has been demonstrated successfully for a number of circumstellar spectra and in particular for three proto-planetary nebulae (Fig. 10), details can be found in Guillois *et al.* [11]. A black-body function has been introduced in the model to account for the reddened photospheric contribution of the central star in the near-IR. All the optical properties of the grains introduced into the calculations are derived directly from laboratory measurements. In each case, the mean steady-state temperature of the grains is deduced from the known properties of the central (F or G type) star and a suitable steady-state dust temperature distribution ranking between 400 K to 180 K is introduced to take into account the decrease of the incident absorbed power with increasing distance from the central star. Note also that the fit in the 10 – 13 μm is able to reproduce the observations in details (Fig. 11).

This mechanism cannot apply to every object in which the UIR bands are observed, and particularly to reflection nebulae or the diffuse ISM. The available exciting fluxes are too weak and the continuum which sustains the features too low (*i.e.*, the contrast of the UIR bands is too high) [30]. In this case, we can consider the contribution of transiently heated very small grains, a mechanism that can dominate the thermal equilibrium mechanism, provided that the grain size is small enough. The difficulty is that, in order to have the right spectroscopic features, carbonaceous materials must have intermediate range order as well as short range order. This sets a lower limit to the size of the grain for the latter to retain the optical properties of the bulk. A limit can be estimated at around 20 Å, *i.e.*, more than ~1000 atoms. This is much larger than the upper limit set for efficient stochastic heating [31].

Hence, we can summarise the dilemma:

- either we consider entities small enough to be easily heated by a stochastic process, but no spectroscopic agreement can be provided up to now;
- or we consider a solid-state model, such as coal, which can provide spectroscopic agreement, but we are confronted with difficulties in simulating the emission spectra in the ISM where stellar photons are scarce and the continuum underlying the UIR bands is very low.

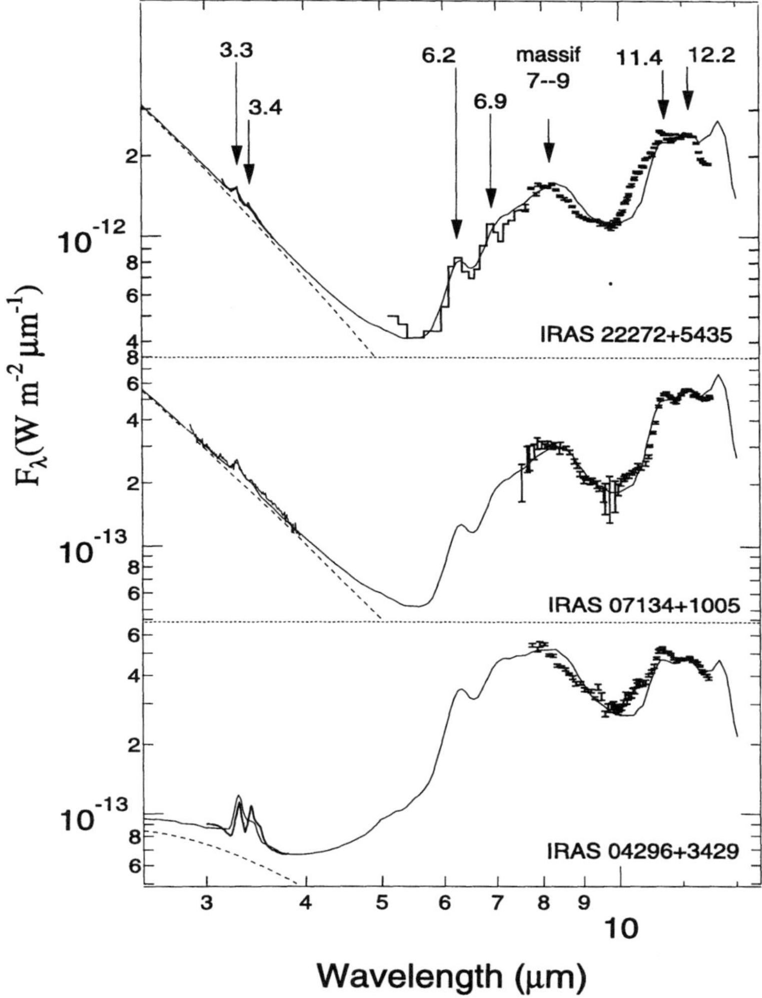

Fig. 10. — Fits of the emission spectra of the PPN IRAS 22272+5435 [21, 23, 28], IRAS 04296+3429 [21, 23], and IRAS 07134+1005 [23, 29] with IR emission of semi-anthracite coal grains in thermal equilibrium with the ambient radiation field [11].

To overcome these difficulties, we think that two directions have to be explored extensively by laboratory experiments:

- exploring PDR physics and chemistry, because the ISO results indicate a dominant contribution of these regions to the source of the UIR emission bands. The role of the interaction of hydrogen atoms with grains, as an additional source of grain heating in these regions, is not well understood [32];

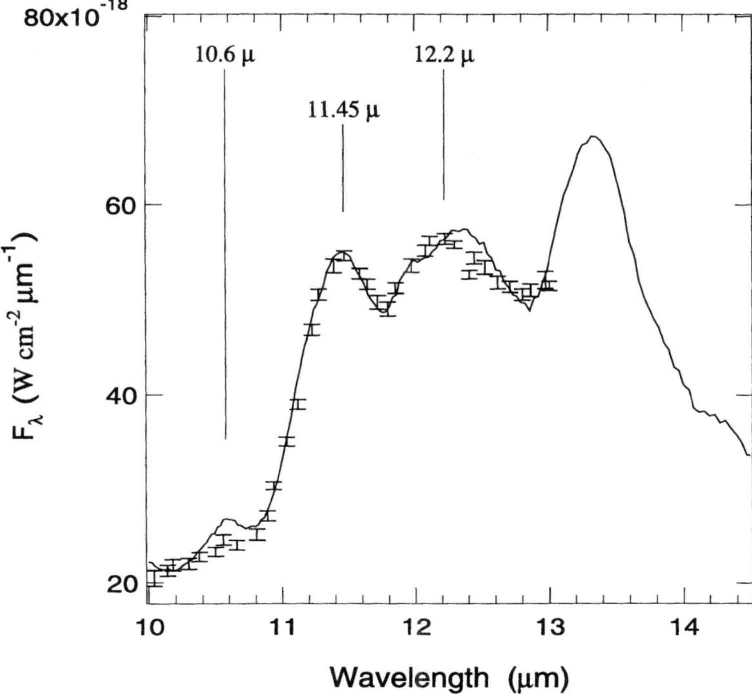

Fig. 11. — Fit of the spectral details in the C–H out of plane deformations region of IRAS 07134+1005 (symbols) [23] with IR emission of semi-anthracite coal grains at 180 K (continuous line) [11].

- measuring the "true" IR spectra of intermediate range aromatic grains (several hundreds of carbon atoms).

6. CONCLUSION

The model of cosmic carbon dust presented here has the following characteristics. It calls for disordered solid state by contrast with molecular state: the width of the IR spectral features is hence better simulated, as is the underlying continuum. It makes use of a continuous family of materials deriving from each other by a progressive natural evolution, involving essentially heat-treatment. Hence, the diversity of the emission spectra recorded in different celestial environments is reproduced, and a link can be established between signatures observed in emission with signatures observed in absorption, or signatures observed in IR and UV. Thus, any member of the family is at a different level the carrier of the features we try to identify.

REFERENCES

[1] Duley W.W. and Williams D.A., *MNRAS* **191** (1981) 701.
[2] First ISO Results, A&A Special Issue, *A&A* **315** (1996) 1.
[3] Uchida K.I., Sellgren K. and Werner M., *ApJ* **493** (1998) L109.
[4] Leger A. and Puget J.L., *A&A* **137** (1984) L5.
[5] Allamandola L.J., Tielens A.G.G.M. and Baker J.R., *ApJ* **290** (1985) L25.
[6] Cook D.J. and Saykally R.J., *ApJ* **493** (1998) 793.
[7] Sakata A., Wada S., Okutsu Y., *et al.*, *Nature* **301** (1983) 493.
[8] Papoular R., Conard J., Guiliano M., *et al.*, *A&A* **217** (1989) 204.
[9] Papoular R., Conard J., Guillois O., *et al.*, *A&A* **315** (1996) 222.
[10] Papoular R., Guillois O., Nenner I., *et al.*, *Planet. Sp. Sc.* **43** (1995) 1287.
[11] Guillois O., Nenner I., Papoular R. and Reynaud C., *ApJ* **464** (1996) 810.
[12] Van Krevelen D.W., Coal (Elsevier, Amsterdam, 1993) p. 1.
[13] Charcosset H. and Donat B.N.P., Advanced Methodologies in Coal Characterisation (Elsevier, New York, 1990) p. 1.
[14] Alvarez D., Borrego A.G., Menendez R. and Bailey J.G., *Energy Fuels* **12** (1998) 849.
[15] Mathis J.S., Rumpl W. and Nordsieck K.H., *ApJ* **217** (1977) 105.
[16] Mathis J.S. and Whiffen G., *ApJ* **341** (1989) 808.
[17] Tokunaga A.T., Diffuse IR Radiation and the IRTS, ASP Conf. Series 124, edited by H. Okuda, T. Matsumoto and T.L. Roellig (ASP, San Francisco, 1996) p. 149.
[18] Buss R.H., Cohen M., Tielens A.G.G.M., *et al.*, *ApJ* **365** (1990) L23.
[19] Pendleton Y.J., Sandford S.A., Allamandola L.J., Tielens A.G.G.M. and Sellgren K., *ApJ* **437** (1994) 683.
[20] Joblin C., Tielens A.G.G.M., Allamandola L.J. and Geballe T.R., *ApJ* **458** (1996) 610.
[21] Geballe T.R., Tielens A.G.G.M. and Hrivnak B.J., *ApJ* **387** (1992) L29.
[22] Sellgren K., *ApJ* **277** (1984) 623.
[23] Justtanont K., Barlow M.J., Skinner C.J., *et al.*, *A&A* **309** (1996) 612.
[24] Witteborn F.C., Sandford S.A., Bregman J.D., *et al.*, *ApJ* **341** (1989) 270.
[25] Waters L.B.F.M., Waelkens C., van Winckel H., *et al.*, *Nature* **391** (1998) 868.
[26] Dan Tran, Ph.D. Thesis, Université Paris XI (1998).
[27] Guillois O., Nenner I., Papoular R. and Reynaud C., *A&A* **285** (1994) 1003.
[28] Buss R.H. Jr., Cohen M., Tielens A.G.G.M., *et al.*, *ApJ* **365** (1990) L23.
[29] Kwok S., Hrivnak B.J. and Geballe T.R., *ApJ* **360** (1990) L23.
[30] Laureijs R., Acosta-Pulido J., Abraham P., *et al.*, *A&A* **315** (1996) L313.
[31] Schutte W., Tielens A.G.G.M. and Allamandola L., *ApJ* **415** (1995) 397.
[32] Guillois O., Ledoux G., Nenner I., Papoular R. and Reynaud C., *Faraday Discuss.* **109** (1998) 335.

LECTURE 8

Organics in the ISM: Where Do They Come from?

Y.J. Pendleton

NASA Ames Research Center,
Mail Stop 245-3, Moffett Field, CA 94035, U.S.A.

Abstract. Spectral comparisions of the 3.4 and 4.62 μm interstellar absorption bands to laboratory spectra of processed ices imply that interstellar organic signatures result from energetic processing of ices on grain mantles in dense molecular clouds. However, in the case of the 3.4 μm aliphatic hydrocarbon band, the interstellar feature has only been observed in the diffuse interstellar medium where no ices exist. The absence of the 3.4 μm band in any dense cloud spectrum observed to date, and the detection of the band in the outflow of an evolved carbon star, suggest the production mechanism for this particular organic species may not involve ice processing after all. The widespread distribution of the 3.4 μm carrier and the striking similarity of the interstellar band to that found in carbonaceous meteorites emphasizes the need to understand the pathway by which interstellar organics evolve. Inside dense clouds, the 4.62 μm absorption feature is seen towards embedded protostellar objects. The carrier of this band is likely the result of interstellar ice processing, and a carbon-nitrogen bearing species has been implicated. However, at least two types of energetic processing (UV photolysis and ion bombardment) can produce spectral features which similarly match the interstellar observations, so at this time the production mechanism is unclear. To produce the 4.62 μm band in the laboratory, the initial starting mixture can contain nitrogen in several forms, but if the nitrogen precursor molecule is in the form of solid state N_2, then only the ion bombardment method can break the strong bond, allowing the nitrogen to interact with other species. The 3.4 and 4.62 μm absorption bands may be the two best indicators of the solid state organic component of the interstellar medium. Therefore, understanding the formation, evolution, and distribution of these components will help to trace the organic material that is available for incorporation into planetary systems.

1. INTRODUCTION

Until relatively recently, crucial portions of the infrared spectrum were unavailable to astronomical spectroscopists due to atmospheric limitations which hinder ground-based efforts, so despite the accumulating body of knowledge, gaps in the spectral observations have remained. The $2 - 30$ μm wavelength region is critically important for the study of organic solids and ices, and with the former Kuiper Airborne Observatory and the Infrared Space Observatory (ISO), the major gaps in the observational picture are now being filled in. The combination of moderate resolution ground-based spectroscopy with the space and airborne observations is providing a more complete inventory of the dust constituents, as well as information regarding the physical location of dust.

Stars and planetary systems form only inside the high density, low temperature environment of dense molecular clouds, therefore the composition of interstellar ices in these clouds and the effects of energetic processing on those ices is important in assessing the inventory of organic material available for incorporation into forming planetary systems. Dust in the diffuse interstellar medium (ISM) that surrounds dense clouds, must also contribute to the material found in dense regions, because new generations of dense clouds form out of swept-up diffuse ISM dust (*e.g.* [1]).

Recent observations from the Short Wavelength Spectrometer [2] on ISO [3] have provided the necessary overview of the entire near and mid-infrared, and as a consequence most of the features due to interstellar ices in the spectrum of a typical embedded protostar have now been identified (*e.g.* [4] and the contributions of Dartois and Schutte in this volume). Details concerning the processing of these ices are not so clear, however, and two absorption bands in particular are serving as probes of the dense and diffuse cloud environments. These two features, the 3.4 and 4.62 μm bands, fall in wavelength regions that can be observed from the ground. Moderate resolution, high signal-to-noise ground based data have proven to be highly complimentary to the essential information provided by ISO.

In this paper, the organic refractory components of the interstellar medium, best revealed through the 3.4 and 4.62 μm solid state absorption features, are discussed as possible tracers of the formation sites and evolution of interstellar dust. These features have been detected in diffuse and dense interstellar clouds, respectively. This paper presents the current state of the field and the puzzling questions that are apparent from the data.

2. THE 3.4 μm BAND

Along lines-of-sight through the diffuse interstellar medium towards bright background sources, a distinct signature of hydrocarbon absorption has been detected near 3.4 μm. Spectroscopy of the 3.4 μm wavelength region reveals sub-features at 3.385, 3.420 and 3.485 μm (2955, 2925, and 2870 cm^{-1}). The positions of the 2955 and 2925 cm^{-1} sub-features are characteristic of the symmetric C–H stretching frequencies of $-CH_3$ (methyl) and $-CH_2-$ (methylene) groups in saturated aliphatic hydrocarbons and the band at 2870 cm^{-1} is characteristic of the asymmetric C–H stretching vibrations of these same functional groups when perturbed by other chemical groups [5]. These distinct sub-features have now been detected along nearly a dozen different sightlines through our galaxy [6] as well as in the spectra of dust embedded Seyfert galaxies [7–9].

Early detections of the diffuse ISM 3.4 μm band [10–14] focused on the line-of-sight towards the galactic center, where the dust extinction is high ($A_v \approx 31$ magnitudes) and the background sources are bright in the near-infrared. Lequeux and Jourdain de Muizon [15] detected the same absorption band towards a nearby, evolved carbon star, CRL 618, and additional lines-of-sight through the galaxy later revealed the presence of the same carrier in the diffuse ISM [5, 6]. There is, however, a distinct absence of the 3.4 μm band in the spectra of dense molecular cloud dust [16, 17]. The difference in the spectra of the refractory component of dense and diffuse cloud dust was first pointed out by Allamandola *et al.* [18,19], and the discrepancy has raised serious questions about the formation site and evolutionary history of this hardy, refractory component which is so widespread throughout our own galaxy and other galaxies. The remarkable similarity of the 3.4 μm band seen in the diffuse interstellar dust to the organic refractory material found in carbonaceous meteorites [6,20] suggest some exchange between the dense and the diffuse medium, since meteorite parent bodies form around stars which arise from dense cloud regions.

Figure 1 displays ground-based observations of the interstellar 3.4 μm absorption band obtained at the NASA Infrared Telescope Facility and the United Kingdom Infrared Telescope. The sub-features of CH_2 and CH_3 groups are seen in absorption in the dust in our galaxy and the dust from the luminous IRAS galaxy 08572+3915 [8]. The extragalactic dust feature has been blue-shifted to account for the distance of the galaxy ($z = 0.05$). The carbonaceous material in the diffuse ISM has an average $-CH_2-/-CH_3$ ratio of $2.0 - 2.5$ and likely contains moderate length aliphatic chains, such as $-CH_2-CH_2-CH_3$ and $-CH_2-CH_2-CH_2-CH_3$, associated with electronegative chemical groups, *i.e.*, moieties like $-OH$, $-C{\equiv}N$, and aromatics [5,6]. These basic organic constituents appear to be distributed throughout galactic and extragalactic dust [7, 8, 21]. The close match in position, strength, and profile between the galactic and extragalactic dust features suggests this type of chemical arrangement may be quite common.

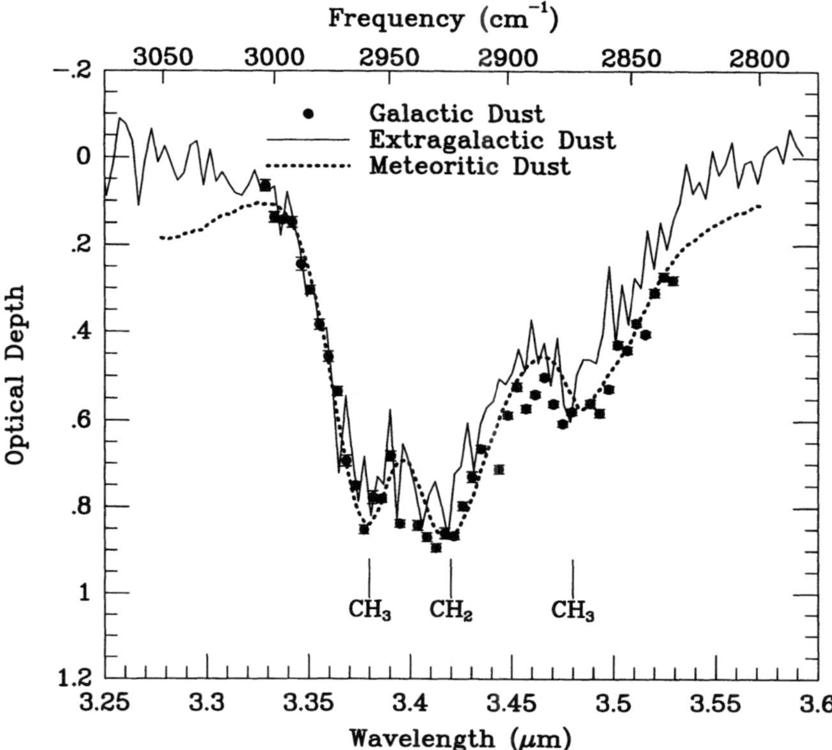

Fig. 1. — A comparison of the aliphatic hydrocarbon feature seen in the diffuse interstellar medium in our galaxy (points), the Murchison meteorite (dashed line) and the dust of a distant galaxy (solid line; redshift of the galaxy has been accounted for). Figure taken from [9]. A similar feature has now been detected in 4 out of 13 galaxies [8].

Figure 1 also displays the infrared spectrum taken from the volatile component sublimed off of the acid insoluble residue of the Murchison meteorite (Murchison data from [22]). Spectral similarity between the light hydrocarbons of carbonaceous meteorites and the diffuse galactic dust [6,20] have suggested that these light hydrocarbons are somehow preserved in the parent body of the meteorite.

The role of ultraviolet photolysis is one of the key questions in the chemistry of interstellar ices. Theoretical and experimental studies have long suggested that FUV photolysis of interstellar ices is important for their composition [23–31]. In the laboratory, the ultra-violet photolysis and subsequent warm-up of interstellar ice analogs produces a hardy, organic refractory residue whose spectrum is similar to the feature observed in the diffuse ISM [5]. A reasonable initial explanation for the production of the interstellar 3.4 μm band was that organic refractory material, processed in the dense

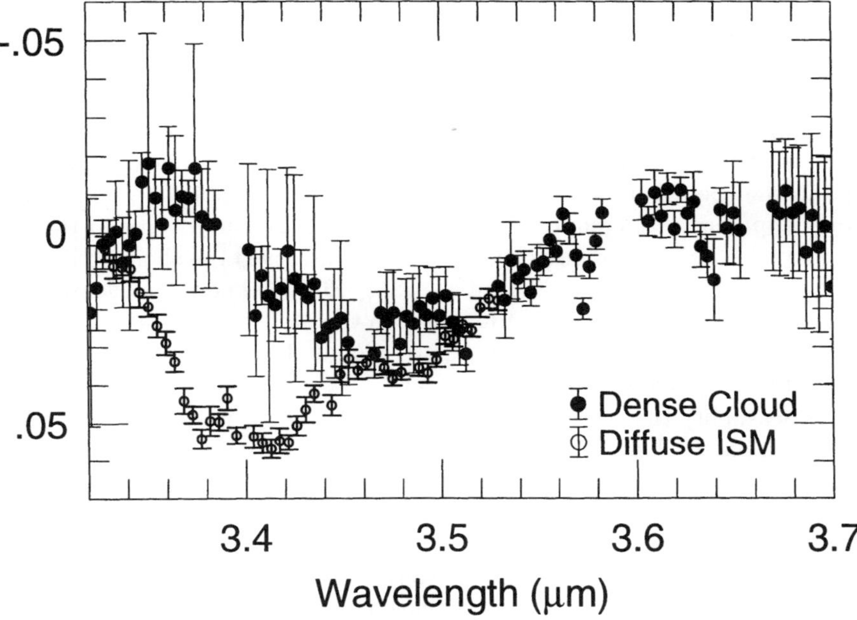

Fig. 2. — A comparison of the absorption bands in the hydrocarbon spectral region for a typical dense molecular cloud line-of-sight (solid points) and the diffuse interstellar medium (open points). Figure taken from [33].

cloud ices during exposure to UV photons and/or upon subsequent heating of the ices, remained in the diffuse ISM after the volatile components of the icy grain mantles were destroyed [32]. Laboratory studies showed that the refractory material could survive the diffuse ISM conditions and the feature was indeed observed in virtually every line-of-sight where there was enough diffuse dust to measure it. However, the apparent absence of the feature in dense cloud spectra raised new questions [9,18,19,33]. Comparisons between the dust absorption features seen in dense and diffuse interstellar clouds, such as that shown in Figure 2, show the absence of the aliphatic diffuse ISM sub-features in the spectra of embedded protostellar objects.

Related to the absence of the aliphatics in the dense clouds is the clear spectral signature of the 3.4 μm band in a nearby, evolved star CRL 618 [15,34,35]. The star is not far away (1.3 kpc; [36]) and the feature is too strong to attribute the absorption to dust along the diffuse ISM line-of-sight [35]. It therefore must be produced close to the evolved star. Interstellar ices cannot exist in this environment, so their processing cannot be the source of the observed aliphatics. It is suspected that the high velocity wind emanating from this source (~200 km/s; [37–39]) plays a role in the formation of the hydrocarbons. Figure 3 is a comparison between the spectrum of the diffuse ISM dust and that of the hydrocarbons produced locally in the outflow of the evolved star CRL 618 [35].

The presence of the 3.4 μm band in the spectrum of CRL 618 suggests the aliphatics seen in the diffuse ISM might form in the outflows of carbon stars. However, even if the aliphatics are made in this way, the absence of the 3.4 μm band in the spectra of dense cloud sources remains a problem. The feature is clearly apparent throughout the diffuse ISM and the carrier should therefore find its way back into a dense cloud at some point, given the short cycling time of grains between the dense and diffuse ISM [40]. We know that dense clouds transfer a substantial amount of material to the diffuse ISM and if the diffuse ISM does not go back into making dense clouds, there simply is not enough material to keep making new stars.

One possibility that readily comes to mind is that perhaps the signature of the 3.4 μm band is hidden under layers of ice on the grains in dense, cold clouds. The problem with this solution is that the aliphatic signature is very strong and laboratory experiments have shown ice layers do not mask the telltale absorption bands [41].

Perhaps the aliphatic material is converted into some other form upon entry into a dense cloud. This possibility should be explored through laboratory experiments to test for intermediary products that we might look for observationally. The destruction of aliphatics must be extremely rapid and very complete, however, because there are currently no dense cloud spectra which reveal the presence of the distinctive aliphatic CH_2 and CH_3 sub-groups as shown in Figure 1 [42].

Carbonaceous meteorites carry a great variety of complex organic materials of varying degrees of volatility, solubility, and structural diversity [43, 44], of which the amino acids and the aliphatic components [45] are especially significant in the context of prebiotic chemistry. A poorly characterized complex of macromolecular material, the kerogen component [46] is of particular interest because it may also occur as a coloring agent on various Solar System bodies [47]. The high degree of deuteration of the complex organic molecules detected in carbonaceous meteorites and in cometary material is strong evidence for the D-fractionation having originally occurred in the interstellar medium. The remarkable similarity of the diffuse ISM aliphatic hydrocarbons to those in carbonaceous meteorites [20] suggests another puzzle that cannot be readily solved at this time. In short, if the aliphatic hydrocarbons detected in the diffuse ISM are truly absent in the dense molecular clouds, the remarkable similarity between the interstellar 3.4 μm band and that found in the meteorites may be deceptive, because it is not at all apparent that one could go from point A (the diffuse ISM) to point C (formation site of the parent body of the meteorite) without passing through point B (the dense cloud stage where the star and planetary system form). However, before disregarding the incredible match shown in Figure 1, it is worth remembering that the Murchison

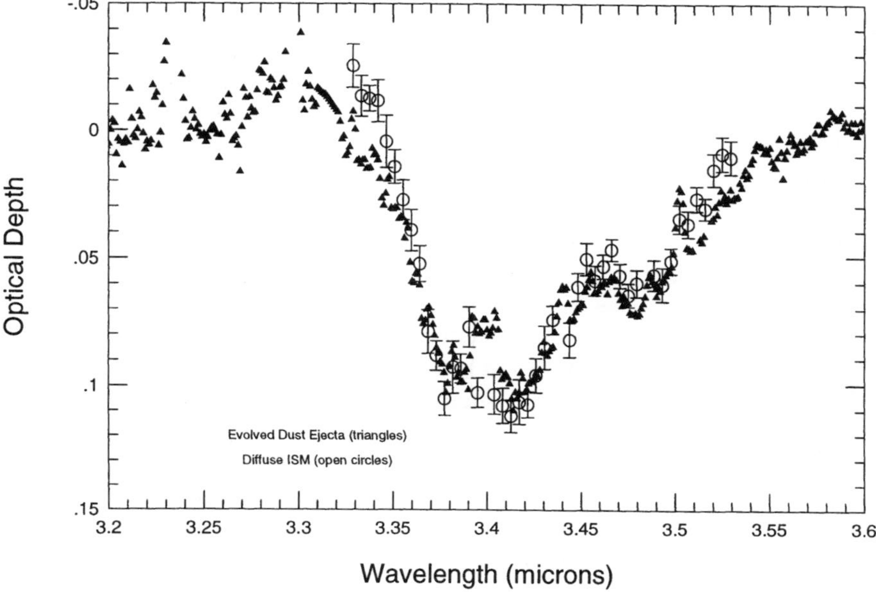

Fig. 3. — An overlay of the diffuse interstellar medium data (open points) and the aliphatic hydrocarbons seen in the outflow of an evolved star (solid points). Figure taken from [35].

meteorite provides a better match to the interstellar feature than do any of the laboratory residues produced from astrophysically plausible starting mixtures. If it is a feature so easily formed that it happened independently in several different places, one has to wonder why it is not easy to duplicate it in the laboratory.

3. THE 4.62 μm BAND

Another solid state absorption feature attributed to an organic carrier is the 4.62 μm interstellar absorption band seen exclusively in the spectra of embedded, dense cloud objects. First reported by Soifer *et al.* [48], this feature has now been observed towards several embedded protostellar sources [49–54]. Laboratory far ultraviolet (FUV) photolysis of CO and NH_3 ices show a prominent absorption feature at 4.62 μm (2165 cm^{-1}) [24, 25, 28, 49, 55–57]. Ion bombardment of ices can also produce a 4.62 μm band [58, 59] as shown in Figure 4. Because different processing mechanisms can reproduce similar bands, the pathway to, as well as the exact identification of, the interstellar 4.62 μm band remains unclear.

The type of process (ultraviolet photolysis or ion bombardment) is directly relevant to the composition of the pre-processed ices on the grain mantles. If the source of the nitrogen comes from solid N_2, then ultraviolet photolysis would not break apart the N_2 bond [60], but ion bombardment could do so [58]. Ice present on bodies in the outer Solar System [61,62] suggests that the infrared inactive N_2 ice may be hiding in the interstellar ices as well. Therefore, studies of the interstellar 4.62 μm (2165 cm^{-1}) band may provide information on both the reservoir of the solid-state nitrogen in interstellar ices and the energetic processing mechanism by which it participates in the chemistry of the X-CN feature. A comparison of the spectra of laboratory residues produced from the UV photolysis of a 12 K ice mixture of H_2O:CH_3OH:CO:NH_3 in the proportions 100:50:10:10 and subsequent warm-up to 100 K (solid line; taken from [63]) with residues produced from the ion bombardment of a mixture of H_2O:N_2:CH_4 in the proportions 1:1:1 (solid line; taken from [58]) is shown in Figure 4.

The OCN^- ion has been identified with the laboratory 4.62 μm band [28,29, 65]. It was suggested that the NCO^- or OCN^- ion is produced through the photolysis of icy mixtures containing CO and NH_3, which would first produce the acid HNCO. In the presence of ammonia, or some other base, an acid-base reaction (the transfer of a proton) then occurs between HNCO and NH_3 to form NCO^- and $NH_4{}^+$. One problem for any ion identification is the balance of charge. For each negative ion made, a positive ion must be produced. In some of these experiments, the proposed counter-ion is NH_4^+ and a band near 6.67 μm (1500 cm^{-1}) is attributed to this ion [28]. Recent laboratory results by Demyk *et al.* [65] also attribute the 4.62 μm absorption in the spectra of UV photolyzed CO:NH_3 mixtures to OCN^-.

As mentioned above, an alternative way to process interstellar ice mixtures is through bombardment with high energy protons. Moore *et al.* ([59] and this volume) have shown that a band is produced at 2170 cm^{-1} (4.61 μm) upon the irradiation of a mixture of H_2O:NH_3:CH_4 (in the proportions $\simeq$ 1:3:2) after irradiation with $\simeq$ MeV protons. They also found that irradiation of mixtures H_2O:N_2:CO_2 (in the proportions $\simeq$ 1:1:1) and H_2O:N_2:CO (in the proportions $\simeq$ 5:1:1) produces a band at about 2180 cm^{-1} (4.59 μm). It is well established (*e.g.* [66–68]) that ion irradiation not only drives the formation of specific molecules, depending on the particular target, but also refractory residues. Recent work by those authors [58] which includes N_2 in the starting mixture produces a band which is similar to the interstellar 4.62 μm band. Ion bombardment, which is analogous to cosmic ray interactions with icy dust grains in dense molecular clouds, has the advantage that the high energy ions can penetrate further into icy grain mantles and can break apart materials such as solid state N_2; the resulting 4.62 μm band is discussed in [58] and is compared to the observations of the protostar W33A in Figure 4b.

Some additional examples of the interstellar 4.62 μm band are shown in Figures 4 and 5 (taken from [54]). The observed near-constant position for the interstellar 4.62 μm band suggests that either the carrier of the band is not very sensitive to the molecular environment, or the grain mantles towards

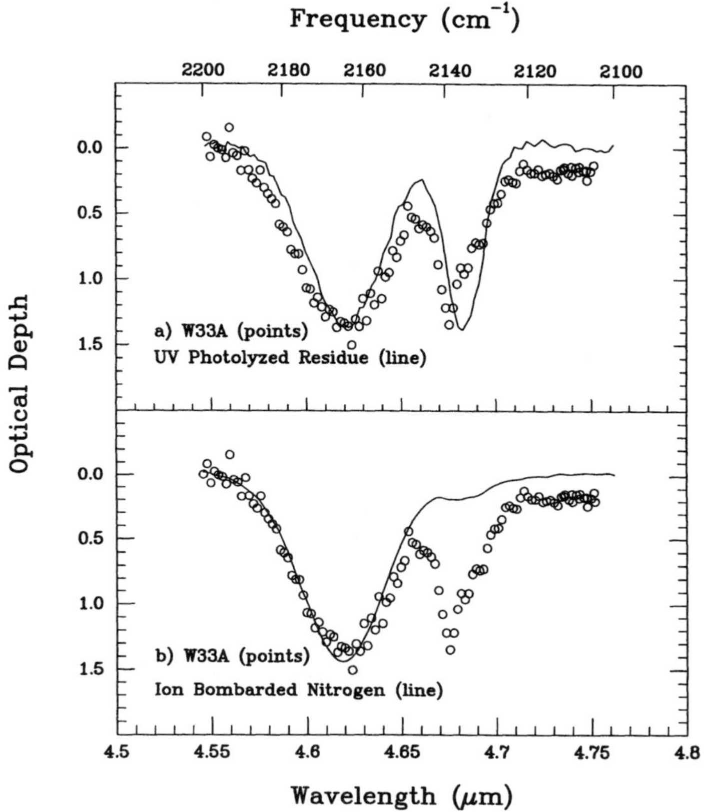

Fig. 4. — A comparison of the 4.62 μm absorption band detected in the dense molecular cloud line-of-sight towards the protostar W33A (points [54]) to a) the organic residue produced via ultraviolet photolysis of an interstellar ice analog mixture containing ammonia (lab data from [64]; line) and b) the organic residue produced through the ion bombardment of an interstellar ice analog mixture containing solid nitrogen (N_2) [58]. The strong absorption band at 4.67 μm in the interstellar data and in lab data in Figure 4a is due to solid state CO. Upon warm-up, CO disappears from the laboratory spectrum, however in this case there was so much CO in the original mixture that some of it remained even at a temperature of 100 K (Bernstein, private communication). A similar 4.62 μm band can be also be produced through the ion bombardment of ammonia containing ice mixtures, however the UV photolysis of solid nitrogen containing mixtures cannot break the strong N_2 bond apart [60] and therefore cannot produce the 4.62 μm band.

these sources reside in very similar conditions. The latter point is unlikely due to the observed variable composition of interstellar ice mantles and the unlikely situation that all the protostars observed are at the same evolutionary stage.

The distribution of the 4.62 μm band within dense clouds may have implications for type of physical environment which favors the production of the band carrier. For instance, Tegler *et al.* [52] found that the 4.62 μm band did not appear in the spectrum of a background field star (Elias 16) seen through the Taurus molecular cloud, while embedded protostars seen through equivalent column densities of dust in the Taurus cloud show the feature. If these results are found to apply in general to molecular clouds, they may suggest that the production of the 4.62 μm band requires close proximity to newly forming stars. This might help constrain the production mechanism.

It is also possible that the 4.62 μm band is produced via simple grain surface chemistry. An example would be a reaction occuring between acids and bases, such as the previously discussed reaction between HNCO and NH_3, since both of these species could be produced through grain surface chemistry. Ammonia can be formed by hydrogenation of accreted atomic N. HNCO occurs through the addition of CO+H to form HCO [69], followed by HCO+N which may rearrange itself into the lowest energy state, HNCO [70,71]. This species was proposed by Charnley [72] as the most chemically plausible carrier of the interstellar 4.62 μm band. Alternatively, the intermediate excited complex NCHO (which first forms from the addition of HCO+N) might transfer energy to the matrix rather than use that energy to rearrange the carbon and nitrogen, and the hydrogenation of NCHO would lead to NH_2CHO [73], a species that has been observed in the gas phase in the ISM [74], but which has no solid state 4.62 μm band.

Energetic processing of interstellar ices, either through ultraviolet photolysis, ion bombardment, or both, and grain suface chemistry are all plausible pathways by which the interstellar 4.62 μm band might arise. Identifying the production mechanism is directly related to identifications of the interstellar ice constituents which reside on the pre-processed grain mantle. The production pathways are described in more detail in [54], but they are summarized here in Table I.

4. CONCLUSIONS

Understanding the production and distribution of organic solids in the ISM aids in our understanding of the initial composition of planetary systems. The recent discovery of organic solid-state material in distant galaxies, as evidenced by the 3.4 μm hydrocarbon absorption in NGC 1068 and IRAS 08572+3915 points out the ubiquitous nature of the carrier, which has now been detected along more than a dozen lines-of-sight through the diffuse interstellar medium in our own Galaxy. Comparisons of the diffuse interstellar medium 3.4 μm hydrocarbon absorption feature with the spectrum of extracts from carbonaceous meteorites has shown a remarkable similarity in profile and relative strength among the sub-features which arise from CH_2 and CH_3 groups.

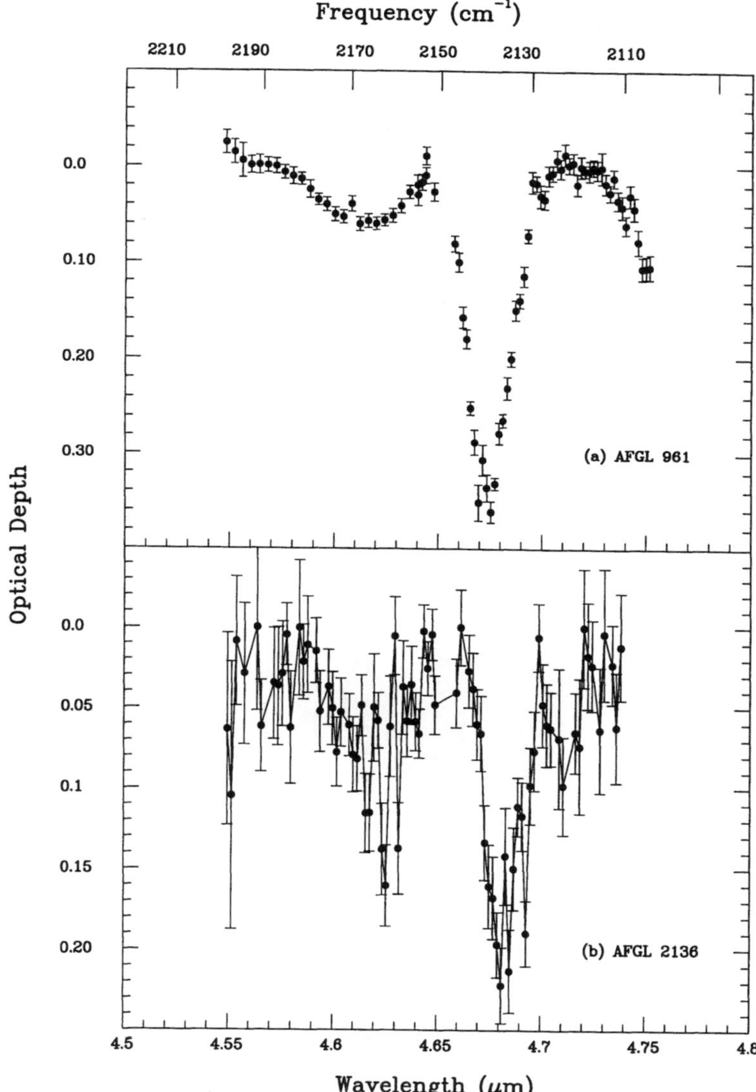

Fig. 5. — Optical depth plot from 4.5 – 4.8 μm (2222 – 2083 cm^{-1}) for embedded protostars (a) AFGL 961 and (b) AFGL 2136. The absorption band at 4.62 μm is likely due to a carbon-nitrogen carrier. Where error bars are not visible, they are smaller than the points.

Table I. — Production pathways.

Starting Molecule	Product	Comments	References
UV Photolysis of Ices			
CO, NH_3 H_2O, CH_3OH ices	XCN probable	Band at 2165 cm^{-1} correlates with band near 1700cm^{-1}, presumably due to a carbonyl stretch which explains shifted peak with respect to typical nitrile; C,N,O are implicated in the carrier.	Hagen 1982; d'Hendecourt et al. 1986 Allamandola et al. 1988 Sandford & Allamandola 1990 Bernstein et al. 1997
N_2-mixtures	no 2165 cm^{-1} band.		Elsilia, Allamandola & Sandford 1997
Grain Surface Reactions			
Hydrogenation/Nitrogen Addition			
H+CO	HCO		van IJzendoorn et al. 1983
N+HCO	NCHO* to HNCO	Stabilization of the excited NCHO* to its lowest energy level form, HNCO. Gas phase HNCO has band at 2269 cm^{-1}, HNCO in solid argon has band at 2259 cm^{-1}.	Jacox & Milligan 1964 Bernstein 1998
	NCHO* to NCHO	If NCHO is the final product, hydrogenation will occur to NH_2CHO. NH_2CHO observed in gas phase in ISM.	Tielens & Hagen 1982 Kuan & Snyder 1996
Acid-Base Chemistry			
CO, NH_3		In conjunction with UV Photolysis: band assigned to OCN^-	Grim 1988; Grim & Greenberg 1987b;1989; Schutte & Greenberg 1997
H_2O,NH_3, HNCO		Without UV. Deposition onto 12K substrate yields OCN^- and NH_4^+, band at 2171 cm^{-1} grows during warm-up.	Keane & Schutte 1998
Ion Bombardment of Ices			
H_2O:NH_3:CH_4	XCN probable	(≃ 1:3:2); MeV protons; produces 2170 cm^{-1} band.	Moore et al. 1983
H_2O:N_2:CO_2	XCN probable	(≃ 5:1:1 and 1:1:1); Band at 2180 cm^{-1}.	Moore et al. 1983
H_2O: CH_4: NH_3 or N_2 ices (2:1:1)	R-O-CN	30-60 keV helium, argon ions; produces 2165 cm^{-1} band. R is a species belonging to the refractory residue.	Palumbo et al. 1998

Ultraviolet photolysis, ion bombardment, and thermal processing all probably play a role in the interstellar chemistry of ices, although each mechanism may have greater relevance under specific astrophysical conditions. Recent studies of the evolution of dust between the dense and diffuse interstellar medium are probing the physical conditions under which refractory species are observed.

Production of the 3.4 μm band in the outflow of the evolved carbon star CRL 618 suggests that the carrier of this band is not formed by the energetic processing of interstellar ices. Even so, the incorporation of the hydrocarbons into the next generation of molecular clouds should occur, and the absence of the 3.4 μm band in dense cloud spectra remains a serious concern.

The remarkable comparison of the diffuse ISM feature to that found in the Murchison meteorite underscores the need to fully understand the nature and evolution of interstellar organic solids. Is the similarity an indicator of unaltered interstellar organics in the meteorite? We must know where these materials are made, how they are created, how they are distributed throughout the interstellar medium, and what type of processing they have endured. Only then will we have a chance to learn about the true availability of these organic constituents for incorporation into planetary systems during the star and planet formation processes.

Laboratory results clearly indicate that ice processing leads to the production of some types of organic material, and the 4.62 μm absorption band observed in dense molecular clouds towards embedded protostellar objects is a likely example. Organic residues produced by energetic processing of astrophysically plausible initial ice mixtures reproduce the 4.62 μm (2165 cm^{-1}) band quite well in both peak position and profile. Ion bombardment has the advantage that it would allow N_2, possibly the dominant nitrogen reservoir in molecular clouds, to participate in the chemistry of the ices. Ions are more energetic than UV photons and can break apart the strong N_2 bond.

The eventual identification of the interstellar 4.62 μm band may reveal an important clue to the production, evolution, and distribution of interstellar organic matter. Whatever identification laboratory studies eventually reveal for the carrier of the interstellar 4.62 μm band, it is now clear that the spectrum of the carrier will not exhibit any other strong bands in the mid-infrared as, ISO has shown there are none in the interstellar data. The formation site of the hydrocarbon 3.4 μm band and the carrier and process which results in the interstellar 4.62 μm band are important to the study of organic material in planetary systems, since these compounds may well have been preserved in some form in meteorites and comets.

REFERENCES

[1] Jones A.P., From Stardust to Planetesimals, edited by Y.J. Pendleton and A.G.G.M. Tielens (ASP, San Francisco, 1997) p. 97.

[2] de Graauw Th., Haser L.N., Beintema D.A., *et al.*, *A&A* **315** (1996) L49.
[3] Kessler M.F., Steinz J.A., Anderegg M.E., *et al.*, *A&A* **315** (1996) L27.
[4] First ISO Results, *A&A* **315** (1996) L27.
[5] Sandford S.A., Allamandola L.J., Tielens A.G.G.M., *et al.*, *ApJ* **371** (1991) 607.
[6] Pendleton Y.J., Sandford S.A., Allamandola L.J., Tielens A.G.G.M. and Sellgren K., *ApJ* **437** (1994) 683.
[7] Bridger A., Wright G. and Geballe T., Infrared Astronomy with Arrays: The Next Generation, edited by I.S. McLean (Kluwer, Dordrecht, 1993) p. 537.
[8] Wright G., Geballe T., Bridger A. and Pendleton Y.P., New Extragalactic Perspectives in the New South Africa, edited by D. Block and J.M. Greenberg (Kluwer, Dordrecht, 1996) p. 143.
[9] Pendleton Y.J., The Cosmic Dust Connection, edited by J.M. Greenberg (Kluwer, Dordrecht, 1996) p. 71.
[10] Soifer B.T., Russell R.W. and Merrill K.M., *ApJ* **207** (1976) L83.
[11] Wickramasinghe D.T. and Allen D.A., *Nature* **287** (1980) 518.
[12] Butchart I., McFadzean A.D., Whittet D.C.B., Geballe T.R. and Greenberg J.M., *A&A* **154** (1986) L5.
[13] McFadzean A.D., Whittet D.C.B., Longmore A.J., Bode M.F. and Adamson A.J., *MNRAS* **241** (1989) 873.
[14] Adamson A.J., Whittet D.C.B. and Duley W.W., *MNRAS* **243** (1990) 400.
[15] Lequeux J. and Jourdain De Muizon M., *A&A* **240** (1990) L19.
[16] Brooke T.Y., Sellgren K. and Smith R.G., *ApJ* **459** (1996) 209.
[17] Chiar J.E., Adamson A.J. and Whittet D.C.B., *ApJ* **472** (1996) 665.
[18] Allamandola L.J., Sandford S.A. and Tielens A.G.G.M., *ApJ* **399** (1992) 134.
[19] Allamandola L.J., Sandford S.A., Tielens A.G.G.M. and Herbst T.M., *Science* **260** (1993) 64.
[20] Ehrenfreund P., Robert F., d'Hendecourt L. and Behar F., *A&A* **252** (1991) 712.
[21] Pendleton Y., New Extragalactic Perspectives in the New South Africa, edited by D. Block and J.M. Greenberg (Kluwer, Dordrecht, 1996) p. 135.
[22] de Vries M.S., Reihs K., Wendt H.R., *et al.*, *Geochim. Cosmochim. Acta* **57** (1993) 933.
[23] Greenberg J.M. and Yencha A.J., Interstellar Dust & Related Topics, I.A.U. Symposium No. 52, edited by J.M. Greenberg and H.C. Van de Hulst (Reidel, Dordrecht, 1973) p. 309.
[24] d'Hendecourt L.B., Allamandola L.J., Grim R.J.A. and Greenberg J.M., *A&A* **158** (1986) 119.
[25] Allamandola L.J. and Sandford S.A., Dust in the Universe, edited by M.E. Bailey and D.A. Williams (Cambridge University Press, Cambridge, 1988) p. 229.
[26] Allamandola L.J., Sandford S.A. and Valero G., *Icarus* **76** (1988) 225.

[27] Grim R.J.A., Greenberg J.M., De Groot M.S., *et al.*, *A&AS* **78** (1989) 161.
[28] Grim R.J.A. and Greenberg J.M., *A&A* **181** (1987) 155.
[29] Schutte W.A., The Cosmic Dust Connection, edited by J.M. Greenberg (Kluwer, Dordrecht, 1996) p. 1.
[30] Schutte W.A., Tielens A.G.G.M., Whittet D.C.B., *et al.*, *A&A* **315** (1996) L333.
[31] Bernstein M.P., Sandford S.A. and Allamandola L.J., *ApJ* **472** (1996) 127.
[32] Greenberg J.M., Comets, edited by L.L. Wilkening (University of Arizona Press, Tucson, 1982) p. 131.
[33] Pendleton Y.J. and Chiar J.E., From Stardust to Planetesimals, ASP 122, edited by Y.J. Pendleton and A.G.G.M. Tielens (ASP, San Francisco, 1997) p. 179.
[34] Geballe T.R., Chiar J.E., Pendleton Y.J. and Tielens A.G.G.M., ISO's View on Stellar Evolution, edited by L.B.F.M. Waters, C. Waelkens, K.A. van der Hucht and P.A. Zaal (Kluwer, Dordrecht, 1998) p. 457.
[35] Chiar J.E., Pendleton Y.J., Geballe T.R. and Tielens A.G.G.M., *ApJ* **507** (1998) 281.
[36] Loup C., Forveille T., Omont A. and Paul J.F., *A&A* **99** (1993) 291.
[37] Gammie C.F., Knapp G.R., Phillips T.G. and Falgarone E., *ApJ* **345** (1989) L87.
[38] Cernicharo J., Guelin M., Martin-Pintado J., Penalver J. and Mauersberger R., *A&A* **22** (1989) L1.
[39] Burton W.B. and Geballe T., *MNRAS* **223** (1986) 13P.
[40] McKee C.F., Interstellar Dust, edited by L.J. Allamandola and A.G.G.M. Tielens (Kluwer, Dordrecht, 1989) p. 431.
[41] Baratta G.A. and Strazzulla G., *A&A* **240** (1990) 429.
[42] Pendleton Y.J., *Orig. Life Evol. Biosphere* **27** (1997) 53.
[43] Cronin J.R., Pizzarello S. and Cruikshank D.P., Meteorites and the Early Solar System, edited by J.F. Kerridge and M.S. Matthews (Univ. of Arizona Press, Tucson, 1988) p. 819.
[44] Cronin J.R. and Chang S., The Chemistry of Life Origin, edited by J.M. Greenberg and V. Pirronello (Kluwer, Dordrecht, 1994) p. 209.
[45] Pizzarello S., Krishnamurthy R.V., Epstein S. and Cronin J.R., *Geochim. Cosmochim. Acta* **55** (1991) 905.
[46] Kerridge J.F., Chang S. and Shipp R., *Geochim. Cosmochim. Acta* **51** (1983) 2527.
[47] Cruikshank D.P., Rouch T.L., Owen T.C., Quirico E. and de Bergh C., Solar System Ices, edited by B. Schmitt, M. Festou and C. de Bergh (Kluwer, Dordrecht, 1997) p. 655.
[48] Soifer B.T., Puetter R.C., Russel R. W., *et al.*, *ApJ* **232** (1979) L53.
[49] Lacy J.H., Baas F., Allamandola L.J., *et al.*, *ApJ* **276** (1984) 533.
[50] Larson H.P., Davis D.S., Black J.H. and Fink U., *ApJ* **299** (1985) 873.
[51] Tegler S.C., Weintraub D.A., Allamandola L.J., *et al.*, *ApJ* **411** (1993) 260.

[52] Tegler S.C., Weintraub D.A., Rettig T.W., *et al.*, *ApJ* **439** (1995) 279.
[53] Weintraub D.A., Tegler S.C., Kastner J.H. and Rettig T.W., *ApJ* **423** (1994) 674.
[54] Pendleton Y.J., Tielens A.G.G.M., Tokunaga A.T. and Bernstein M., *ApJ* **513** (1999) in press.
[55] Hagen W., Ph.D. Thesis (Rijks Universiteit, Leiden, 1982) p. 1.
[56] Grim R.J.A., Ph.D. Thesis (Rijks Universiteit, Leiden, 1988) p. 1.
[57] Schutte W.A. and Greenberg J.M., *A&A* **317** (1997) L43.
[58] Palumbo M.E., Strazzulla G., Pendleton Y.J. and Tielens A.G.G.M., *ApJ* (1999) in press.
[59] Moore M.H., Bertram D., Khanne R. and A'Hearn M.F., *Icarus* **54** (1983) 388.
[60] Elsilia J., Allamandola L.J. and Sandford S.A., *ApJ* **479** (1997) 818.
[61] Cruikshank D.P., Roush T.L., Owen T.C., *et al.*, *Science* **261** (1993) 742.
[62] Cruikshank D.P., Roush T.L., Bartholomew M.J., *et al.*, *Icarus* **135** (1998) 389.
[63] Bernstein M.P., Sandford S.A., Allamandola L.J., Chang S. and Scharberg M.A., *ApJ* **454** (1995) 327.
[64] Bernstein M.P. (1998) in preparation.
[65] Demyk K., Dartois E., d'Hendecourt L., *et al.*, *A&A* **339** (1998) 553.
[66] Strazzulla G. and Johnson R.E., Comets in the Post-Halley Era, edited by R.L. Newburn Jr., M. Neugebauer and J. Rahe (Kluwer, Dordrecht, 1991) p. 243.
[67] Strazzulla G., From Stardust to Planetesimals, ASP Conf. Series 122, edited by Y.J. Pendleton and A.G.G.M. Tielens (ASP, San Francisco, 1997) p. 423.
[68] Strazzulla G., Solar System Ices, edited by B. Schmitt, C. de Bergh and M. Festou (Kluwer, Dordrecht, 1997) p. 281.
[69] van IJzendoorn L., Allamondola L.J., Baas F. and Greenberg J.M., *J. Chem. Phys.* **78** (1983) 7019.
[70] Charnley S.B. *MNRAS* **291** (1997) 455.
[71] Charnley S.B., Astronomical and Biochemical Origins & the Search for Life in the Universe, edited by C.B. Cosmovici, S. Bowyer and D. Wertheimer (Editrice Compositori, Bologna, 1997) p. 89.
[72] Charnley S.B. (1998) submitted.
[73] Tielens A.G.G.M. and Hagen W., *A&A* **114** (1982) 245.
[74] Kuan Y. and Snyder L., *ApJ* **470** (1996) 981.

LECTURE 9

Laboratory Analogues for Interstellar Carbon Dust

L. Colangeli[1], V. Mennella[1],
E. Bussoletti[2], P. Palumbo[2] and A. Rotundi[2]

[1] *Osservatorio Astronomico di Capodimonte, Via Moiariello 16, 80131 Napoli, Italy*
[2] *Istituto Universitario Navale, Via A. De Gasperi 5, 80133 Napoli, Italy*

Abstract. Carbon grains are, together with silicates, the major component of cosmic dust. Various forms of carbon–based particles are formed in the envelope of C–rich stars and populate the interstellar medium, either as isolated grains or as components of mixed and/or core–mantle particles, including silicates. The characterisation of laboratory analogues provides a fundamental contribution to the interpretation of astronomical observations concerning cosmic dust. In the recent years, various experimental groups have started a systematic quantitative program of laboratory analyses aimed at studying the structure, chemistry and optical properties of carbon materials. The results obtained contribute to form a sort of reference frame in which astronomical data can have a clearer interpretation. The final goal is to discriminate the properties of carbon in different space environments and to identify the driving processes which determine their evolution in space. In this paper a summary of the most recent results concerning the simulation of cosmic carbon analogues is reported, with the aim of offering an updated view of the present understanding of cosmic carbon properties.

1. INTRODUCTION

The presence of various forms of carbon dust in interstellar and circumstellar space is evidenced by several spectroscopic fingerprints. In particular, the ubiquitous interstellar extinction band in the UV, falling at 4.60 μm^{-1} (220 nm), has been attributed to various carbon based materials, such as graphite, amorphous carbon, coals, polycyclic aromatic hydrocarbons or composite grains [1–5]. However, so far, no definite identification of the actual nature of the carrier(s) has been possible, taking into account the observational constraints along different lines of sight: stability of the peak position within 1% and variations of the band width by 25% [6]. An extinction bump due to carbon grains is also observed in the circumstellar environment of some carbon stars, centered around 4.17 μm^{-1} (240 nm) [7]. The infrared spectral range is rich of features due to typical C–C and C–H vibrational modes [8]. In this case, however, the identification of the actual carrier is even more difficult as several different hydrogen–rich carbon–based molecular or solid materials could potentially produce such features.

Observed features may be diagnostic of the actual physical, chemical and structural properties of carbon in different space environments. However, their interpretation must rely on a deep and systematic comparison with data obtained, under carefully controlled conditions, in laboratory experiments. A proper description of cosmic carbon compounds must account both for the elemental abundance constraints [9] and for the chemico–physical processes active in space, that may favour the formation (or the evolution) of some specific molecular or solid compounds [10]. It is thus clear, that observations alone are not sufficient to derive a full description of cosmic carbon evolution. Input must come from laboratory experiments aimed at simulating the formation and evolution of materials in space.

2. THE LABORATORY APPROACH

Several research groups are nowadays involved in laboratory experiments aimed at simulating the formation and evolution of carbon dust, accounting for the actual processes active in space. The common aim is to reach a clear description of cosmic carbon properties and their evolution. To achieve this goal it is necessary to correlate extinction, absorption and emission spectral features with the actual carbon form. Moreover, the formation mechanisms and the ability of species to survive in space conditions must be accounted for.

Laboratory experiments must then be oriented to: a) produce various forms of carbon grains under well controlled boundary conditions, b) link their physical, chemical, structural and optical properties; c) analyse the evolution due to mechanisms such as thermal, UV and ion processing.

The production methods typically used to condense carbon–based submicron grains are:

- arc discharge between carbon/graphite electrodes in an inert (Ar, He) or reducing (H_2) atmosphere [11];
- burning of hydrocarbons, such as benzene [11];
- high power laser bombardment of bulk targets in an inert or reducing atmosphere [12];
- resistive heating of graphite rods followed by molecular beam extraction [13, 14];
- IR laser pyrolysis of gas phase molecular species [15];
- quenching of hydrocarbon plasmic gas in vacuum [4].

It must be stressed that very often these methods are not able to simulate actual formation mechanisms active in space, but are used to produce grains with well controlled structure and morphology. They are appropriate for studying the influence on the final grain properties of boundary conditions during production, even if they are rather different from actual synthesis methods expected in space. The produced grains are useful for testing effects induced by further processing. In this way, the laboratory results have a strong impact on the interpretation of astronomical observations.

Of course, to have a full characterisation of laboratory samples, several methods of analysis must be applied to them. A typical set of tools that can be used to monitor structure, morphology, chemistry and optical behaviour of grains is summarised in Table I.

3. RESULTS

Examples of carbon grain spectra from the vacuum UV to the mm range are shown in Figure 1. The following typical spectral features are well evidenced: a large bump at $80 - 90$ nm, due to $\sigma - \sigma^*$ electronic transitions; a bump, more or less pronounced depending on the sample, at $200 - 260$ nm, due to $\pi - \pi^*$ electronic transitions. The hydrogen content in the production chamber has a relevant impact on the structure of grains. In fact, the $\pi - \pi^*$ bump varies in intensity and shape according to the relative amount of H_2 and Ar (see Fig. 2). The observed behaviour can be interpreted by considering that hydrogen mainly promotes sp^3 bond formation and localisation of the π bonds. A progressive increase in number of sp^2 bonds in the carbon structure is obtained as the hydrogen content is diminished.

Vibration modes of C–C and C–H bonds fall in the mid-IR and are evident in the hydrogen–rich sample ACH2 (Fig. 3). In the C–H stretching mode region, $3.3 - 3.4$ μm (*i.e.*, $3000 - 2900$ cm^{-1}), the ACH2 grains show a typical aliphatic character, *i.e.*, dominated by sp^3 bonds. Interestingly enough, grains produced by IR laser pyrolysis of gas phase molecular species show a prevalent 3.28 μm band, *i.e.*, a dominant sp^2 aromatic character in the as-produced material [15]. A similar result is obtained after thermal annealing of ACH2 in

Table I. — Tools for characterising cosmic dust analogues.

Technique	Information
Scanning electron microscopy (SEM)	3-D aggregation, surface status, grain shape and size
Transmission electron microscopy (TEM)	2-D grain distribution, internal structure
Electron diffraction	Crystalline degree
Energy dispersive X-ray analysis (EDX)	Elemental distribution
Far UV - UV spectroscopy	Electronic transitions
Visible spectroscopy	Optical gap
IR spectroscopy	C–X resonances
FIR spectroscopy	Structure
Electron energy loss spectroscopy (EELS)	Electronic properties

vacuum up to 500 °C (Fig. 3), even if, in this case, the band intensity is reduced by the progressive hydrogen effusion. These differences clearly indicate that caution must be applied in comparing properties of carbon grains produced by different techniques: their nature may be strongly affected by the condensation environment. Thus, a fruitful comparison of different materials must be based on a careful identification of structural, morphological and spectral properties.

The complexity of the carbon grain structures condensed in the laboratory has been recently evidenced [18]. Actually, carbon grains produced by arc discharge present various forms. The most abundant chain–like aggregates (CLA) are an amorphous carbon smoke formed by spherical grains with size between 7 and 15 nm. Other more organised structures are present: poorly graphitised carbons, bucky–onions, graphitic ribbons, single crystal carbon platelets. Most of these components derive from an auto–annealing process during grain coagulation. It must however be stressed that they are sufficiently rare, with respect to the CLA, that they cannot be considered responsible for the spectral behaviour observed for carbons. This implies that it is the intimate structure of the CLA that determines the main optical properties of synthesised carbons.

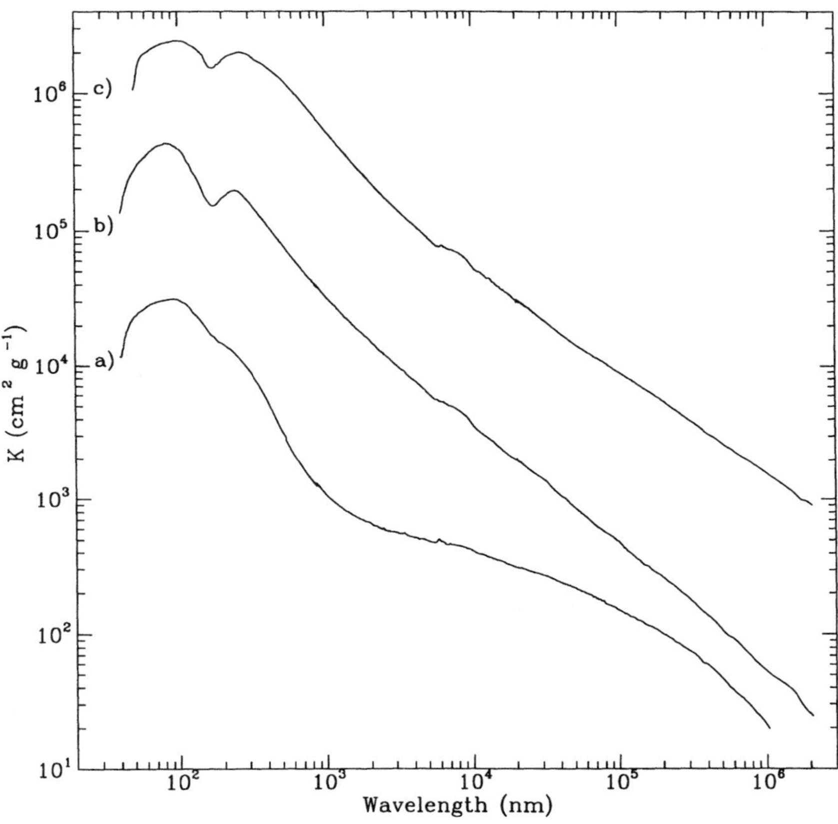

Fig. 1. — Mass extinction efficiency of carbon grains produced in a 10 mbar H_2 atmosphere (ACH2 – curve a), by arc discharge in a 10 mbar Ar atmosphere (ACAR – curve b) and by burning benzene in air (BE – curve c). The BE and ACH2 spectra are shifted down and up by one order of magnitude, respectively [11].

4. THE PROBLEM OF CLUMPING

Before considering the data obtained from laboratory simulations applicable to the interpretation of astronomical observations, a careful check is necessary to identify potential systematic errors that can affect their validity. In this respect, a typical problem that must be faced in laboratory experiments is the clumping of grains. Actually, as already mentioned in the previous section, grains collected in condensation experiments usually occur in aggregated forms. Therefore, the results obtained for the optical behaviour must be properly

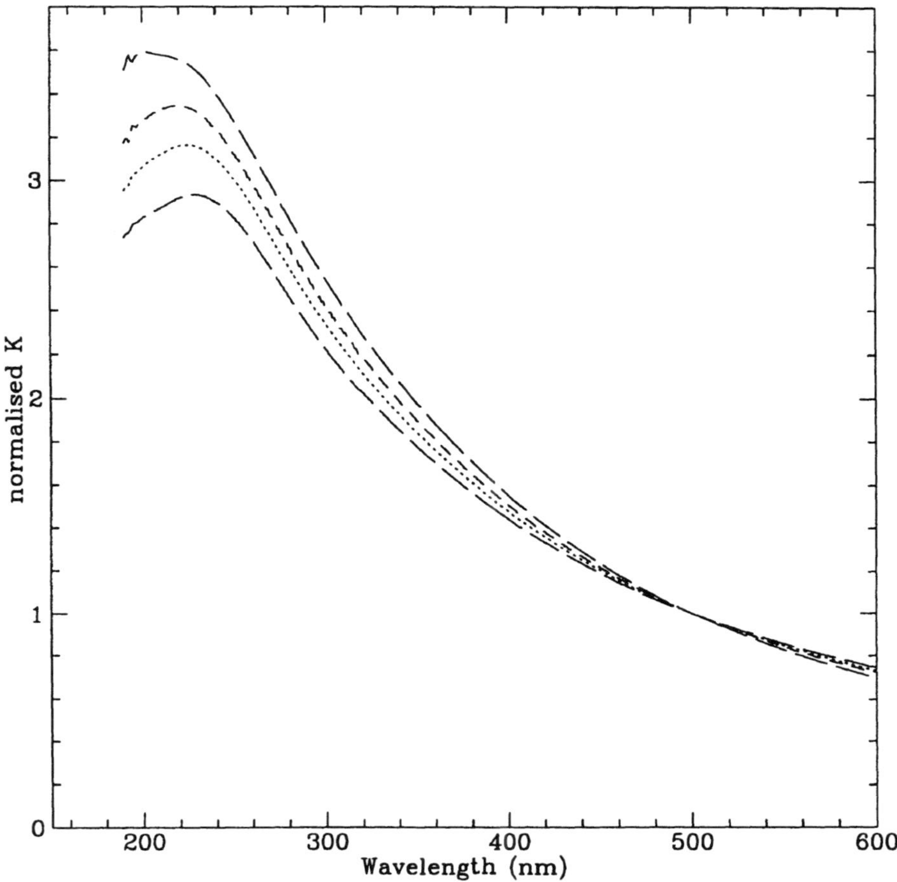

Fig. 2. — Normalised mass extinction coefficient for carbon grains produced in mixed H_2 – Ar atmosphere. Curves from bottom to top refer to 0.4, 0.6, 0.8 and 1% of H_2 in Ar [16].

interpreted in order to be applicable to cosmic dust because, depending on the actual environment, grains in space are expected to occur either as isolated entities or in clumped structures [19].

Appropriate theories have been elaborated to describe the aggregate formation in terms of ballistic particle cluster aggregates or ballistic cluster-cluster aggregates [20]. Various models are also available to compute the optical properties of clumped grains, such as the continuous distribution of ellipsoids, the effective medium theories, the discrete dipole approximation or the discrete multipole approximation [21–23]. This network of theoretical tools has been applied to laboratory data to derive the optical properties of isolated grains.

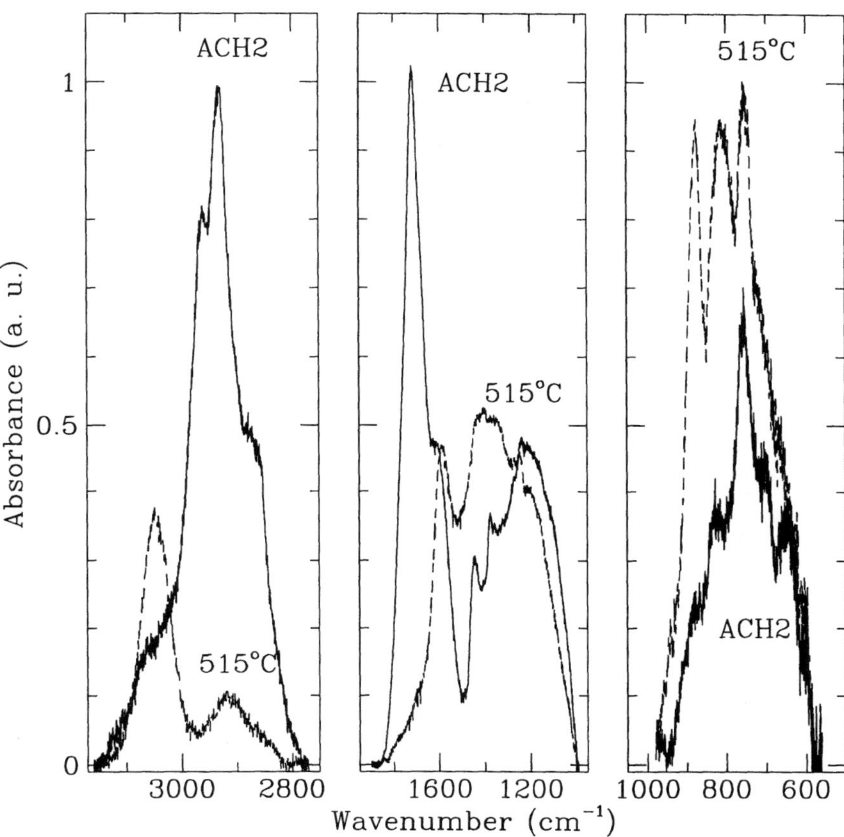

Fig. 3. — Absorbance, normalised to the continuum, of carbon grains produced by arc discharge in a 10 mbar H_2 atmosphere (ACH2). The absorbance spectrum of the sample after thermal annealing in vacuum at 515 °C is also shown [17].

Another approach to obtain the optical properties of single grains is based on the use of experimental techniques able to isolate particles. In this case, the grains are trapped in solid matrices for spectroscopic measurements. In Figure 4 an example of extinction obtained for carbon grains "trapped" in solid boron oxide matrix is shown. Only carbon grains produced at 1 mbar by arc discharge in Ar and trapped in boron oxide display a sharp UV feature falling at around 200 nm. This result is in line with previous findings [13, 14] for carbon grains isolated in solid argon cold matrix. Even if this approach is certainly promising, we have to recall that also in this case appropriate models have to be applied to render the spectroscopic data free from the matrix effects.

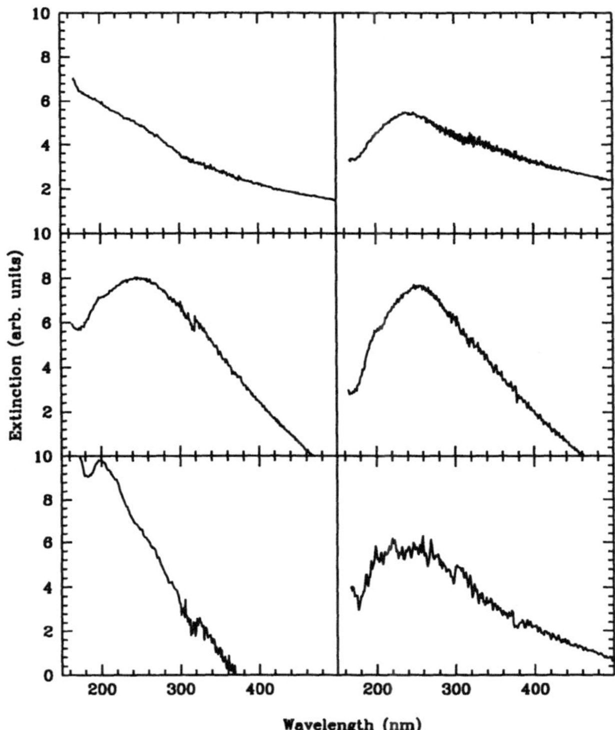

Fig. 4. — Extinction of carbon grains produced by arc discharge in Ar atmosphere at 1 mbar (left panels) and at 10 mbar (right panels). The top panels refer to grains collected on a quartz substrate, the middle panels refer to grain embedded in a boron oxide matrix, while the bottom panels refer to grains "trapped" in the boron oxide matrix. Only grains produced at 1 mbar and trapped in boron oxide occur as isolated particles and display a sharp UV feature at 200 nm (left bottom panel).

Finally, a precise identification of the actual isolation degree of grains is very difficult.

5. PROCESSING

The use of various production techniques allows us to synthesise grains characterised by different physical and optical properties. However, in order to investigate the evolution of cosmic carbon dust, laboratory experiments must be also oriented to study the modifications induced by processes similar to those active in space.

Experiments on thermal annealing can be useful in identifying possible evolutionary processes acting on carbon grains, even if other mechanisms are expected to be more efficient in space environment. In Figure 3 we showed that the thermal annealing in vacuum of hydrogenated carbon grains produces modifications in the IR spectral properties, compatible with a progressive aromatisation of the carbon structure. This behaviour is confirmed by the analysis of the UV spectrum (Fig. 5). The appearance of a UV feature that progressively moves towards longer wavelengths (up to 260 nm at 800 °C) indicates that the structure of grains changes from a mainly aliphatic to a dominant aromatic status, *i.e.*, the carbon network tends to arrange in an increasing number of larger sp^2 clusters.

Processes closer to actual mechanisms active in space are UV irradiation and ion bombardment. The UV irradiation of hydrogenated carbon with doses up to 4×10^{22} eV cm^{-2} produces effects similar to the thermal annealing (Fig. 6). The maximum applied dose must be compared with the expected UV dose of 3×10^{23} eV cm^{-2}, deposited in carbonaceuos grains on a diffuse cloud time–scale (3×10^7 yr) [24]. The laboratory results show again the appearance of a bump near 220 nm. However, the shape results in a sharper band than that obtained by thermal annealing. This result could indicate that the sp^2 clusters produced by UV irradiation have a narrower size distribution.

Ion bombardment of hydrogenated carbon at energy doses up to 660 eV/C–atom induces again the formation of a UV bump (Fig. 7). However, on a diffuse interstellar medium time–scale, the expected dose for carbon grains should not exceed about 30 eV/C–atom [25]. Thus, cosmic ray bombardment should not be able to compete with effects produced by UV irradiation. It is interesting to note the behaviour of dehydrogenated carbon grains under the same ion treatment (Fig. 8). It seems that the aromatic structure of these grains is destroyed to some extent by the processing. In conclusion, ion bombardment seems able to aromatize disordered grains and disorder aromatic structures.

6. FINAL CONSIDERATIONS

Laboratory experiments on carbon dust are aimed at producing particles characterised by different properties, in terms of morphology, structure, chemistry and optical behaviour. The processing of samples can show how sensitive various kinds of carbon grains are to mechanisms that are expected to drive the evolution of cosmic dust in the interstellar medium. The results obtained so far, even if not conclusive, allow us to infer some important properties of carbon, that can be valid for cosmic grains.

In particular, it is nowadays evident that the intimate structure of solid particles, more than their morphology, can drive the optical behaviour.

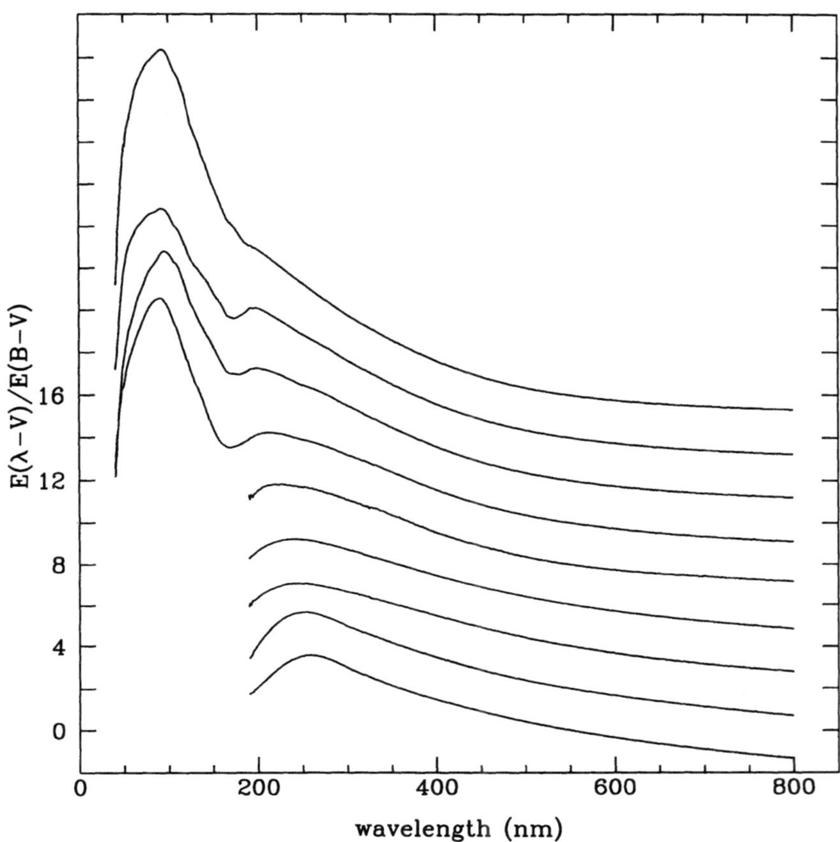

Fig. 5. — Extinction, normalised to $B-V$, of carbon grains produced by arc discharge in H_2 atmosphere at 10 mbar (top curve) and after thermal annealing for 3 hours in vacuum. The curves are arbitrarily shifted in ordinate and refer (from top to bottom) to different temperatures up to 800 °C (bottom curve). The appearance of a bump at around 210 nm and its progressive shift to 260 nm testifies to the structural evolution of the material [26].

The presence of hetero–atoms, hydrogen primarily, in the network may explain the dominance of aliphatic structures. On the other hand, the processing of grains in the interstellar medium, mainly by UV irradiation, may produce an aromatisation of disordered grain networks, with a consequent modification of the optical properties.

If we apply the laboratory data available so far to interpret main features observed in the interstellar medium, we may conclude that a form of aromatic

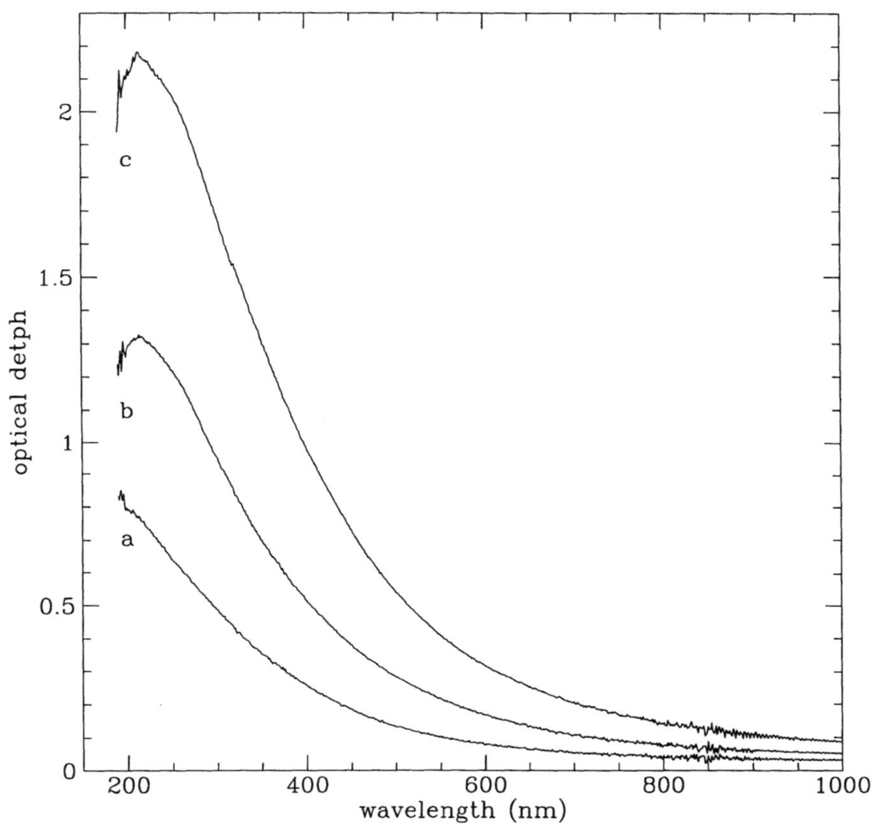

Fig. 6. — Optical depth of hydrogenated carbon grains, as produced (a) and after irradiation in vacuum with UV doses of 1×10^{22} eV cm^{-2} (b) and 4×10^{22} eV cm^{-2} (c) [27].

carbon must be the carrier of the ubiquitous UV extinction band, while the 3.4 μm absorption feature, observed towards several galactic center sources (see article by Pendleton this volume), could be associated with carbon particles dominated by an aliphatic character. This discrepancy could be reconciled if we admit the presence of different carbon grain populations in the interstellar medium: small aromatic grains could be active in the UV, while larger (or agglomerated?) aliphatic grains could contribute to the IR absorption features.

It is, however, clear that further investigation in laboratory is needed before solving the puzzles that remain open on the nature of carbon dust in space.

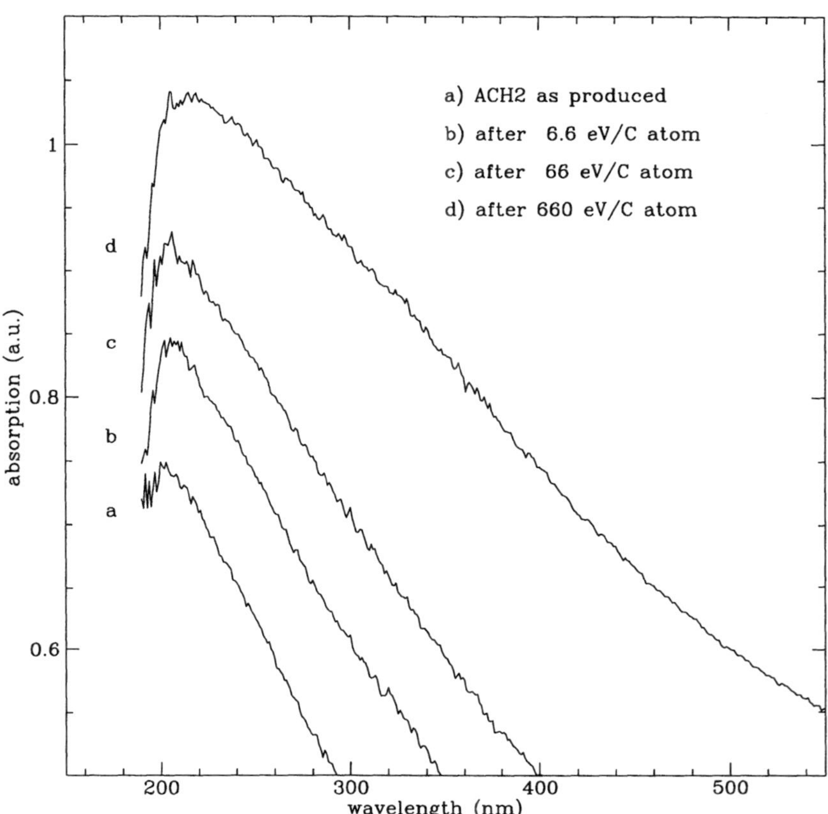

Fig. 7. — Absorption of hydrogenated carbon grains, as produced (a) and after 6.6 eV/C–atom (b), 66 eV/C–atom (c) and 660 eV/C–atom (d). The ion bombardment has been performed by using a beam of 3 keV He^+ ions at a current of 1.5 μA cm^{-2}, to minimise heating [27].

ACKNOWLEDGMENT

This work has been supported by MURST, ASI and CNR contracts.

REFERENCES

[1] Draine B.T. and Lee H.M., *ApJ* **285** (1984) 89.
[2] Bussoletti E., Colangeli L. and Orofino V., *ApJ* **321** (1987) L87.

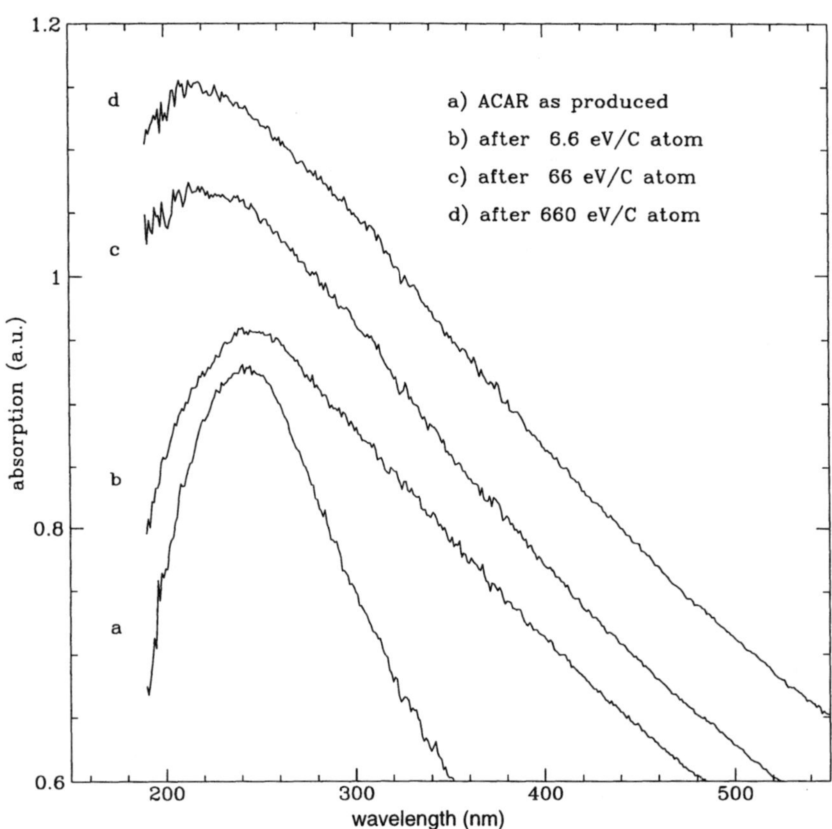

Fig. 8. — Absorption of carbon grains produced by arc discharge in Ar atmosphere, as produced (a) and after 6.6 eV/C–atom (b), 66 eV/C–atom (c) and 660 eV/C–atom (d). The UV peak tends to shift towards shorter wavelengths [27].

[3] Papoular R., Guillois O., Nenner I., *et al.*, *Planet. Space Sci.* **43** (1995) 1287.

[4] Sakata A., Wada S., Tokunaga A.T., *et al.*, *ApJ* **430** (1994) 311.

[5] Joblin C., Léger A. and Martin P., *ApJ* **393** (1992) L79.

[6] Fitzpatrick E.L. and Massa D., *ApJ* **307** (1986) 286.

[7] Buss R.H., Lamers H. and Snow T.P., *ApJ* **347** (1989) 977.

[8] Pendleton Y.J., Sandford S.A., Allamandola L.J., Tielens A.G.G.M. and Sellgren K., *ApJ* **437** (1994) 683.

[9] Snow T.P. and Witt A.N., *Science* **270** (1995) 1455.

[10] Tielens A.G.G.M. and Allamandola L.J., Interstellar Processes, edited by D.J. Hollenbach and H.A. Thronson (Reidel, Dordrecht, 1987) p. 397.
[11] Colangeli L., Mennella V., Palumbo P., Rotundi A. and Bussoletti E., *A&AS* **113** (1995) 561.
[12] Stephens J.R., *ApJ* **237** (1980) 450.
[13] Schnaiter M., Mutschke H., Henning Th., *et al.*, *ApJ* **464** (1996) L187.
[14] Schnaiter M., Mutschke H., Dorschner J., Henning Th. and Salama F., *ApJ* **498** (1998) 486.
[15] Herlin N., Bohn I., Reynaud C., *et al.*, *A&A* **330** (1998) 1127.
[16] Mennella V., Brucato J.R., Bussoletti E., *et al.* Cosmic Physics in the Year 2000, edited by S. Aiello, N. Iucci, G. Sironi, A. Treves and U. Villante (Editrice Compositori, Bologna, 1997) p. 225.
[17] Mennella V., Colangeli L., Cecchi Pestellini C., *et al.*, From Stardust to Planetesimals, edited by M.E. Kress, A.G.G.M. Tielens and Y.J. Pendleton (NASA CP–3343, Moffett Field, 1997) p. 109.
[18] Rotundi A., Rietmeijer F.J.M., Colangeli L., *et al.*, *A&A* **329** (1998) 1087.
[19] Mathis J.S. and Whiffen G., *ApJ* **341** (1989) 808.
[20] Stognienko R., Henning Th. and Ossenkopf V., *A&A* **296** (1995) 797.
[21] Bazell D. and Dwek E., *ApJ* **360** (1990) 142.
[22] Ossenkopf V., *A&A* **251** (1991) 210.
[23] Kozasa T., Blum J. and Mukai T., *A&A* **263** (1992) 423.
[24] Jenniskens P., *A&A* **274** (1993) 653.
[25] Mennella V., Baratta G.A., Colangeli L., *et al.*, *ApJ* **481** (1997) 545.
[26] Mennella V., Colangeli L., Blanco A., *et al.*, *ApJ* **444** (1995) 288.
[27] Mennella V., Colangeli L., Palumbo P., *et al.*, *ApJ* **464** (1996) L191.

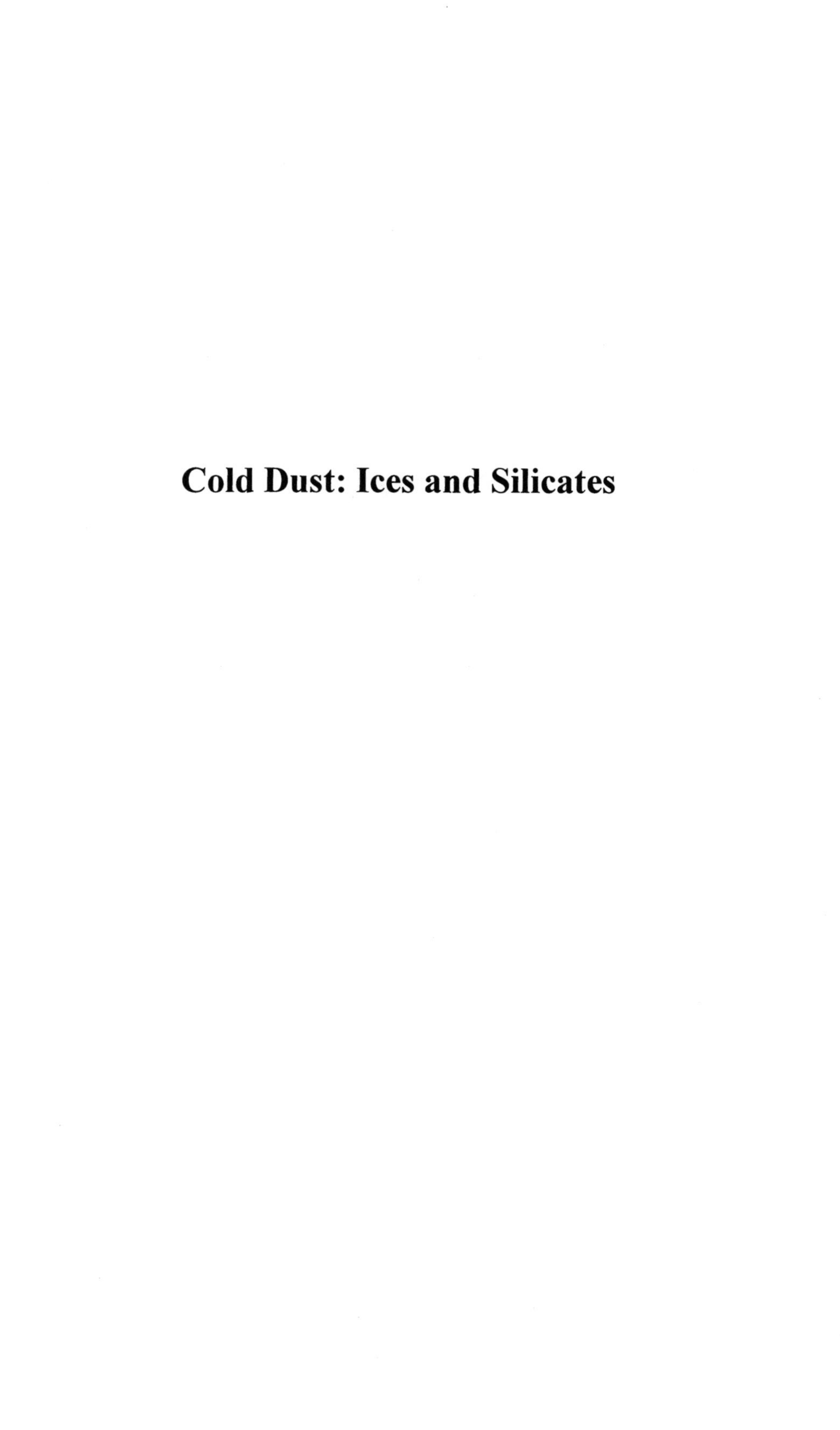

Cold Dust: Ices and Silicates

LECTURE 10

Combined SWS-LWS Spectra of Solid State Features

P. Cox[1] and P.R. Roelfsema[2,*]

[1] *Institut d'Astrophysique Spatiale, Université Paris XI, 91405 Orsay, France*
[2] *SRON Laboratory for Space Research Groningen, P.O. Box 800, 9700 AV Groningen, The Netherlands*

Abstract. The spectrometers on board ISO have for the first time allowed the measurement of complete infrared spectra from 2.5 to 196 μm of many astronomical sources revealing the richness of the signatures of the dust across the entire wavelength range. In this paper, we present combined SWS and LWS spectra which have been obtained towards compact H II regions where a variety of emission and absorption features are observed, and towards embedded infrared sources whose spectra are dominated by absorption bands.

1. INTRODUCTION

The launch in 1995 of the Infrared Space Observatory (ISO) [1] has had a profound impact on the study of interstellar dust. It opened the *entire* 2.5 $-$ 196 μm wavelength region for spectroscopic studies at spectral resolutions of $\frac{\lambda}{\Delta\lambda}$ from $\sim$ 100 to $\sim$ 500, and was ideally adapted for the study of the broad emission and absorption bands of interstellar dust grains and ices. The availability of complete infrared spectra up to 196 μm for a great number of astronomical sources allows the study of the dust bands together with the other components of the infrared spectrum: the infrared continuum, the atomic fine

* Presented on behalf of members of the SWS and LWS Teams.

structure lines and molecular gas bands. This richness of information is one of the greatest achievements of ISO and part of what has been called the *ISO revolution*.

This paper summarises some of the results obtained with combined SWS and LWS spectra for compact galactic H II regions and embedded infrared sources. The emphasis of this paper is on the importance of the complete wavelength coverage which was made possible with ISO. All the data presented in this paper have been reduced using the procedures described in the Appendix.

Remarkable results have also been obtained when combining SWS and LWS spectra for young pre-main sequence stars and evolved (AGB and post-AGB) stars. Examples can be found in the contributions in this volume by Waters *et al.* for Herbig Ae/Be stars and oxygen-rich circumstellar dust shells and by Sylvester for carbon-rich stellar envelopes. In the case of the oxygen-rich environments, a very rich series of new solid state features due to crystalline silicates was discovered.

2. COMPACT H II REGIONS

In order to study the distribution of elemental abundances in the Galaxy, infrared spectra from 2.5 to 196 μm have been measured for a sample of 46 compact H II regions in a series of ISO guaranteed time programs (*e.g.*, [2]). The wide spectral coverage allows the contemporaneous measurement of the major atomic lines (from O, N, S, Ne, Ar, Si and C) and the continuum and spectral signatures of dust. The sources are uniformly distributed in galacto-centric distance ($0 \leq R_{\rm gal} \leq 25\,{\rm kpc}$) enabling a derivation of the galactic abundance variation over larger size scales than in previous studies (*e.g.*, [3,4]). This data set will also show how the dust properties vary with the elemental abundances.

The compact nature of the H II regions which were measured does not allow an analysis of the distribution of the dust bands and continuum across the source, or their possible variation with distance from the exciting star. However, the large data set and the variety in the physical properties of the sources, in particular the spectral type of the exciting star(s), offers the possibility of studying global trends, such as the variation of the shapes of the dust bands with the hardness of the radiation field. The first results of this study have already been reported [5].

In Figure 1, the combined SWS-LWS spectra of four compact H II regions are shown to illustrate the spectral diversity of the sample. The spectrum of IRAS 18162−2048 presents a smooth continuum with prominent dust features in emission and in absorption, molecular absorption bands, but weak fine structure lines. In contrast, IRAS 21190+5140 has a very rich emission line spectrum, weak PAH bands and no indication of molecular features. IRAS 12073−6233 shows no broad features but a series of fine structure lines on top of a strong continuum. In all cases, strong continuum emission is present which is also detected at mid-infrared wavelengths ($\sim 10\,\mu$m). Note that

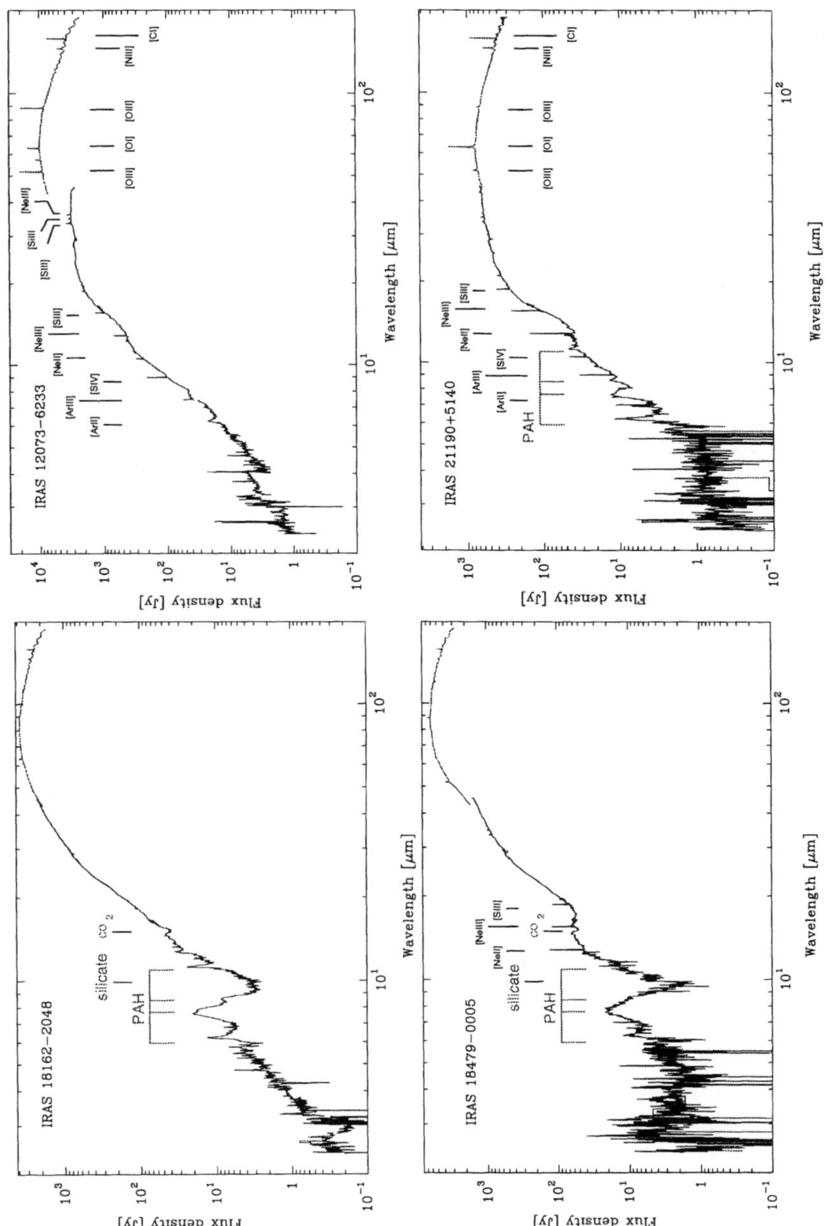

Fig. 1. — The combined SWS and LWS 2.5 – 196 μm spectra of four compact H II regions. The the atomic fine structures lines, and the dust and ice bands are labelled in the figure.

for IRAS 18162−2048 and IRAS 21190+5140 which are compact far-infrared sources, *i.e.*, point sources for both spectrometers, the match of the SWS and LWS spectra (around 40 μm) is nearly perfect. The two other sources in Figure 1 are more extended at mid- and far-infrared causing the match between the SWS and LWS spectra to be poorer: the jump is due to the fact that the beam of the LWS ($\sim 70''$) is larger than the SWS aperture ($\sim 20'' \times 33''$) in the overlapping wavelength region.

In the mid-infrared, the spectra of compact H II regions display the series of well-known emission features at 6.2, 7.7, 8.6 and 11.3 μm which are commonly attributed to Polycyclic Aromatic Hydrocarbons (PAHs) – see the contributions by Boulanger and Salama in this volume on small dust particles. Figure 2 presents the spectra between 5 and 13 μm of 18 H II regions. No continuum was subtracted from the data and the spectra are ordered by increasing mid-infrared continuum from bottom to top and from the left to the right panel. The spectra clearly show the characteristic C–C stretch (at 6.2 and 7.7 μm) and the C–H bending modes (8.6 and 11.3 μm) features of small aromatic compounds. The relative strengths of the bands vary significantly from source to source, in particular the 7.7/8.6 band ratio with values ranging from 0.6 to ~ 6. In some cases the 6.2 μm band is very strong, such as for IRAS 23133−6050 - see spectrum (j) in Figure 2. These variations are possibly due to different mixtures of neutral/ionized and compact/non-compact PAHs resulting from the different physical conditions in the sources, such as the hardness of the radiation field (*e.g.*, [2,5]). As can be seen in Figure 2, the strongest 8.6 μm bands are present in the H II regions with the strongest mid-infrared continua and high excitation atomic fine structure lines ([SIV] and [ArIII]). Similar results are derived in the case of the H II region/Photodissociation interface in M 17 [6, 7].

3. EMBEDDED INFRARED SOURCES

Depending on the geometry and/or the degree of evolution of the sources, the infrared spectra of compact H II regions can be entirely dominated by absorption bands. In one case, the source may be located behind a massive molecular cloud and observed through a large column density of molecular gas and dust, in another case, the young exciting star(s) is still embedded in its native environment and the small ionized region is surrounded by a large volume of dense molecular gas. The infrared spectra of embedded infrared sources have been important in the analysis of the refractory cores of dust grains (mainly the silicates) and have been the prime sources of information in deriving the chemical composition of the ice mantles which form around the dust cores in the cold regions deep inside molecular clouds. Infrared absorption spectra of sources embedded in dense molecular clouds have been obtained during the last few decades from the ground, airborne platforms (mainly the KAO), and the IRAS *LRS*, allowing major progress in our understanding of the composition of ices in the interstellar medium. In particular, these studies established the

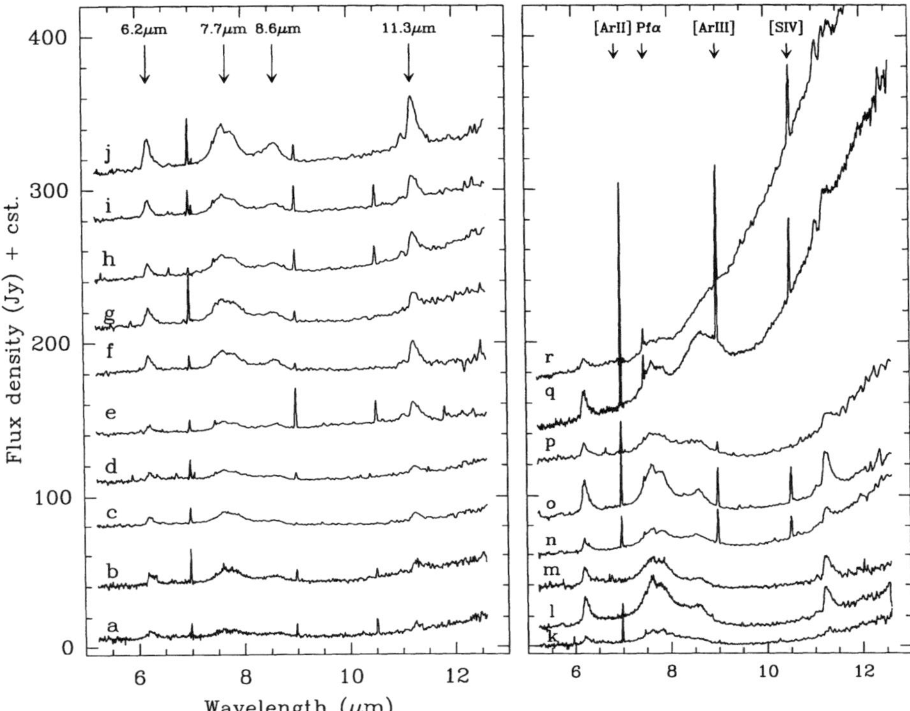

Fig. 2. — The 5 – 13 μm SWS spectra of compact H II regions showing the shapes of the PAH features (identified in the figure) and variation of the relative intensities from source to source. No continuum has been subtracted. The sources are as follows: (a) IRAS 19207+1410 (5), (b) IRAS 18502+0051 (40), (c) IRAS 17279−3350 (80), (d) IRAS 17221−3619 (110), (e) IRAS 23030+5958 (140), (f) IRAS 22308+5812 (180), (g) IRAS 18116−1646 (210), (h) IRAS 12063−6259 (240), (i) IRAS 10589−6034 (280), (j) IRAS 23133+6050 (310), (k) DR 21, (l) IRAS 19442+2427 (10), (m) IRAS 18162−2048 (35), (n) IRAS 17455−2800 (60), (o) IRAS 15384−5348 (85), (p) IRAS 15502−5302, (q) IRAS 18434−0242 (140), and (r) IRAS 12073−6233 (170). The numbers given in brackets are the additive factors (in Jy) which are used for display purposes. The spectra are ordered by increasing strength of the continuum from bottom to top and from the left to the right panel.

dominant role of H_2O ice in molecular clouds. The spectrometers on board the ISO satellite have opened new windows in the infrared and enabled us to study for the first time major ice constituents such as CO_2 and CH_4 in a great number of sources – see the contributions of Dartois *et al.*, Ehrenfreund and Schutte in this volume.

The complete spectra of two embedded infrared sources, IRAS 18316−0602 and IRAS 19110+1045 have been presented [8], both are luminous protostars with luminosities of $10^{4.3}$ $L_\odot$ and $10^{5.8}$ $L_\odot$, respectively. IRAS 18316−0602 (also known as RAFGL 7009S) has been studied in detail [9, 10] where the exceptional richness of its SWS spectrum has been reported – see the contribution of Dartois in this volume. IRAS 19110+1045 is a young protostar associated with a dense, compact submillimeter source with evidence of infalling gas [11].

Figure 3 shows the SWS and LWS spectra of the sources IRAS 18316−0602 and IRAS 19110+1045. No correction was applied to match the SWS and LWS portions of the spectrum. The good agreement in the overlap region of the two spectrometers (41−43 μm) confirms the high quality of the flux scale calibration of both instruments. Both spectra are dominated by a series of absorption bands which are identified in Figure 3 – see the contribution of Dartois *et al.* in this volume for a description of the bands shortwards of 25 μm. Centred on the overlap region between the SWS and the LWS spectra, there is a broad absorption band at $\sim$ 44 μm. This wide and deep band cannot be explained by an artefact of the calibration since there are infrared point sources with comparable flux densities at 40 μm (*i.e.*, 10^3 Jy) which show a flat spectrum without any evidence of a broad absorption band at 44 μm (Fig. 1). This band has been identified with the transverse optical (TO) vibrational band of H_2O ice [8] detected for the first time in absorption. It is important to stress that the detection of this prominent band, which extends from 30 to 80 μm, would not have been possible without spectroscopy over the wide spectral range of the combined SWS and LWS spectra.

The 44 μm absorption band is not the same in the two sources (Fig. 3). In IRAS 18316−0602, the band is strong and structureless, indicating that the absorption is dominated by amorphous water ice. In the case of IRAS 19110+1045, the absorption band is weaker and shows a small dip at 43 μm, suggesting a higher relative abundance of crystalline H_2O ice. A comparison with the laboratory transmittance data shows that the observed bands are quite different and that the 62 μm band is not detected (Fig. 3). These differences are entirely due to radiative transfer effects which for absorption bands become critical at mid- and far-infrared wavelengths. The results of a simple model which takes into account the temperature dependence of the ice properties have been presented [8]. The model predictions agree fairly well with the observations (see dash-dotted lines in Fig. 3). Most of the bands are reproduced both in shape and in width. In particular, the H_2O ice bands are well explained: the lack of a 62 μm band is predicted by the model, as observed; the depth of the 44 μm band is compatible with the strengths of *all* the other H_2O ice bands in the spectra; the shapes of the bands are reproduced, structureless in the case of IRAS 18316−0602 and with a small dip in the case of IRAS 19110+1045. These facts strongly support the identification of the 44 μm band with the TO vibrational band of H_2O ice.

The column densities of H_2O in ice are derived for IRAS 18316−0602 and IRAS 19110+1045, $\sim 1.2 \times 10^{19}$ and $\sim 5 - 6 \times 10^{18}\,\mathrm{cm}^{-2}$, respectively [8]. The relative abundances of the ice constituents H_2O:CO_2:CO are 1:0.2:0.15 for IRAS 18316−0602 [9] and 1:0.05:$<$0.015 for IRAS 19110+1045.

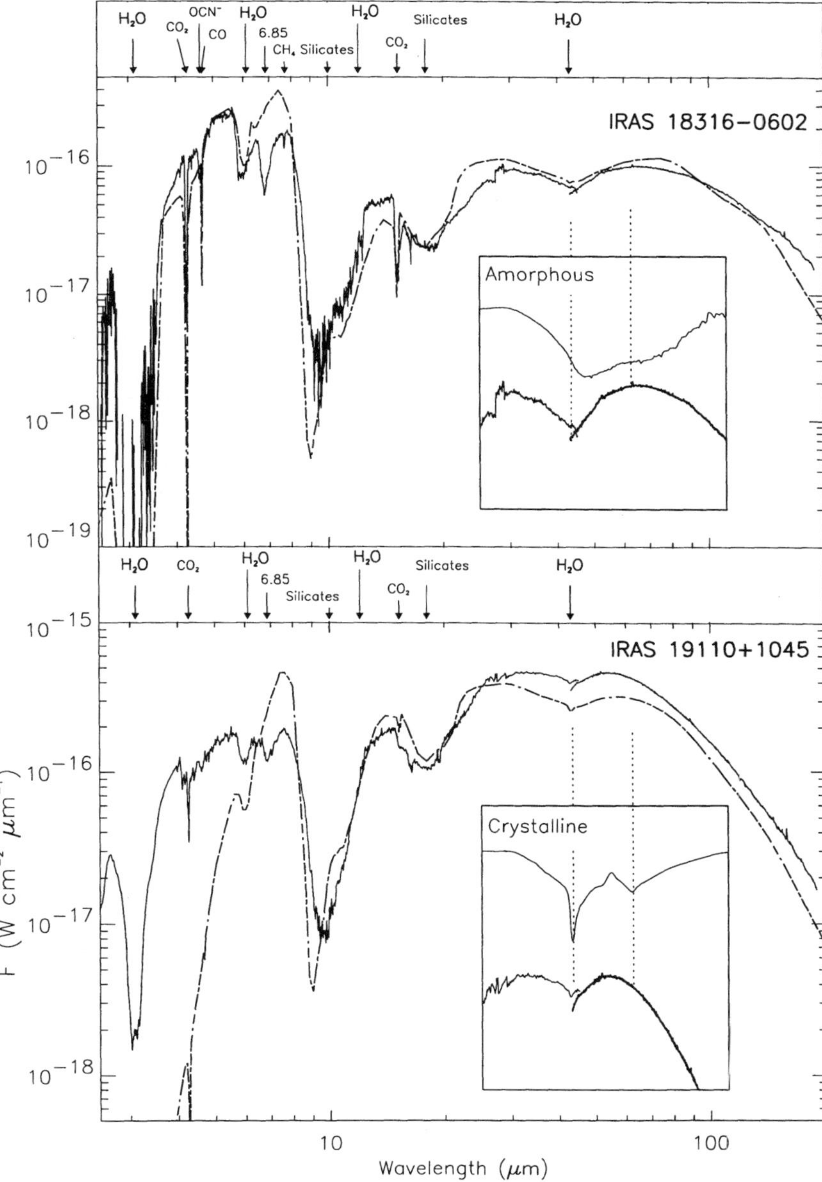

Fig. 3. — Combined SWS and LWS spectra of the embedded infrared sources IRAS 18316−0602 (RAFGL 7009) and IRAS 19110+1045. The position and the identification of the main absorption bands are given for each source. Note that the SWS and LWS spectra were calibrated independently and that no corrections were made to improve the match between the two instruments. The insert boxes display the transmittances of amorphous and crystalline water ices (taken from [12]) together with the source spectra - the intensity axis has been expanded for clarity purposes. The results of radiative transfer calculations [8] are shown as dash-dotted lines.

These numbers underscore the differences in the degree of evolution between the sources. IRAS 19110+1045 appears to be a more evolved protostar where significant ice mantle evaporation has already occured, consistent with the absence of CO and CH_4 ices, the weakness of the CO_2 band and the structure seen in the 43 μm band.

The above results on the two massive protostars IRAS 18316−0602 and IRAS 19110+1045 illustrate the importance of wide wavelength range spectroscopy for the study of interstellar ices. All the well-known infrared embedded sources have been observed with ISO and it would be of great interest to make a systematic study of the combined spectra of these sources in order to investigate the changes in the shape of the 44 μm H_2O ice band as a function of the evolutionary state of the object. For the most obscured sources less abundant components of the interstellar ices, such as NH_3, could be searched for at longer wavelengths although their detection could be very difficult. The long wavelengths modes are lattice modes which will weaken or vanish as soon as the material is diluted in another matrix. Nevertheless, the detection of additional modes at longer wavelengths would provide invaluable additional information on the chemical composition of interstellar ices.

4. CONCLUSIONS

The availability of complete spectra from 2.5 to 196 μm has opened up an entirely new field in the investigation of the dust properties in astronomical sources. This paper illustrates the potential of the combined SWS and LWS data. It is clear that the rich harvest of spectra obtained during the ISO mission will provide for the years to come an important database to further study the nature of the interstellar dust grains and ice mantles. More work, both observational and in the laboratory, will be needed to build on the ISO spectroscopic legacy. Also, models should be improved to enable a reproduction of the details of the entire infrared spectrum including the emission and absorption bands, instead of explaining broad-band photometric measurements as in the past.

Looking ahead, two major space missions will allow us to obtain complete spectra at infrared wavelengths. The Infrared Spectrograph (IRS) for the Space Infrared Telescope Facility (SIRTF) which enables low and moderate-spectral resolution spectroscopy from 4 to 40 μm [13]. On board the Far-Infrared Space Telescope (FIRST), the heterodyne receivers, HIFI, which will cover at high spectral resolution the wavelength range from 200 to $\sim$ 600 μm [14], and the FTS imager of SPIRE which will allow uninterrupted spectro-imagery at variable spectral resolution between 20 and $\sim$ 1000 in the range 200 to 650 μm [15].

ACKNOWLEDGMENTS

We wish to acknowledge the work of the SWS and LWS Instrument dedicated teams in VILSPA for the establishment and monitoring of the calibration. We thank E. Dartois and A.G.G.M. Tielens for valuable discussions. Finally, the infinite patience of the editors of this book is gratefully acknowledged.

Appendix A

All the spectra described in this paper were observed with the LWS and SWS full AOT 01 scans yielding spectra with a resolution $\frac{\lambda}{\Delta\lambda}$ from ~100 to ~500. The 2.5 – 45 μm SWS grating spectra were taken in the full scan AOT1 mode at speed 2 (or 3) [16]. The data were processed using standard SWS Interactive Analysis procedures. Version 7.0 of the SWS off line processing (OLP) system flux and wavelength calibration was applied for the 12 AOT bands independently [17, 18]. The final SWS spectra have a resolution of $\frac{\lambda}{\Delta\lambda} \sim 400$ and a wavelength accuracy better than one tenth of a resolution element. The absolute fluxes have an estimated accuracy of about 7% at the shortest wavelength to about 15% at ~ 40 μm, with a significantly better relative spectral accuracy. Residual calibration errors are present only in the 27 – 30 μm range (SWS AOT band 3E) where the relative spectral responsivity is not yet fully calibrated. The 43 – 196.5 μm LWS spectra were measured using the LWS AOT01 [19]. Each observation consists of at least six fast grating scans with 0.5 s integration ramps at each commanded grating position. The spectra were taken with four samples per resolution element. The absolute and relative flux calibration of the spectra is relative to Uranus [20]. The data were reduced using the LWS off-line processing software (version 7.0). The ten sub-spectra, one for each detector, were then rescaled (with scaling factors not exceeding 15%) to the SW4 detector, considered to be one of the most photometrically reliable detectors.

REFERENCES

[1] Kessler M.F., Steinz J.A., Anderegg M.E., *et al.*, *A&A* **315** (1996) L27.

[2] Roelfsema P.R., Cox P., Kessler M.F. and Baluteau J.P., Star Formation with the Infrared Space Observatory, ASP Conf. Series 132, edited by J.L. Yun and R. Liseau (ASP, San Francisco, 1998) p. 76.

[3] Shaver P.A., McGee R.X., Newton L.M., Danks A.C. and Pottasch S.R., *MNRAS* **204** (1983) 53.

[4] Afflerbach A., Churchwell E. and Werner M.W., *ApJ* **478** (1997) 190.

[5] Roelfsema P.R., Cox P., Tielens A.G.G.M., *et al.*, *A&A* **315** (1996) L289.

[6] Verstraete L., Puget J.L., Falgarone E., *et al.*, *A&A* **315** (1996) L337.

[7] Cesarsky D., Lequeux J., Abergel A., *et al.*, *A&A* **315** (1996) L309.

[8] Dartois E., Cox P., Roelfsema P.R., *et al.*, *A&A* **338** (1998) L21.
[9] d'Hendecourt L., Jourdain de Muizon M., Dartois E., *et al.*, *A&A* **315** (1996) L365.
[10] Dartois E., d'Hendecourt L., Boulanger F., *et al.*, *A&A* **331** (1998) 651.
[11] Hunter T.R., Phillips T.G. and Menten K.M., *ApJ* **478** (1997) 283.
[12] Trotta F. and Schmitt B., *Applied Spectroscopy* (1999) in preparation.
[13] Houck J.R., *et al.*, 1999, Astrophysics with Infrared Surveys: A Prelude to SIRTF, ASP Conf. Series, edited by M.D. Bicay, R.M. Cutri and B.F. Madore (ASP, San Francisco, 1999) in press.
[14] Whyborn N.D., The Far-Infrared and Submillimeter Universe, ESA SP-401, edited by M. Rowan-Robinson and G. Pilbratt (ESA Pubs. Div. ESTEC, Noordwijk, 1997) p. 19.
[15] Griffin M.J., The Far-Infrared and Submillimeter Universe, ESA SP-401, edited by M. Rowan-Robinson and G. Pilbratt (ESA Pubs. Div. ESTEC, Noordwijk, 1997) p. 31.
[16] de Graauw T., Haser L.N., Beintema D.A., *et al.*, *A&A* **315** (1996) L49.
[17] Schaeidt S.G., Morris P.W., Salama A., *et al.*, *A&A* **315** (1996) L55.
[18] Valentijn E.A., Feuchtgruber H., Kester D.J.M., *et al.*, *A&A* **315** (1996) L60.
[19] Clegg P., Ade P.A.R., Armand C., *et al.*, *A&A* **315** (1996) L38.
[20] Swinyard B.M., Clegg P.E., Ade P.A.R., *et al.*, *A&A* **315** (1996) L43.

LECTURE 11

Molecules in the Gas and Solid State Towards the Young Massive Protostar RAFGL7009S

E. Dartois[1,*], K. Demyk[1], M. Gerin[2] and L. d'Hendecourt[1]

[1] *Institut d'Astrophysique Spatiale, Bât. 121, Campus d'Orsay, Université Paris XI, 91405 Orsay Cedex, France*
[2] *Radioastronomie Millimétrique, Laboratoire de Physique de l'ENS, 24 rue Lhomond, 75231 Paris Cedex 05, France*

Abstract. With its large spectral coverage and resolution, the SWS instrument on board ISO is well-adapted to the study of solid state features arising from icy interstellar grain mantles present in dense molecular clouds. It can thus be used to probe the molecular material along lines of sight toward bright infrared protostellar sources. An example is presented here with observations of the protostellar object RAFGL7009S. Deep solid state features are detected in this heavily extinguished source. These features are compared with laboratory data which allow the identification of ice mixtures composed of H_2O, CH_4, CO, CO_2 and CH_3OH. Quantitative estimates of the abundances of these molecules are derived. Superimposed on these solid state features, we observe the ro-vibrationnal lines of the gas phase species H_2O, CO, CO_2 and CH_4. Analysis of these lines helps us to derive some physical parameters, such as the gas temperature and the gas-to-solid ratios for these molecules. These results may shed light on the relationship between the gas and the grains with possible implication for interstellar chemistry around these objects. We show that careful laboratory experiments involving matrix isolation techniques allow us to describe the physical nature of the observed ices as well as their photochemical evolution. Two characteristic absorption bands seen in the laboratory are tentatively attributed to the molecular ions OCN^-

* *Now at*: IRAM, 300 rue de la Piscine, 38406 Saint-Martin-d'Hères, France.

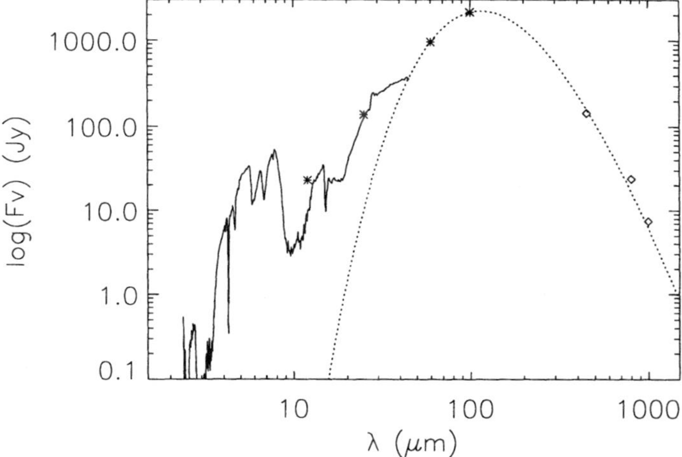

Fig. 1. — Global spectral energy distribution of RAFGL7009S. The solid line is the ISO-SWS data. The stars represent the IRAS filter values. The diamonds are long wavelength measurements from the ground by McCutcheon *et al.* [1]. The dotted line is a fit with a function, used in the inversion derivation discussed in the text.

and NH_4^+, the first of these band is also seen in the source. Together with the CO_2 and possibly H_2CO, these ionic species represent the only hint today, that UV photolysis is an important mechanism for the evolution of interstellar ices.

1. INTRODUCTION

The primary astronomical targets for the infrared spectroscopy of ices are protostars that display very high column densities of cold foreground material [2]. Such high column densities are required for the detection not only of major species but also of minor ones. The latter may be used as indicators of a certain degree of chemical evolution generated by energetic UV and/or cosmic-ray processing. In addition, temperature variations may stimulate chemical reactions which are initiated by reactive species stored in the cold icy mantle. These processes are described elsewhere in this volume, mainly by Schutte and Moore. We focus here on one source, RAFGL7009S, pointing out the real advantages of using ISO to obtain a global image of the infrared spectroscopy of protostellar objects.

2. GLOBAL PROPERTIES OF RAFGL7009S

In order to derive the global physical conditions present in the observed source, one can look at the source spectral energy distribution (SED), as shown in Figure 1. We want to investigate the proportion of cold grains present in the cloud surrounding the protostar RAFGL7009S. These grains emit in the far-infrared (FIR) and in this spectral region we are generally in the optically thin limit. Assuming this is the case, the flux continuum emitted by the source is related to the mean mass absorption coefficient of the grains κ_ν by:

$$F(\nu) = \frac{\kappa_\nu}{D^2} \int_0^\infty W(T) B_\nu(T) \mathrm{d}T \tag{1}$$

where D is the source distance, $W(T)$ is the dust mass distribution at a given temperature T and $B_\nu(T)$ is the Planck function. This equation can be inverted to obtain $W(T)$. To do this we cover a reasonable grain emissivity range with two frequency dependances for the absorption coefficient. The first given by κ_ν (cm^2 g^{-1}) $= \kappa_0 \nu^{3/2} = 7.9\times10^{-18}\ \nu^{3/2}$ [3] and the second, derived more recently [4] gives $\tau/N_{\rm H} \sim 10^{-25} \times (\lambda/250\ \mu\mathrm{m})^{-2}$. Assuming that the gas to dust densities are related by $\rho_{\rm gaz}/\rho_{\rm dust} \sim 100$, we derive $\kappa_\nu(\mathrm{cm}^2\mathrm{g}^{-1}) = \kappa_0 \nu^2 = 4.15\times10^{-24} \nu^2$. For the long wavelength part of the spectrum we use an analytical function shown in Figure 1 by the dotted line. We then make use of the modified Mobius theorem, as previously [3], to invert equation (1). The resultant dust mass distribution is shown in Figure 2. This distribution indicates very cold temperatures. Integrating over the mass at temperatures below 30 K we derive, assuming a source distance of 3 kpc, a mass of 20 to 40 $M_\odot$ of grains.

To investigate the global properties of the gas phase, we observed this source with the IRAM 30M telescope. We chose to observe the molecules CS, $C^{34}S$ and H_2CO to probe the mean hydrogen density of the cloud. Indeed, in 1994, Jansen *et al.* [5] published calculations that predicted the intensity ratios expected for some transitions of these molecules as a function of the kinetic temperature and the density. Our observations for the CS molecule, shown on the abacuses in Figure 3 assuming $^{34}\mathrm{S}/^{32}\mathrm{S} = 24.4 \pm 5$ [6], point to a high mean density over the IRAM beam of about 10^6 cm^{-3}. Furthermore, studying the transitions of CH_3C_2H, CH_3CN and CH_3OH, we derive cold rotational temperatures, ranging from 10 to 50 K.

All our measurements point to a very dense, cold cloud where we expect to detect strong absorption features arising from ice mantles onto the more refractory grains.

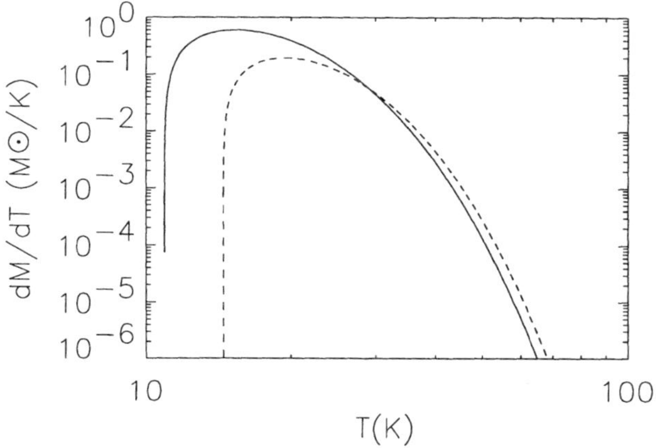

Fig. 2. — Dust mass distribution at a given temperature resulting from the inversion of the long wavelength portion of equation (1) with $\kappa_\nu \propto \nu^\alpha$: line $\alpha = 2$, dotted line $\alpha = 1.5$ (see text).

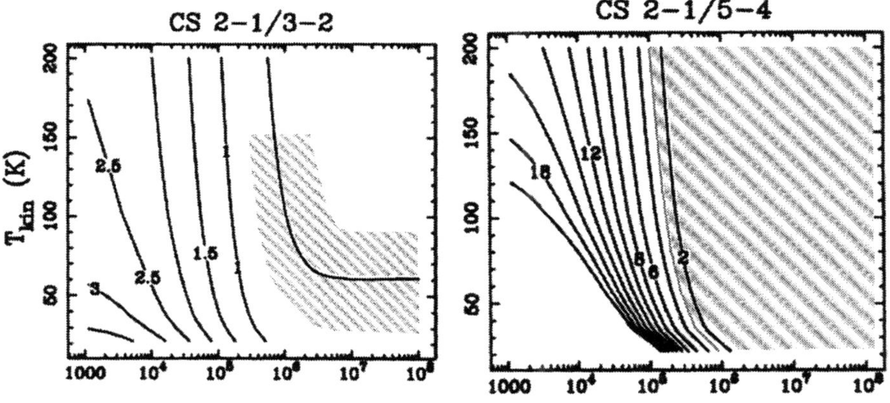

Fig. 3. — Abacuses from Jansen *et al.* [5] on which are reported the CS ratios measurements performed at IRAM (dashed region).

3. THE GAS AND SOLID PHASES

3.1. Solid phase

The ISO 2.5 – 18 μm spectrum of RAFGL7009S, obtained with the Short Wavelength Spectrometer instrument (SWS) is shown in Figure 4. The extinction towards the source is so large that the water ice stretching mode at 3 μm, the CO_2 antisymetric stretching mode around 4.27 μm and the

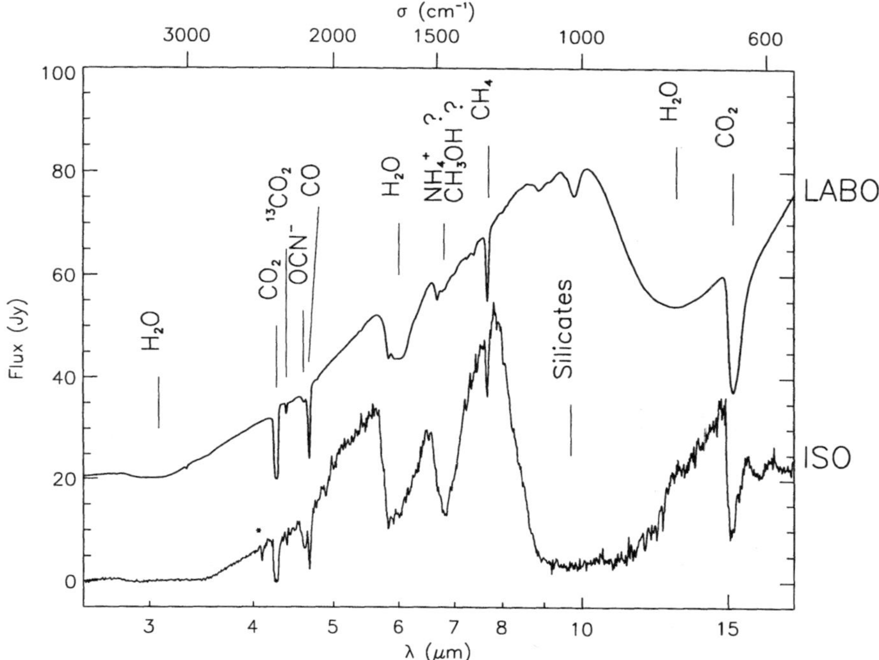

Fig. 4. — The 2.5 – 18 μm SWS spectrum of RAFGL7009S. The identified molecular transitions are labeled above the spectra.

9.7 μm silicate SiO stretching absorption bands are saturated. This spectrum shows prominently the ice absorptions arising from the CO_2, CH_4, CO and "XCN" molecules, and also a band at 4.9 μm, attributable to OCS. Above the spectrum, and shifted for clarity, is superimposed a transmittance spectrum multiplied by an estimated source continuum. The spectrum was recorded in the laboratory, after UV photolysis of an initial ice mixture of H_2O:CO:CH_4:NH_3:$O_2$$\sim$10:2:1:1:1 deposited at 10 K. AT about 4.62 μm, on the short wavelength side of the CO feature, the so-called "XCN" absorption band is present in both the ISO and laboratory spectra. However, the optical depth in the astrophysical case is much larger than the one produced in the experiment.

The large 6 μm (water ice bending mode) and 6.8 μm (at this stage still unidentified) bands dominate the spectrum as in many other protostar spectra. The water ice bending mode line shape is considerably modified by photolysis, which simultaneously induces a reorganisation of the ice matrix structure, following the energy input, and also creates new molecules such as formaldehyde,

Table I. — Solid phase column densities toward RAFGL7009S.

Molecule	Mode	Wavelength (μm)	N (cm^{-2})	% relative to H_2O
H_2O	ν_2	6.0	1.2×10^{19}	100
CO	ν_1	4.67	1.8×10^{18}	15
CO_2	ν_2	15.2	2.5×10^{18}	21
$^{13}CO_2$	ν_3	4.39	4.0×10^{16}	0.33
CH_4	ν_4	7.7	4.3×10^{17}	3.6
OCN^-	ν_2	4.62	4.4×10^{17}	3.7
H_2CO	ν_2	5.81	3.3×10^{17}	3.0
OCS	ν_1	4.9	2.0×10^{16}	0.17

H_2CO, and others (see also Schutte, this volume).

This source is interesting because it is deeply embedded, but this implies that the silicate absorption is saturated. This saturation effect prevents us from observing other molecular "fingerprints" as can be seen by comparing the laboratory and astronomical spectra shown in Figure 4. In the laboratory spectrum, in which we did not attempted to reproduce the silicate absorption, one can see features between 9 and 11 μm due to the presence of NH_3 and CH_3OH, the latter being a product of the photolysis. The efforts that must be developed in order to extract these modes frim the silicate absorption wing are well illustrated by recent solid NH_3 observations [7]. The CO_2 molecule, unobservable from the ground, is a major constituant of the ice mantles.

At this stage of the interpretation, laboratory spectroscopy succeeds in reproducing the observed ice spectrum and identifying most of the bands. This gives access to the relative abundances of the molecules in the mantles (see Tab. I) but this is only a line of sight integrated information.

3.2. Gas phase

Higher resolution observations of RAFGL7009S obtained with the SWS06 observing template ($R \sim 2000$) revealed sharper transitions superimposed on the broad solid phase absorptions. These lines are the gas phase counterparts to the ice constituents. In spite of the fact the SWS resolution is not fully adapted to the gas phase rovibrationnal line analysis, we can neverthless obtain a reasonable estimate for the gas excitation temperatures and column densities. This allows us to probe, with the same observationnal parameters, the gas and solid phases and to derive gas-to-solid ratios for many molecules. Furthermore, most of the observed molecules (H_2O, CH_4 and CO_2), with the exception of CO, are unobservable from the ground due to telluric water and carbon dioxide absorptions or are inactive in the submillimeter wavelength region (CH_4 and CO_2). The comparison between gas and solid phase permits us to constrain the relative importance of each phase and allows us to draw some conclusions about the desorption mechanisms and the molecular "reservoir" contained in the icy grain mantles. This is a step toward the understanding of the grains role in cloud chemistry.

3.2.1. *Gas phase CO band*

We observed the rovibrational transitions of CO in RAFGL 7009S between 4.5 and 5.15 μm as well as the solid state features from CO, OCS, and "XCN" (see Tab. I). We constructed gas phase absorption models which were used, in conjonction with experiments to measure the solid phase spectra, to interpret the ISO data. Using rotational diagrams, *i.e.* plotting $\ln(N_J/2J+1)$ versus E_J/k allows us to derive the excitation temperature of the gas (see Mitchell *et al.* [8] for a detailed description of the rotation diagrams for diatomic molecules such as CO). We observe two temperature components. To measure the high temperature excitation we used the high rotational levels of the ^{12}CO and this led to a temperature of about 740 $\pm$ 255 K (3σ). The low temperature component was determined using the ^{13}CO lines not blended with the main isotope. The scatter in values from the two different scans is quite large but gives a temperature of 40 $\pm$ 27 K (3σ). Assuming that low J levels are dominated by the low temperature component and high levels by the high temperature component, we estimate the column densities involved. Table II gives a summary of the ^{13}CO and ^{12}CO column densities deduced, assuming a $^{12}CO/^{13}CO$ ratio of 60 because some of the ^{12}CO lines are saturated. Figures 5 and 6 show the composite spectra constructed using solid phase absorption spectra measured in the laboratory combined with the gas phase models 1 and 2.

The high temperature is probably generated in a region close to the central object and may be related to an outflow or to interactions between the outflow and circumstellar matter [9].

3.2.2. *Gas phase CO_2 band*

We also observe the ν_2 bending mode of CO_2 in the ISO spectrum of RAFGL 7009S. In addition to the P and R branches, the CO_2 ν_2 mode presents a Q branch which shows transitions in the range 667.39 cm^{-1} to 672.18 cm^{-1} from $J=2$ up to $J=68$, which are present as a single feature at the SWS resolution (~1500), and has a fairly strong integrated absorption coefficient. A model was constructed using the HITRAN 96 database constants for the ν_2 bending mode line positions [10] and Reichle and Young's integrated absorption coefficient [11]. We use a Voigt profile for the lines with an adjustable Doppler FWHM given by the relation $\Delta\lambda = 2\sqrt{ln2}\lambda v/c$ where the turbulent velocity v is given

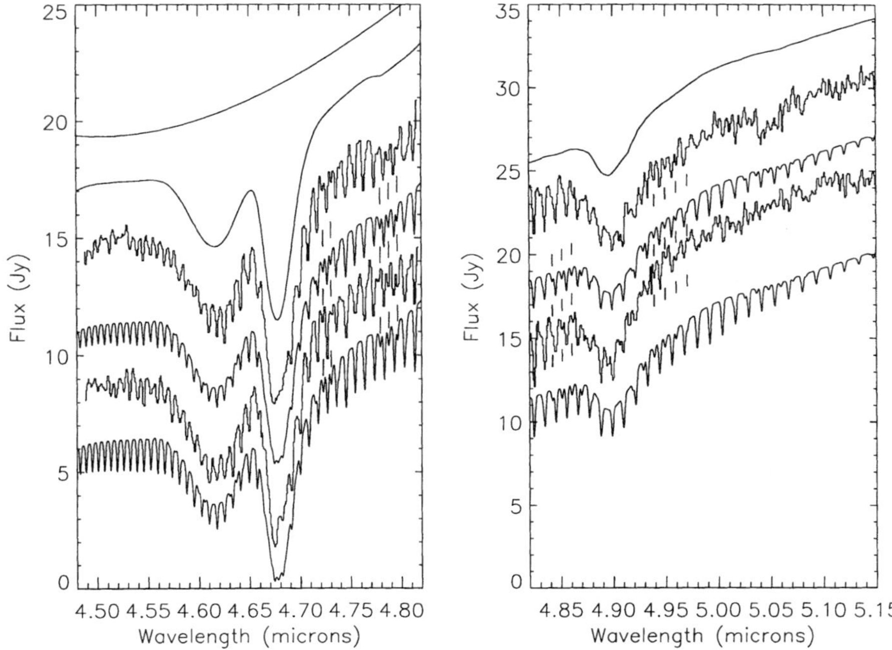

Fig. 5. — Gas phase CO rovibrational transitions and ice features at 4.9 μm (OCS), 4.62 μm ("XCN") and solid CO at 4.67 μm. Left panel, from upper to lower curve: adopted continuum; solid state spectrum obtained in the laboratory from an ice deposited at 13 K containing CO and NH_3 and UV irradiated (multiplied by the adopted continuum); upward scans from ISO-SWS06 observation; composite spectrum using the adopted continuum, gas phase Model 1 described in Table II (including ^{12}CO and ^{13}CO components at $T = 40$ K and 740 K assuming a ^{12}CO/^{13}CO ratio of 60) and solid state spectrum from the laboratory; downward scans from ISO-SWS06 observation; composite spectrum built using the adopted continuum, gas phase Model 2 (Tab. II) and solid state spectrum from the laboratory. Right panel, from upper to lower curve: adopted continuum (the laboratory data do not fully reproduce this solid feature); upward scans from ISO-SWS06 observation; gas phase Model 1 as presented above multiplied by the adopted continuum; downward scans from ISO-SWS06 observation; gas phase Model 1 as presented above multiplied by the adopted continuum. Straight lines denote ^{13}CO transitions that are seen at a good separation from the ^{12}CO transitions. The models correspond to the high column density cases.

in km s^{-1}, and the Lorentz profile is given by the lifetime of the considered level, assuming an excitation temperature, $T_{\rm ex}$, and LTE conditions in the gas. We convolve the calculaterd spectrum with a gaussian having the ISO-SWS06 resolution to allow comparison with observations. The resultant spectrum is then compared with the observations after extraction of the continuum level and assuming a temperature (40 K). We have few constrains on the exact

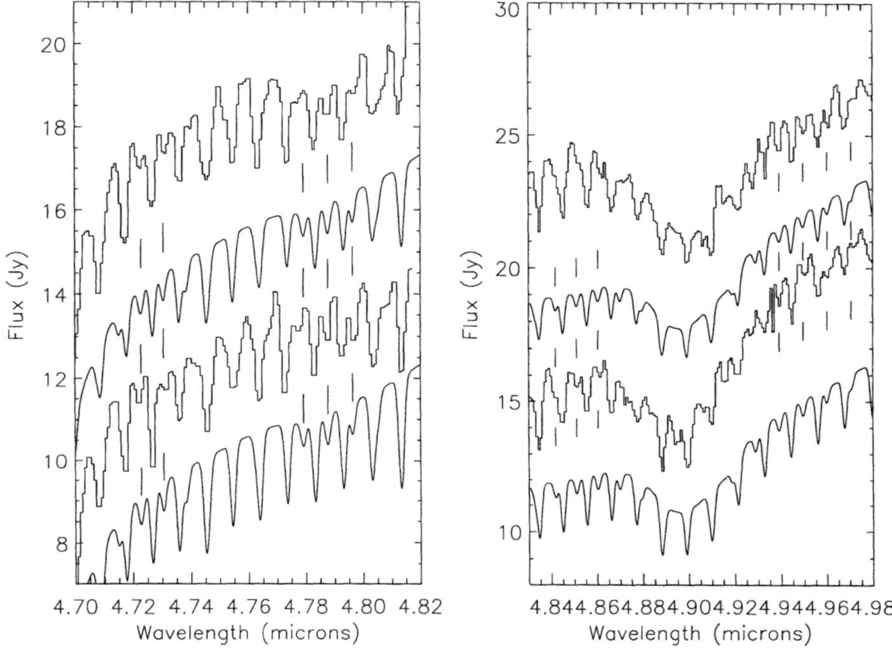

Fig. 6. — A close-up of part of Figure 5 showing the ^{13}CO sub-structures.

temperature of the gas but we can exclude a 200 K excitation temperature because of the asymmetry it would produce in the Q branch. We assumed a turbulent velocity of v = 3 km s^{-1} based on CO isotope radio lines ($^{13}C^{16}O$, $^{12}C^{18}O$, [12]).

The gas column density is determined by computing gas transmittance spectra with increasing number of CO_2 molecules along the line of sight and dividing the ISO reduced spectrum by the expected transmittance from the model. We calculate the variance of this division, which diminishes until it reaches a minimum governed by the signal-to-noise, column density and model uncertainties. The best fit was obtained with $N_{CO_2} \approx 1^{+1}_{-0.5}.10^{17}$ molecules cm^{-2}. To derive the CO_2 ice column density we use the band strength given by Gerakines *et al.* [13] and derive 2.5×10^{18} cm^{-2} as already reported in d'Hendecourt *et al.* [14]. The CO_2 gas to solid ratio we determine is 5×10^{-2}, which is similar to other protostellar objects [15]. CO_2 is clearly much more abundant in the solid than in the gas phase.

3.2.3. Gas phase H_2O band

The H_2O ν_2 rotation-vibration bands are seen in absorption from $\sim$ 5.5 μm to 6.9 μm. The ortho and para H_2O lines are observed, some of which are

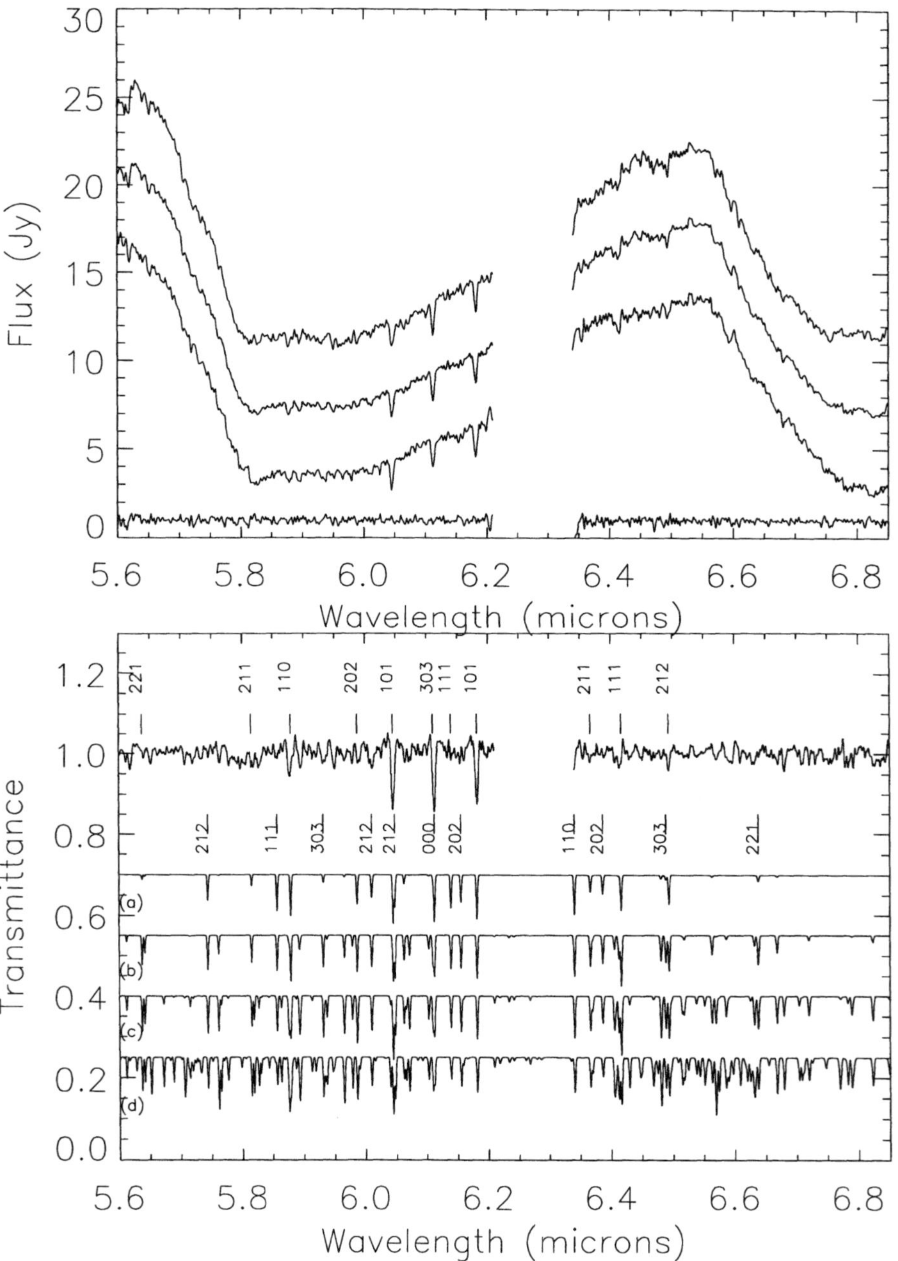

Fig. 7. — Upper panel (top to bottom): upward scans, all scans, downward scans and the difference between upward and downward scans divided by $\sqrt{2}$ to estimate the noise on the spectra. Lower panel: Gas phase H_2O ν_2 band Observed in RAFGL 7009S and model spectra with $v = 3$ km s^{-1} and 2×10^{18} cm^{-2} at temperature (a) $T = 25$ K (b) $T = 50$ K (c) $T = 100$ K (d) $T = 300$ K. The assignment of the pure rotationnal levels from which the transitions arises are given for the strongest bands occurring at $T = 50$ K, showing that we mainly see a cold gas component.

Table II. — Gas phase CO estimates and models used in Figures 5 and 6.

Molecule	^{12}CO mol cm^{-2}	^{13}CO mol cm^{-2}
Gas estimates T estimates	 740 K ± 255 K	$2.1 \times 10^{17} \pm 0.9(1\sigma)$
T used Model 1 Model 2 Model 3 Model 4	740 K 1.8×10^{19}(3 km s^{-1}) 1.8×10^{19}(5 km s^{-1}) 7.2×10^{18}(3 km s^{-1}) 7.2×10^{18}(5 km s^{-1})	740 K 3.0×10^{17}(3 km s^{-1}) 3.0×10^{17}(5 km s^{-1}) 1.2×10^{17}(3 km s^{-1}) 1.2×10^{17}(5 km s^{-1})
Molecule	^{12}CO molec cm^{-2}	^{13}CO molec cm^{-2}
Gas estimates T estimates		$9.7 \times 10^{16} \pm 2.4(1\sigma)$ 40 K ± 27 K
T used Model 1 Model 2 Model 3 Model 4	40 K 7.2×10^{18}(3 km s^{-1}) 7.2×10^{18}(3 km s^{-1}) 4.4×10^{18}(3 km s^{-1}) 4.4×10^{18}(3 km s^{-1})	40 K 1.2×10^{17}(3 km s^{-1}) 1.2×10^{17}(3 km s^{-1}) 7.3×10^{16}(3 km s^{-1}) 7.3×10^{16}(3 km s^{-1})

saturated. The temperature estimate from individual lines seems rather difficult to obtain because the strongest lines suffer from saturation effects and the non-saturated ones are almost lost in the noise. A tentative assignment of the observed lines can give information on the range of gas temperatures encountered. In Figure 7 one can see that the strongest observed transitions coincide with low level excitation lines (ground, $J = 1$ or $J = 2$ levels). We can exclude a dominant absorption from gas much hotter than about 50 K because the spectrum between *e.g.* 5.65 and 5.75 or 5.9 and 6 μm would display high temperature lines, as evidenced in the recent results of Helmich *et al.* [16].

Most of the water vapor lies at temperatures between 20 and 50 K in what we call the cold component. We have detected over a restricted wavelength range, with higher signal-to-noise ratio, gas at higher temperature (around 300 K). However, the gas phase water is dominated by the cold component.

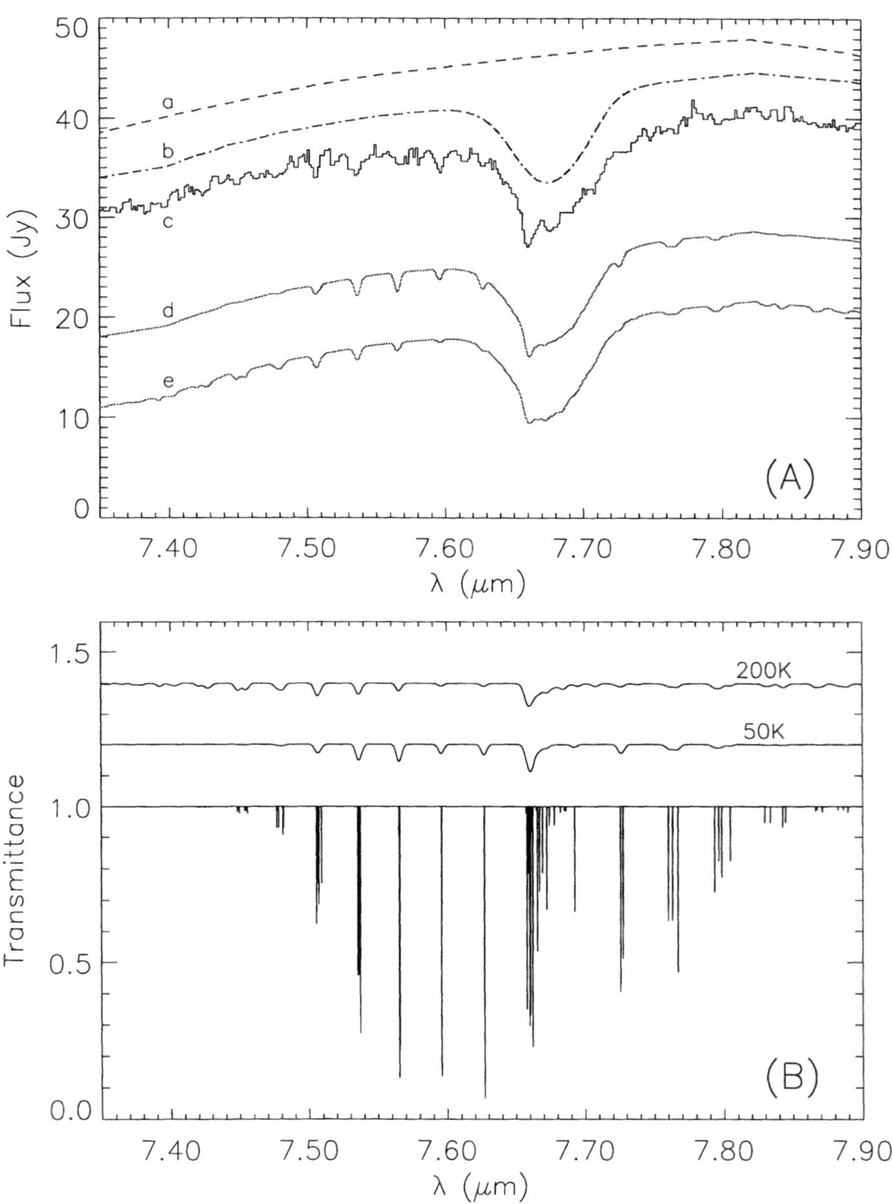

Fig. 8. — Gas phase CH_4 in the ν_4 region. Panel (A): a) adopted continuum, b) resultant spectrum with the CH_4 laboratory ice feature, c) ISO spectrum of RAFGL 7009S, d) and e) spectra when the gas phase model (d) 50 K and e) 200 K) and laboratory ice spectrum are both taken into account. Even the broad absorption around 7.4 μm is well reproduced by laboratory experiment. Panel (B) (top to bottom): model spectrum with T = 200 K, doppler parameter b = 3 km s^{-1} and a column density of 10^{17}cm^{-2} at R = 1800. Model spectrum with T = 50 K, doppler parameter b = 3 km s^{-1} and a column density of 10^{17}cm^{-2} at R = 1800. Full resolution for the second spectrum is shown.

3.2.4. Gas phase CH_4 in the ν_4 region

Methane is inactive in its rotational ground state and is thus not observable in the radio range. CH_4 can only be seen through the rovibrational modes [17] such as the in the ν_4 region falling in the 7.7 μm range. The CH_4 analysis is shown Figure 8, using a slighlty different approach from the one adopted for H_2O. Indeed, we use a laboratory photolysed ice spectrum obtained from a mixture (H_2O:CO:CH_4:NH_3:O_2–15:8:3:2.75:5) to provide the synthetic spectrum (see Fig. 8A). The gas phase model spectrum is generated using the extensive list of weighted squared transition moments for the ν_2/ν_4 dyad given in the HITRAN database. We deduce the oscillator strength of each line from these data. The population of the ground state is given in LTE by the formula:

$$N_{\rm J} \ = g_{\rm N} \ g_{\rm I} \ {\rm e}^{-hc\sigma_{\rm J}/kT}/Q_{\rm r}(T).$$

Where $g_{\rm N} = 5, 3, 2$ for symetry A, E and F and $Q_{\rm r}$ is the rotational partition function given by:

$$Q_{\rm r}(T) \ = \ \Sigma_{\rm J} \ g_{\rm N} \ g_{\rm I} \ {\rm e}^{-hc\sigma_{\rm J}/kT}.$$

Assuming a Doppler parameter and a line profile, as given for H_2O data, a model can be made and compared to observations. The best fit is obtained with a column density of $1.2^{+0.5}_{-0.3}\times10^{17}$ cm^{-2} with a temperature of about 50 K and is shown in Figure 8B. The methane gas/solid ratio is 0.23, similar to that for H_2O but very different from CO_2. A quasi-exact match between the laboratory data, to which were added the gas phase model and the observed feature, is obtained showing that grain size effects or shapes are unimportant for this molecule at these wavelengths ($a \ll \lambda$).

3.2.5. Discussion

From the analysis presented above we derived the gas to solid ratios of the studied molecules. The one we derive for CO (see Tab. III) is the only one above unity and might simply reflect the fact that CO is mainly produced in the gas phase and is a very stable and volatile molecule. Looking at the ratios we clearly see that the solid phase dominates the gas phase in this source.

A striking aspect is that the CO_2 seems extremely depleted in the gas compared to the other abundant molecules present in the solid. This low gas-to-solid ratio, combined with the laboratory evidence that CO_2 is produced very efficiently from the photolysis of H_2O and CO containing ices, two major components of the mantles, tend to indicate this molecule is mainly produced in the solid phase and reacts or is destroyed when released in the gas [18].

Another important question is to know whether water forms on grains or in the gas. The formation of the H_2O molecule in the gas phase involves reactions with activation energies. The high abundance of water ice in dark clouds such as Taurus (10% of the cosmic oxygen abundance in H_2O, [19]) is not reproduced by chemical models assuming formation in the gas followed by accretion on dust. Formation on grain surfaces therefore seems required. Most

Table III. — Gas to solid ratios in RAFGL 7009S (Solid phase values are taken from d'Hendecourt *et al.* 1996 [14]).

Molecule	Gas (cm^{-2})	Solid (cm^{-2})	Gas/Solid
H_2O	$> 2 \times 10^{18} < 1 \times 10^{19}$(*cold comp.*)	1.1×10^{19}	≥ 0.18
CO_2	$1.0^{+1}_{-0.5} \times 10^{17}$	2.5×10^{18}	~ 0.04
CH_4	$1.2^{+0.5}_{-0.3} \times 10^{17}$	4.3×10^{17}	~ 0.28
CO	$6.1^{+1.7}_{-1.7} \times 10^{18}$ (*cold component*)	1.8×10^{18}	~ 3.4

models including water formation on grain surfaces predict H_2O rich mantles [20-24], water being in most cases the dominant solid state species. The main result concerning water from our observations is that most of the water is in the solid phase. This strongly argues for formation on dust and we believe that the cold gas phase water is desorbed from grains. Other scenarii have been proposed, *e.g.* Bergin *et al.* [25], where, in a three stage model including preshock, shock, and postshock evolutions for the gas, they produce water rich ice mantles by subsequent accretion, thus providing an alternative for the creation of high water abundance such as in protostars.

Finally, the comparison between the water and methane gas-to-solid ratios raises the question of the desorption mechanism. Indeed, despite the fact that the vapor pressures for these two species are very different, the ratios derived are similar. If we interpret these ratios as tracing the evaporation of these molecules from the icy surfaces, that is to say if they are linked to the desorption mechanism, either we are dealing with an out of equilibrium process or the desorption is not correlated to the vapor pressures of these species. The last point could be interpreted in terms of the CH_4 molecules being trapped in a water ice dominated matrix which governs the desorption temperature. It has been shown that, to a certain degree, up to 10% by number, of volatile species such as CO and CH_4 could be retained in a water ice matrix almost until the ice entirely sublimates [26].

4. MOLECULAR IONS AND INTERMOLECULAR INTERACTIONS

4.1. OCN^- and NH_4^+

The 4.62 μm band has been observed for almost twenty years in protostars [27, 28] but remains unidentified although this band was present in early laboratory experiments of UV photolysis of astrophysically relevant ices.

The position of the band indicates that the molecule responsible for this feature contains a C≡N bond, numerous molecules such as nitriles or isonitriles were proposed to be the carrier of the interstellar band [29]. In 1987 Grim and Greenberg [30] proposed the negative ion OCN^- as an alternative assignment. They proposed that OCN^- is produced simultaneously with the positive ion NH_4^+ by acid-base reactions. The main vibration of NH_4^+ falls at 6.85 μm and they argued that this molecule could be the carrier of the unidentified 6.85 μm interstellar band [31]. This strong feature, usually observed in infrared sources where the 4.62 μm band (so-called the XCN band) is present [32,33], peaks in a region where all molecules containing $-CH_3$ and $-CH_2$ groups absorb making its identification, on the basis of only one band, ambiguous.

Laboratory work using electron donnor and acceptor dopants in addition to astrophysically relevant molecules allows us to establish the ionic character of the carrier of the laboratory band. It confirms, by another experimental method, the assignment of the interstellar feature to OCN^- given by Grim and Greenberg [30] and more recently by Schutte and Greenberg [34] , using isotopic substitution and the detection in their experiments of three other weak bands of OCN^- at 7.72, 8.29 and 15.87 μm (1296, 1206 and 630 cm^{-1} respectively).

Electron acceptors slow down the production of negative ions while favouring the formation of positive ions. Conversely, electron donors favour the production of negative ions at the expense of positives ions. To CO:NH_3 (1:1) mixtures we added carbon tetrachloride (CCl_4) or sulfur hexafluoride (SF_6) as electron acceptors and 2-methyltetrahydrofuran (C_5OH_3) as an electron donor. Photolysis of CO:NH_3 ices with UV photons led to the appearance bands at 4.62 μm (2165 cm^{-1}), 6.85 μm (1460 cm^{-1}) and two bands at 4.27 μm (2342 cm^{-1}) and 15.2 μm (658 cm^{-1}), the latter two indicating the production of CO_2. When an electron donor or acceptor is added to the mixture the optical depths of the 4.62 and 6.85 μm bands vary (Fig. 9). The 4.62 μm band is deeper when the electron donnor is present in the mixture whereas the 6.85 μm band is enhanced when an acceptor is used indicating that the carriers of these two bands are of ionic nature, a negative ion for the carrier of the 4.62 μm band and a positive ion for the 6.85 μm band.

In presence of an electron acceptor, the production of the 4.62 μm band is prevented and a band at 4.43 μm (2255 cm^{-1}) appears, attributed to HNCO [35] and produced from CO and the decomposition products of NH_3 under photolysis. The annealing of this mixture shows a transfer of the band to the 4.62 μm band and a simultaneous increase in the optical depth of the 6.85 μm band. This supports the identification of the 4.62 and 6.85 μm bands with OCN^- and NH_4^+, respectively, assuming that OCN^- is produced via an acid-base reaction where HNCO is the acid and NH_3 is the base. HNCO transfers a proton to NH_3 leading to the production of the two ions.

Following the evolution of molecules containing oxygen during the UV photolysis of the CO:NH_3 mixture we have calculated the integrated cross section of OCN^-. During photolysis, the CO molecules disappear and CO_2, OCN^-, $HCONH_2$ are produced. The integrated cross-section of $HCONH_2$

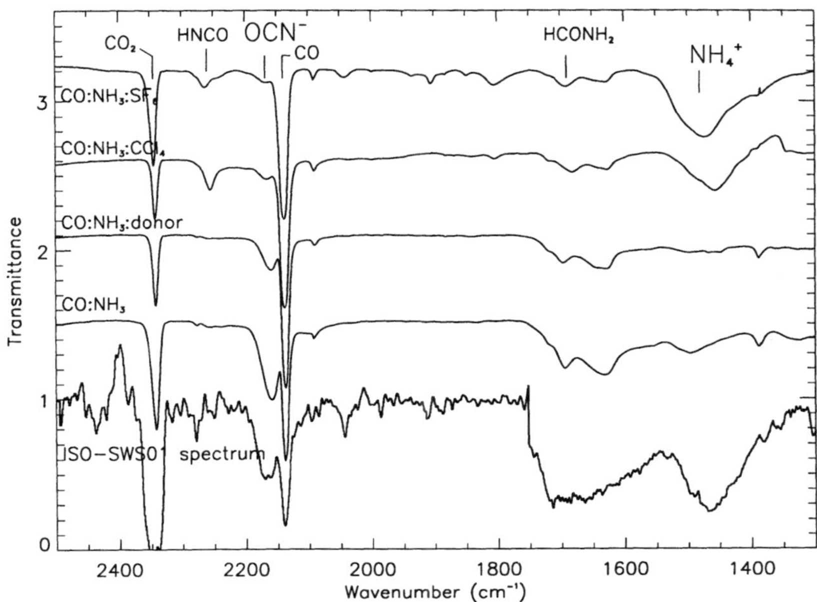

Fig. 9. — Comparison of the ISO-SWS01 spectrum of the protostellar object RAFGL7009S with spectra of the photolysed laboratory mixtures $CO:NH_3$ (1:1), $CO:NH_3:e^-$ donor (10:10:1), $CO:NH_3:SF6$ (10:10:1) and $CO:NH_3:CCl_4$ (10:10:2). Position of the OCN^- and NH_4^+ bands in the laboratory mixtures have been indicated.

was determined using an $HCONH_2$ photolysis experiment. As we neglect, in our estimate, some minor oxygen-bearing species (such as HNCO), we determine a lower limit for the integrated cross section of OCN^- of $\mathcal{A}_{OCN^-} \geq 4.28 \times 10^{-17}$ cm/molecule.

From this we derive the value of $\mathcal{A}_{NH_4^+}$ by assuming that in the $CO:NH_3$ mixture, both ions are produced in the same amount. We can thus express $\mathcal{A}_{NH_4^+}$ in term of $\mathcal{A}_{OCN^-}$ and the ratio of the area of the 6.85 to the 4.62 μm band. This gives us the lower limit for the absortpion cross-section of NH_4^+ of $\mathcal{A}_{NH_4^+} \geq 1.3 \times 10^{-17}$ cm/molecule.

The position of the OCN^- band in the experiments is in very good agreement with the interstellar spectrum of RAFGL7009S. As no band in the $CO:NH_3$ experiment contradicts the ISO spectrum, OCN^- is a good candidate for the interstellar 4.62 μm band. Furthermore the maximum abundance derived for OCN^- in the source is 4.75×10^{17} cm^{-2}, *i.e.* about 4% of the water abundance. This reasonable abundance further reinforces the assignment of the interstellar band with OCN^-. Because the mechanism of production of OCN^-

is an equilibrium mechanism that conserves the charge, this assignment implies that a positive ion is present in the grain mantles towards RAFGL7009S with the same abundance as OCN^-. As no other cation has been identified in the source, NH_4^+ should be present in the ice. NH_4^+ participates in the interstellar 6.85 μm band (Fig. 9) but cannot reproduce the whole band. Assuming the same abundance as OCN^-, NH_4^+ account for about 6 – 7% of the interstellar feature and is clearly not the only carrier of this band. A complete description of the experiments involved and their interpretation can be found in Demyk *et al.* [36].

4.2. Methanol-CO_2 complexes

The resolution achieved by the SWS instrument on board ISO allows us to distinguish the substructures present in the solid absorption bands. Specific attention must be paid on the carbon dioxide bending mode around 15.2 μm, which is deformed and varies with different lines of sight (see [37], and Ehrenfreund, this volume). This particular shape can be interpreted from at least three aspects. At first glance, a simple reason for this shape change could be due to the presence of additional molecules on the line of sight, absorbing in this wavelength range. However, thanks to the ISO wavelength coverage, we can reject a lot of candidates because we did not observe the numerous other infrared transitions associated with the molecules proposed. A second aspect is the possible grain shape effect on the absorption pattern. The fact that we do not detect this effect for any other modes (except for some scattering in the red wing of the water ice stretching mode around 3 μm) and that the grain size distribution in interstellar space is generally dominated by grains very small compared to infrared wavelengths weakens this hypothesis. A third approach we tested in the laboratory was the posibility of modifying the absorption lineshape by intermolecular interactions of the carbon dioxide with another ice constituant.

To search for the reasons of the CO_2 bending mode triple structure we recorded the infrared spectrum of co-deposited mixtures of (CO_2:X–1:1), X being a molecule from the alcohol, ketone, ether group or H_2O. For most of these experiments (except for formic acid and water), at low temperature, just after deposition, we obtained a doubly peaked CO_2 bending mode. When we warmed up the sample, we where able to produce the interstellar triple peak structure in the 50 – 80 K temperature range, depending on the molecules deposited with CO_2. We interpret this as the combination of two effects. The first one is the creation in the ice of an intermolecular interaction of the EDA type (Electron Donor Acceptor), where the lone eletron pairs of the molecule in interaction with the carbon dioxide transfer part of their electronic density onto the CO_2 carbon atom. This interaction is responsible for a slight bent of the CO_2 molecule. The bending mode, initially doubly degenerated, has now two independant bending vibrations, in-plane and out-of-plane of the bent CO_2 molecule. This is the origin of the two peaks seen at low temperature.

We do not see this double peak when water is added in large proportion to the mixtures. The EDA interaction is ineffective if water (or formic acid) dominate the mantle because they can hydrogen bond. The hydrogen bonding then totally dominate the ice structure and the EDA complex formation is no longer possible. When we warm-up the sample, some of the CO_2 molecules in the interaction are freed and migrate in the mantle to form either a separate layer or pure CO_2 "bubbles". We believe this process is responsible for the additional substructure seen in the interstellar spectra. The details of this process are currently being developed [38].

Among the possible candidates to fulfill the interaction requirements, we need to distinguish the ones that are astrophysically relevant in terms of abundance and which can fit with the overall infrared spectrum measured with ISO. The methanol molecule is thus a reasonable candidate. In the short wavelength atmospheric windows, we have observed RAFGL7009S in collaboration with Schutte and Geballe and detected very specific absorptions of the solid methanol molecule [39]. The wavelength range covered in this observation is represented in Figure 10 by a small rectangular box. This identification leads us to derive a solid methanol to water ice ratio of ~ 0.3, making it the second most abundant molecule in the mantles along this line of sight.

Finally, what would be the impact of such a high abundance of methanol relative to water ice on the spectrum of RAFG7009S? Methanol possesses numerous infrared transitions (see Fig. 10). We can then predict and compare the relative importance of the other modes. In particular we can evaluate the methanol contribution to the so-called 6.85 μm band. To do this we invert the equation:

$$\int \tau(\bar{\nu})\mathrm{d}\bar{\nu} = \mathcal{A} \times N \tag{2}$$

where $\mathcal{A}$ is the integrated absorption cross section of the transition under consideration and N the column density of molecules. The CH_3 deformation modes and the OH bending mode around 6.85 μm possess an integrated cross-section estimated between 1.1 and 1.2×10^{-17} cm mol^{-1} for the proportions found in the interstellar mixtures. From this and the water ice to solid methanol ratio derived we can explain from 30 to 60% of the 6.85 μm interstellar band (taking into account the baseline derivation and cross section uncertainties). Part of the band (of the order of 10%) could be also explained by the ammonium (NH_4^+) ion but another unidentified carrier may exist.

The $2\nu_8$ mode falls at 2040 cm^{-1} (4.9 μm), right at the location of a band attibuted to OCS. Using the same reasoning, we deduce that 25% to 35% of this band can be accounted for by the overtone of the strongest methanol mode. The bandwidth of this overtone transition explains rather well the observed 4.9 μm profile with quite a large base and narrower center. This fact also explains why Palumbo *et al.* [40] obtained a better fit to this with a CH_3OH-OCS mixture. This is not necesarily an interaction effect between the two molecules but simply an addition effect along the line of sight.

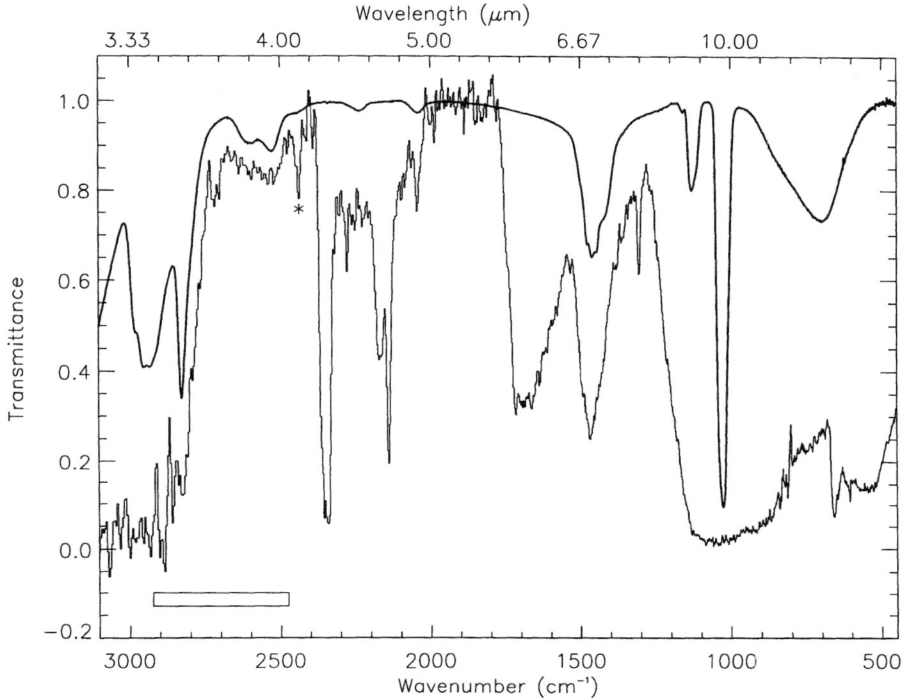

Fig. 10. — Pure solid methanol spectrum (upper solid line) deposited at 10 K superimposed on the transmittance of the RAFGL7009S ISO observation. The spectrum is normalised to the abundance derived from ground based observations in the wavelength region represented by the small rectangular box.

5. DISCUSSION

The ISO spectral data analysis of a deeply embedded object like RAFGL7009S has proven to be extremely successful in the detection of abundant solid state molecules that are unobservable from the ground and that are in some cases radio-quiet. Thus, in the future, models pertaining to the general chemistry of cold molecular clouds must take these molecules (especially CO_2 and CH_4) into account. Moreover, as we show in this paper, the solid phase molecular abundances dominate those of the gas phase. Most of the ongoing chemistry may take place in this solid phase where complex molecules will be synthesised, thanks to the protection of the grain and surface surface which acts as a "catalyst". The depth of our interpretation would not have been possible without the great advantage of the SWS instrument, namely the large wavelength coverage. Such coverage allows the detection of molecules through their various vibrational modes. Combined with ground based observations, especially in the short wavelength region, we can attribute part of the 6.85 μm band to methanol. This paradox, that although clearly seen in the very first analysis of the data, this band was not immediatly attributed to methanol stresses the importance in observing the entire infrared spectrum. One must keep in mind that this infrared wavelength region is not fully diagnostic and that numerous aliphatic molecules posses transitions in this region. Only the combination of the analysis of the peculiar shape of the CO_2 bending mode in the laboratory and the detection of very specific methanol modes from ground based observations allows a firm identification of this transition to methanol.

This is especially clear for the methanol molecule. We must also emphasize that this in-depth analysis would not have been possible without the association of dedicated laboratory experiments. These experiments have allowed us not only to identify and quantify mixtures of simple molecules (H_2O, CO, CO_2 and CH_4) but also to obtain information on the ice structures, the role of molecular interactions and the formation of complexes. The methanol molecule has been positively identified and is the second most abundant ice species in this source. In itself, such an abundance represents a challenge for the chemical models of such environments. The experiments performed have also allowed us to highlight the presence of charged species and the process of charge exchange in the ice matrix. Finally, in addition to the solid phase species, and although its resolution is not well-adapted for gas phase studies, the SWS instrument has observed gas phase ro-vibrational transitions from the same molecules detected in the solid. Gas to solid-state ratios are derived that may allow us to gain insight into the mechanism of grain evaporation for which out of equilibrium processes have been proposed but never fully tested. Such studies are in the range of possibilities with ISO, which has opened a new way to look at the spectra, as now we must fulfill a global reproduction over the entire wavelength range to establish an identification.

REFERENCES

[1] McCutcheon W.H., Sato T., Purton C.R., Matthews H.E. and Dewdney P.E., *ApJ* **110** (1995) 1762.

[2] Whittet D.C.B., Dust and Chemistry in Astronomy, edited by T.J. Millar and D.A. Williams (IOP, Bristol, 1993).

[3] Xie T., Goldsmith P.F., Snell R.L. and Zhou W., *ApJ* **402** (1993) 216.

[4] Boulanger F., Abergel A., Bernard J.P., Burton W.B., Desert F.-X., Hartmann D., Lagache G. and Puget J.-L., *A&A* **312** (1996) 256.

[5] Jansen D.J., van Dishoeck E.F. and Black J.H., *A&A* **282** (1994) 605.

[6] Chin Y.N., Henkel C., Whiteoak J.B., Langer N. and Churchwell E.B., *ApJ* **305** (1996) 960.

[7] Lacy J.H., Faraji H., Sandford S.A. and Allamandola L.J., *ApJL* **501** (1998) L105.

[8] Mitchell G.F., Maillard J.-P., Allen M., Beer R. and Belcourt K., *ApJ* **363** (1990) 554.

[9] Sheperd D.S. and Churchwell E., *ApJ* **457** (1996) 267.
[10] Rothman L.S., Gamache R.R. and Tipping R.H., *JQSRT* **48** (1996) 469.
[11] Reichle H.G. and Young C., *Can. J. Phys.* **50** (1972) 2662.
[12] McCutcheon W.H., Sato T., Dewdney P.E. and Purton C.R., *AJ* **10** (1991) 1435.
[13] Gerakines P.A., Schutte W.A., Greenberg J.M. and van Dishoeck E.F., *A&A* **296** (1995) 810.
[14] D'Hendecourt L.B., *et al.*, *A&A* **315** (1996) L365.
[15] Van Dishoeck E.F. and Helmich F.P., *A&A* **315** (1996) L177.
[16] Helmich F.P., *et al.*, *A&A* **315** (1996) L173.
[17] Boogert A.C.A., Schutte W.A., Helmich F.P., Tielens A.G.G.M. and Wooden D.H., *A&A* **317** (1997) 929.
[18] D'Hendecourt L.B., Allamandola L.J., Grim R.J.A. and Greenberg J.M., *A&A* **158** (1986) 119.
[19] Whittet D.C.B., Bode M.F., Longmore A.J., Admason A.J., McFadzean A.D., Aitken D.K. and Roche P.F., *MNRAS* **233** (1988) 321.
[20] Jones A.P. and Williams D.A., *MNRAS* **209** (1984) 955.
[21] Tielens A.G.G.M. and Hagen W., *A&A* **114** (1982) 245.
[22] d'Hendecourt L., Allamandola L.J., Baas F. and Greenberg J.M., *A&A* **152** (1985) 130.
[23] Hasegawa T.I., Herbst E. and Leung C.M., *ApJS* **82** (1992) 167.
[24] Bergin E.A., Langer W.D. and Goldsmith P.F., *ApJ* **441** (1995) 222.
[25] Bergin E.A., Neufeld D.A. and Melnick G.J., *ApJ* **499** (1998) 777.
[26] Schmitt B., Grim R. and Greenberg J.M., Interstellar Dust: Contributed Papers (1989) p. 265.
[27] Soifer B.T., Puetter R.C., Russel R.W., Willner S.P., Harvey P.M. and Gillet F.C., *ApJ* **232** (1979) L53.
[28] Lacy J.H, Baas F., Allamandola L.J., *et al.*, *ApJ* **276** (1984) 533.
[29] Larson H.P., David D.S., Black J.H. and Fink U., *ApJ* **299** (1985) 873.
[30] Grim R.J.A. and Greenberg J.M., *ApJ* **321** (1987) L91.
[31] Grim R.J.A., Greenberg J.M., Schutte W.A. and Schmitt B., *ApJ* **341** (1989) L87.
[32] Tielens A.G.G.M. and Allamandola L.J., Physical Processes in Interstellar Clouds (Dordrecht, Reidel, 1987) p. 333.
[33] Tielens A.G.G.M., Allamandola L.J., Bregman J., *et al.*, *ApJ* **287** (1984) 697.
[34] Schutte W.A. and Greenberg J.M., 1997, *A&A* **317** (1997) L43.
[35] Bondybey V.E. and English J.H., *J. Mol. Spec.* **92** (1982) 431.
[36] Demyk K., Dartois E., d'Hendecourt L.B., Jourdain de Muizon M., Heras A. and Breitfellner M., *A&A* **339** (1998) 553.
[37] de Graauw T., *et al.*, *A&A* **315** (1996) L345.
[38] Dartois E., *et al.*, in preparation.
[39] Dartois E., Schutte W., Geballe T.R., Demyk K., Ehrenfreund P. and d'Hendecourt L., *ApJ*, 1999, accepted.
[40] Palumbo M.E., Tielens A.G.G.M. and Tokunaga A.T., *ApJ* **449** (1995) 674.

LECTURE 12

Ice Evolution in the Interstellar Medium

W.A. Schutte

Raymond and Beverly Sackler Laboratory for Astrophysics at Leiden Observatory, P.O. Box 9513, 2300 RA Leiden, The Netherlands

Abstract. The composition of ices in dense interstellar clouds as revealed by the infrared absorption bands in the spectra of obscured embedded or background stars is reviewed. We summarise the observational evidence for the presence of a number of ice phases of distinct composition, and discuss how this segregation may have been caused. Possible evidence for energetic processing of the ices is indicated.

1. INTRODUCTION

At the densities of dense interstellar clouds ($10^3 - 10^5$ H atoms cm^{-3} [1]) the timescale for the accretion of molecules on interstellar dust particles is small ($\sim 10^5$ yr; *e.g.* [2]). A decisive aspect of dense cloud chemistry therefore is the condensation of a large fraction of the volatile molecules into icy dust mantles. While H, H_2 and He are hardly depleted, all other species can be indefinitely retained at the temperature of dense cloud grains ($\sim 10 - 15$ K [2]). The presence of gas phase species heavier than He can thus only be explained by the existence of one or more desorption mechanisms which are effective for, at least, part of the frozen species. It is clear that understanding the solid state processes in dense clouds, *i.e.*, modification of species in the ice mantles and the desorption mechanisms, is essential for a better understanding of the chemistry of these objects.

The frozen molecules cause characteristic mid-infrared ($2.5 - 20$ μm) absorption bands in the spectra of sources obscured by dense cloud extinction (*i.e.*, embedded young stellar objects or background field stars).

These features correspond to transitions between molecular vibrational modes. Due to the low temperature of interstellar dust and the very low accretion rates, ices in dense clouds will be amorphous. Since the profile and position of an infrared band of a frozen molecule depends on the structure and composition of the ice, identification of the features requires comparison with appropriate laboratory analogs prepared under simulated interstellar conditions.

The structure and composition of the icy mantles on dense cloud grains will reflect their history. Aspects of this are the initial condensation which is immediately followed by grain surface chemistry, chemical processing of the ice by UV photons and energetic ions (*e.g.*, cosmic rays), and the "distillation" of the ice by selective desorption events, *e.g.*, heating by cosmic rays. Therefore, studying the ices in various environments (obscured or translucent regions, near embedded stellar objects) will give information on how physical and chemical processes interact while the cloud evolves.

This paper is intended to review the current state of knowledge of interstellar ice. It will attempt to interpret this information in the context of the ambient physical and chemical conditions in dense clouds. Section 2 gives a description of the laboratory techniques used to produce interstellar ice analogs. Section 3 quantitatively reviews the observational information on the composition of interstellar ices. Section 4 discusses the segregation of interstellar ice in separate phases. Section 5 discusses energetic processing of the ices from a theoretical, experimental, and observational point of view. Section 6, finally, summarises the main insights.

2. LABORATORY TECHNIQUES

A schematic view of a basic laboratory set-up for the production of interstellar ice analogs is shown in Figure 1. It involves a high vacuum system equipped with a cold finger which is cooled to ~ 10 K by means of a cryostat. A deposition system allows the slow condensation of mixtures of gases on the cold finger resulting in the formation of a layer of amorphous ice. In Figure 1, the set-up is located in the sample chamber of an infrared spectrometer allowing in-situ spectroscopy. Infrared spectroscopy is the most applied method of analysis, since the spectra can be directly compared to astronomical observations to deduce the composition of interstellar ices (Sect. 3). The CsI window is turned to face the deposition tube during sample condensation, or perpendicular to the infrared beam during data acquisition. For further details on the experimental techniques we refer the reader to [3,4]. Table I gives an overview of laboratory infrared spectroscopy of interstellar ice analogs published to date.

The basic set-up of Figure 1 can be extended to simulate specific interstellar processes, or to implement various ways of analysis, *e.g.*, addition of a mass spectrometer [6,7], or UV/visual spectrometry (*e.g.*, [8]). A hydrogen discharge UV lamp can be mounted to simulate photo-processing [5,9]. Cosmic Ray bombardment can be simulated by adding a source of keV/MeV ions [6,7,10,

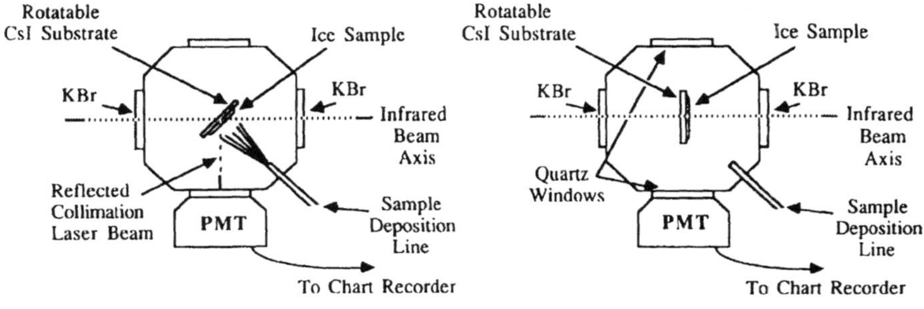

Fig. 1. — Schematic of the vacuum chamber of a basic set-up to produce interstellar ice analogs. The pictures show the system in the configurations used (a) during sample deposition and (b) during infrared spectral measurements [3].

Table I. — Infrared spectroscopy of interstellar ice analogs.

Reference	Species	Comment
[18]	H_2O	pure
[19]	numerous	mixtures with H_2O, low resolution
[20]	numerous	pure ices, low resolution
[64]	CO	
[21]	CO_2	
[22]	CH_3OH	only in H_2O, only CO stretching mode
[23]	O_2	
[3]	numerous	simple mixtures
[24]	CO	ion irradiated ices
[25]	OCS	
[4]	CO, CO_2	band strengths
[57]	CH_3OH, H_2CO	only CH stretch features
[26][a]	H_2O	in apolar ice mixtures
[27]	(iso-)nitriles	numerous species, simple mixtures
[28]	CH_4, SO_2	
[60][a]	CO, CO_2	apolar ice mixture
[68]	CO	apolar ice mixtures
[29]	ethanol, ethane, H_2O_2	simple mixtures

[a] Available on the internet at: http://www.strw.leidenuniv.nl/ehrenfreund/isodb

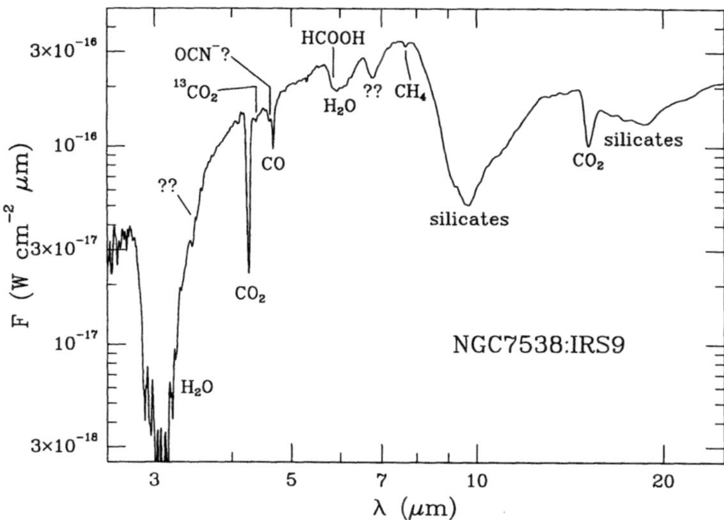

Fig. 2. — Infrared spectrum of the high mass embedded young stellar object NGC 7538:IRS9 as observed by ISO (from [13]).

11]. For simulating grain surface chemistry between molecules and atoms, the sample can be exposed to a flow of atomic hydrogen or oxygen by dissociating H_2 or O_2 in a discharge [12].

3. ICES IN DENSE CLOUDS: OBSERVATIONAL CONSTRAINTS

Figure 2 shows the mid-infrared spectrum of the high mass embedded young stellar object (YSO) NGC 7538:IRS9 as was recently obtained by the Infrared Space Observatory (ISO) [13]. With ISO it has, for the first time, become possible to obtain a high resolution spectrum over the entire mid-infrared range. The spectrum of NGC 7538:IRS9 is characterised by a broad background emission caused by the thermal radiation of hot dust (200 – 1000 K) in the vicinity of the star, with superimposed absorption bands due to cold dust further out ($T < 100$ K). As indicated, apart from two strong silicate bands, most features can be identified with ice components. We note that the CO_2, CH_4 and HCOOH features were first clearly observed by ISO, although indications of the presence of CO_2 and CH_4 were earlier seen in IRAS [14] and ground-based data [15].

The infrared ice absorption bands can be identified by matching them with the spectra of amorphous laboratory samples. An example of this is given in Figure 3, which shows the 5 to 8.5 μm region of NGC 7538:IRS9. Features are present at 6.0, 6.8 and 7.7 μm. The figure shows the laboratory match to the 6.0 μm band. It uses two components, *i.e.*, the OH bending mode of H_2O,

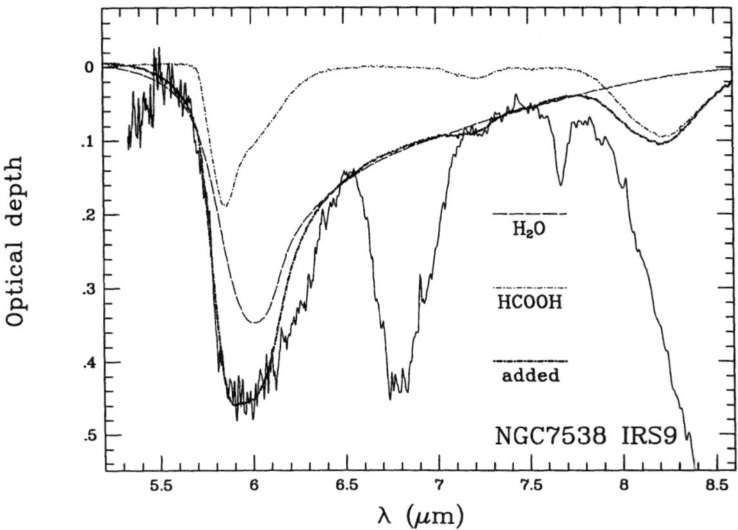

Fig. 3. — ISO spectroscopy from 5 to 8.5 μm towards the high mass YSO NGC 7538:IRS9. The 6.0 μm band is matched with a combination of the OH bending mode of amorphous H_2O ice and amorphous HCOOH ice [16].

and the CO stretching mode of formic acid (HCOOH). While the composite laboratory spectrum gives a good match to the centre and blue wing of the 6 μm feature, some excess absorption remains near 6.2 μm. This absorption probably originates in refractory grain material [16].

The line of sight towards embedded sources generally probes material that has been processed to some extent by the nearby source [17]. Therefore, observations of unassociated background field stars are essential to study dense cloud dust. Suitable sources are luminous red giants, where the cool stellar photosphere provides the infrared background.

3.1. Inventory of solid and gaseous volatile species

Table II compares the ice abundances derived from the mid-infrared absorption bands for the "typical" high–mass YSO NGC 7538:IRS9 and for the field star Elias 16 behind the Taurus dense cloud. Column densities are obtained by dividing the integrated optical depth of the features by the band strength determined from laboratory experiments (Sect. 2). Dividing by the column density of H_2O yields the relative ice abundances given in Table II. The absolute abundance X is obtained relative to the column density of hydrogen atoms. The column density, $N_{\rm H}$, along a given line-of-sight can be determined

Table II. — Ice and gas abundances towards the high mass YSO NGC 7538:IRS9 and the background star Elias 16.

Species	Position (μm)	Abundances with respect to: H_2O ice		Hydrogen	Ref.
		Elias 16	NGC 7538	Elias 16	
Ices					
H_2O	3.05,6.0	100	100	1.25×10^{-4}	[16,17,56]
H_2O_2	3.5	< 5	< 5	$< 6.3 \times 10^{-6}$	[29]
CO in ...					
polar ice	4.67	6.5	7	8.0×10^{-6}	[45,54,66]
apolar ice	4.67	27	8	3.3×10^{-5}	[45,54]
CO_2	4.27,15.2	14	12	2.1×10^{-5}	[59,61]
CH_3OH	3.54,9.75	< 3.4	6.3	$< 4.3 \times 10^{-6}$	[56,58]
H_2CO	5.81	–	< 3	–	[16,84]
CH_4	7.68	–	1.6	–	[15,46]
HCOOH	5.85	–	3.0	–	[16]
OCN^- (?)	4.62	< 0.4	0.5	$< 6 \times 10^{-7}$	[35,38,40]
NH_3	2.97	< 6	< 6	$< 8 \times 10^{-6}$	[17]
unid.	6.8	–	–	–	[16]
Gas phase					
CO	IR lines	60	120	7.4×10^{-5}	[47,48]
O_2	IR lines	< 4	–	$< 5 \times 10^{-6}$	[49,50]
H_2O	IR lines	–	< 4	–	[51]

from the measurement of the optical depth of the silicate feature at 9.7 μm:

$$N_{\rm H} = \{\tau(9.7) - \tau_{\rm thresh.}(9.7)\} \times \left\{ \left[\frac{N_{\rm H}}{A_{\rm V}}\right] \left[\frac{A_{\rm V}}{\tau(9.7)}\right] \right\} \tag{1}$$

using $N_{\rm H}/A_{\rm V} = 1.9 \times 10^{21}$ cm^{-2} mag^{-1} [30], and $A_{\rm V}/\tau(9.7) = 18.9$ mag [31,32]. Towards Elias 16 $\tau(9.7) = 0.57$ and $\tau_{\rm thresh.}(9.7) = 0.1$ [33]. The threshold optical depth corresponds to the outer, translucent, regions of the Taurus cloud where no ices are present.

The assignment of the 4.62 μm band to OCN^- should be regarded as still tentative. The strong OCN^- ν_3 feature falls very close to the interstellar 4.62 μm band [34–37]. In the laboratory, OCN^- can be produced both by photolysis of ices containing CO and NH_3, as well as by warm-up or photolysis of ices containing iso-cyanic acid (HNCO) with NH_3 [35,37–39]. While a thorough overview of alternative candidates was recently given by [40], no other candidate seems at this point able to satisfy the observational constraints.

3.2. The O, C, and N budget in dense clouds

Table II shows that the total amount of oxygen and carbon in identified gaseous and icy material towards Elias 16 equals $X(\mathrm{O}) = 2.8 \times 10^{-4}$ and $X(\mathrm{C}) = 1.4 \times 10^{-4}$. This can be compared to the total amount of condensible material, *i.e.*, all matter not locked in the dust. Under the assumption that no dust material is converted to volatile species in the dense cloud environment, this quantity is equal to the abundance of gaseous material in the diffuse medium, since in such lines of sight no ices are present on the grains [32, 41, 42]. This yields 3.3×10^{-4} and 1.4×10^{-4} for $X(\mathrm{O})$ and $X(\mathrm{C})$, respectively [43, 44]. The uncertainty in these numbers is about 5% for $X(\mathrm{O})$ and 15% for $X(\mathrm{C})$ [43]. Comparison with the abundances in Table II shows that the observed species towards Elias 16 account for 86% of the condensible oxygen and for 100% of the condensible carbon. We conclude that the majority of the oxygen and carbon in dense clouds can be accounted for using ISO observations.

The amount of condensible nitrogen as determined from diffuse lines of sight equals $X(\mathrm{N}) = 7.8 \times 10^{-5}$ [44]. No nitrogen containing species, except for the minor OCN^- component, have yet been observed in the icy grain mantles. Nitrogen containing molecules observed in the gas phase account for only a minor fraction [52, 79]. This suggests that most nitrogen resides in molecules with weak infrared features. Indeed, models of dense clouds predict that most nitrogen will be in the form of the infrared inactive and radio quiet species N_2. This very stable molecule can be formed efficiently in the gas phase [53].

3.3. Source to source variation

H_2O is the dominant component of ices in all known lines of sight. Its absolute abundance (*i.e.*, relative to hydrogen) is considerably lower towards YSO's than towards field stars (*cf.*, [54]). This difference is plausibly caused by sublimation of the ices close to the protostar.

The abundance of apolar CO is highly variable, ranging from close to 0 (for the high mass YSO GL2136) up to ~50% relative to H_2O towards field stars and some low mass protostars. This variation correlates with the degree of thermal annealing of the ices caused by the YSO, consistent with the high volatility of CO [17].

The 3.54 μm feature of solid methanol has so far only been detected towards three high mass YSO's [55–58]. These data show that the methanol abundance is quite variable, from less than 3% up to 8%. The observations suggest that the abundance may be considerably lower in low mass protostars and field stars relative to high mass protostars [58].

The abundance of CO_2 derived from the ISO data shows remarkably little variation, *i.e.*, abundances range from 12 – 16% relative to H_2O, with no systematic difference between YSO's and field stars [59–61].

Recent surveys show that the 4.62 μm band, which is possibly due to OCN^-, is generally present in the spectra of embedded YSO's [36, 40]. The derived abundance is quite variable, from 0.1 – 0.8% relative to H_2O (deduced from the observations reported by [40], with a band strength of 1.0×10^{-16} cm/mol [37]). The feature was not detected towards the field star Elias 16, suggesting that it may be confined to protostellar regions [36]. A claim that the 4.62 μm feature anti-correlates with the apolar CO feature [36] was not confirmed by a more recent study [40].

4. ICE PHASES

Detailed comparison between the profile of the interstellar ice features and those of laboratory analogs gives information on the composition of the ice matrix in which the molecule is embedded. Such comparisons have revealed that interstellar ice is segregated into a number of distinct phases. In this section, the observational evidence for the existence of ice phases dominated by H_2O, and by apolar molecules is reviewed. Subsequently, mechanisms that could be responsible for the segregation are considered. Methanol-rich ice, which is seen towards high mass YSO's [62] has been discussed elsewhere [63].

4.1. Observations

The 4.67 μm feature of solid CO generally consists of two components, a broad feature ($FWHM \approx 12$ cm^{-1}) centred at 2136 cm^{-1}, and a narrower band ($3 - 11$ cm^{-1}) near 2140 cm^{-1}. Matching these features by laboratory spectra reveals that these components originate in ices of distinctly different compositions. The broad feature requires CO diluted in an H_2O-dominated ice, while the narrow feature is matched with CO in ices of apolar molecules, *i.e.*, O_2, N_2, CO_2, and CO itself, with at most only traces of H_2O [54, 64–67]. The presence of the apolar component shows that interstellar ices are segregated into phases of distinctly different composition. Following [2], we will name the apolar ice phase "type III" ice, while we designate the H_2O-dominated ice "type I" ice. Type II ice, which is rich in CH_3OH, is not discussed in this paper.

4.2. Origin

In the following, we discuss only the origin of the ice phases observed in the relatively simple situation of the general dense cloud medium, *i.e.*, towards field stars. For a more extended discussion we refer to [63]. Two mechanisms have been proposed to differentiate interstellar ices in such regions. First, chemically and physically different regions along the line of sight could produce various condensates (*e.g.*, [68], and references therein). Second, sporadic ice desorption events, *e.g.*, warm-up by a cosmic ray, could lead to the desorption of the most volatile ice components. Subsequent re-condensation would form a separate layer of volatile species on top of the refractory ice material [69]. We will henceforth denote these segregation mechanisms *Diverse Condensation* (DC) and *Diverse Desorption* (DD).

According to the DC model, the apolar and polar components of interstellar ices are associated with dense regions of different density and extinction. This is based on models of the chemical evolution of dense clouds [70–72]. Such models assume either that condensation takes place following chemical equilibration in the gas phase, or that an indiscriminate desorption mechanism maintains an equilibrium between gas and ice. According to the models, condensation at densities between 2×10^3 cm^{-3} $< n_H <$ 2×10^4 cm^{-3} leads to mantles dominated by H_2O with important inclusions of CO, CO_2, N_2, O_2 and possibly CH_3OH. Condensation at $n_H >\sim 10^5$ cm^{-3} leads to mantles consisting of CO, O_2 and N_2. This difference is driven by the increasing conversion of atoms to molecules in the gas phase in more opaque, higher density regions. In the lower density regions, abundant atomic hydrogen and oxygen is able to hydrogenate and oxidise the condensing molecular material (CO, N_2, O_2) on the grain surface [63, 70, 73]. At high densities, the atomic content of the gas becomes very low, and the condensing species are incorporated into the ice unchanged.

An important implication of the DC model is that the elemental composition of the various ice phases should all be equal to the overall composition of the condensible material. This gives O:C:N = 2.4:1:0.56 (Sect. 3.1). For comparison, the observed composition of type I (H_2O-dominated) ice towards the background star Elias 16 is $O/C \approx 6$ (Tab. II). While the composition of the type III (apolar) ice is not so well constrained, the best fits of the apolar CO band towards Elias 16 indicate that the CO matrix contains only a limited quantity of impurities, implying $O/C < 2$ [54, 66, 74]. Thus, observations contradict the prediction by the DC model of a fixed $O/C \approx 2.4$ for all ice phases. We therefore conclude that the DC mechanism cannot be responsible for the production of the types I and III ices towards Elias 16.

Various mechanisms have been proposed that could produce the selective desorption required by the DD model, such as Cosmic Ray spot heating, Cosmic Ray whole grain heating and explosive desorption by release of stored chemical energy ([2], and references therein). The DD model predicts that the composition of the type I and type III ices reflects a difference in volatility. In this scenario, the initial condensation produces an homogeneous ice of composition as predicted by grain models for 2×10^3 cm^{-3} $< n_H < 2 \times 10^4$ cm^{-3}, *i.e.*, H_2O-rich, with abundant inclusions of CO, CO_2, N_2, O_2 and some CH_4 and NH_3. Subsequently, the selective desorption would deplete, over time, the volatiles from this mantle, which re-condense on top of the original ice. The processed original ice constitutes the type I ice, while the re-condensed volatile layer gives the type III ice.

The DD scenario may allow for the different elemental composition of types I and III ices towards Elias 16. The low carbon content of the type I ice could be explained by the gradual depletion of CO and CH_4, along with O_2 and N_2 by selective desorption. In a few desorption and re-condensation cycles the O_2 could gradually convert to H_2O by grain surface reactions as long as atomic hydrogen is still abundant in the gas phase, *i.e.*, up to densities of

$\sim 10^5$ H atoms cm^{-3} [70,71]. This "new" H_2O would be purified by the next desorption event and become part of the type I ice mantle. CH_4 is likely to be quickly converted to CO in the gas phase [71]. Thus, the final type III ice would consist primarily of CO and N_2. Assuming that most of the nitrogen in dense clouds resides in N_2 (Sect. 3.1) and that the distribution gas/ice is similar for N_2 and CO (consistent with the close volatility of these species), the composition of type III ice would be associated with a molecular ratio N_2/CO of ≈ 0.3. Such an inclusion of N_2 would hardly influence the CO band profile [60] and would therefore be undetectable. This model predicts that the gas phase in dense clouds will be dominated by CO and N_2 with similar relative abundances.

5. ENERGETIC PROCESSING

A possibly important aspect of the evolution of interstellar ice is the chemical modification caused by energetic processing. This involves UV irradiation or irradiation by keV/MeV ions, *e.g.*, cosmic rays. Several sources for energetic processing are predicted by theory. The galactic UV field may penetrate deep into interstellar clouds due to the inhomogeneous, clumpy, structure of such objects [75]. UV photons from embedded YSO's and their hot accretion disks could illuminate the cloud environment [76]. Ionisation of molecular hydrogen by keV-MeV ions or X rays of cosmic or stellar origin produces secondary and tertiary free electrons which are able to excite molecular hydrogen and produce UV [77,78]. X-rays and ions will also process the grains directly.

In the following, we study the importance of energetic processing of interstellar ices. First, the results of laboratory simulations are reviewed. We will focus here on the effects of UV processing, referring for a review on the effects of ion processing to ([63], see also Moore this volume). Second, we describe in detail the observational data in support of evidence for processing.

5.1. Laboratory simulation of energetic processing

The laboratory studies reviewed here were performed at a temperature of ~ 10 K, representative of the grain temperature in the general dense cloud medium. An extensive database of infrared spectroscopy of UV irradiated ices is available on the internet (http://www.strw.leidenuniv.nl/lab).

5.1.1. H_2O dominated ices

UV photolysis of pure H_2O leads to the production of the first order product OH, and higher order products HO_2 and H_2O_2 [5]. Addition of CO at an abundance level of a few tenths quenches the formation of these species, with instead production of CO_2, HCO, H_2CO and HCOOH [80].

Adding ammonia has a strong influence on the photochemistry, since its presence enables reactions between NH_3, a base, and photochemically produced acids. This leads to the production of ions in the ice matrix. In photolysed $H_2O/CO/NH_3$ mixtures, OCN^- and $HCOO^-$ are produced, as well as NH_4^+. In addition, formamide ($HCONH_2$) is a typical photoproduct [35,38,81].

The UV photolysis of H_2O/CO_2 ices has not been studied yet. However, since CO is the prime photoproduct in the irradiation of pure CO_2 ice, while CO_2 is the main product of the irradiation of H_2O/CO ice, the result obtained with both mixtures should be qualitatively similar.

5.1.2. *CH_3OH containing ices*

Ices with CH_3OH as the dominant carbon containing molecule show very different photochemical properties as compared to CO and CO_2. Photolysis of pure CH_3OH produces H_2CO and CH_4 as first order products [5]. Further irradiation produces CO, CO_2, HCO and methyl formate (H_3C-O-HCO). Photolysis of CH_3OH/H_2O at comparable abundance gives identical products as photolysis of pure methanol [82,83].

Again, including NH_3 has a dramatic effect on the photochemistry. However, the effect is quite different from that obtained for H_2O/CO ices, because the presence of NH_3 in CH_3OH/H_2O mixtures primarily promotes formaldehyde reactions rather than acid-base reactions. Formaldehyde is the prime dissociation product of CH_3OH, and the presence of NH_3 strongly enhances its ability to react with itself and other molecules in the ice matrix [84]. In $CH_3OH/H_2O/NH_3$ mixtures thus large structures are efficiently produced, *i.e.*, hexamethylene tetramine (HMT; $C_6H_{12}N_4$), and formaldehyde polymers like polyoxymethylene [83].

5.1.3. *Apolar ices*

Apolar ices of astrophysical relevance consist of the molecules CO, O_2, N_2 and CO_2. The photochemistry of single component ices was studied by Gerakines *et al.* [5]. Photolysis of CO gives only a very small yield of new products (CO_2 and some C_3O), due to the high stability of this molecule. Likewise, UV radiation has very little effect on N_2 ice. Irradiation of CO_2 produces CO and the CO_3 radical. Irradiation of O_2 produces ozone, O_3, at high yields.

Mixtures of CO and N_2 do not show strong photochemical effects, as expected from the results obtained by irradiating these molecules individually. Upon addition of O_2 a more rich photochemistry ensues, showing clear features of CO_2, N_2O, O_3, and weak bands of CO_3 [68].

5.2. Evidence for processing of interstellar ices

When searching for evidence of energetic processing of interstellar ices, a problem which has to be faced is that molecules which are photochemically produced in the laboratory may, in space, be produced by other processes as well, in particular grain surface chemistry. For example, CO_2, a ubiquitous constituent of interstellar ices (Sect. 3.1), can be produced by irradiating H_2O/CO ices, or possibly by grain surface reactions between CO and O [70,72]. Another product of photolysis of H_2O/CO ices is H_2CO. A detection of the formaldehyde CO stretching mode at 5.81 μm was reported towards the highly obscured high mass YSO RAFGL7009S [60]. However, CH_3OH is generally observed towards YSO's as well, at considerably higher abundances than H_2CO (Sect. 3). CH_3OH is produced very inefficiently by UV processing [57]. It therefore seems plausible that the radiationless processes producing CH_3OH (presumably grain surface chemistry [72]) may produce the related species H_2CO as well.

Thus, while interstellar ices may well be energetically processed, in the absence of molecules which can be uniquely identified with photoproducts, the effects of such processing would be hard to distinguish if independent information on the composition of the original, unprocessed, ice is lacking. However, some products may be unique to photoprocessing. Two such species are HMT and the OCN^- ion. HMT has only weak infrared spectral signature [83] and would probably be very hard to observe. OCN^- on the other hand produces a strong feature near 2165 cm^{-1} (4.62 μm). A similar band is observed towards embedded YSO's (Sect. 3). While laboratory studies produce OCN^- by photolysis of ices containing CO and NH_3 [35,38], in space iso-cyanic acid may be directly produced by grain surface chemistry [85]. Some OCN^- could then be produced by acid-base reactions on the grain surface, or close to the protostar where the ices are heated and molecules can diffuse and react inside the ice. However, laboratory simulation of these processes indicate that by themselves they cannot cause sufficient conversion of HNCO to OCN^- to reproduce the observations, and that some kind of energetic processing is required [37]. The 2165 cm^{-1} band may thus be a key indicator of the energetic processing of ices near YSO's. Further laboratory work is needed to explore this possibility.

6. SUMMARY

A review has been given of the current knowledge on the physical and chemical evolution of ices in dense interstellar clouds. For quiescent general dense cloud regions, almost the entire budget of condensible atoms (O, C, N), *i.e.*, all material not locked up in the dust, appears to be accounted for by the observed ice and gas-phase molecules now that the entire infrared spectrum has become accessible with ISO. About 2/3 of the condensible material is frozen on the grains, while 1/3 is gaseous. The quiescent cloud chemistry can be qualitatively described in the form of the initial condensation of a gas consisting of molecular (CO, O_2, N_2) as well as atomic (O, H) material, with oxidising and reducing grain surface chemistry, followed by the gradual diversification of the ices by selective desorption events. Such an evolutionary scheme can well account for the presence of separate polar, H_2O-dominated, and apolar, CO-dominated ices (types I and III ice, respectively) in quiescent lines of sight.

While a variety of sources may be able to energetically process ices in dense clouds, especially near YSO's, observational evidence for processing is still scarce. Although extensive processing may have taken place, it is hard to identify key indicators which cannot be produced in any other way, *e.g.*, grain surface chemistry. A promising candidate in this respect is the OCN^- ion, which may account for the 4.62 μm feature observed towards many YSO's. In the laboratory this ion is produced by photolysis of ices containing CO and NH_3. Precise laboratory simulation studies are necessary to clarify whether production of interstellar OCN^- requires energetic processing, or may also be accomplished by grain surface chemistry or warm-up of ices near the YSO.

REFERENCES

[1] Goldsmith P.F., Interstellar Processes (Reidel, Dordrecht, 1987) p. 51.

[2] Schutte, W.A., The Cosmic Dust Connection, edited by J.M. Greenberg (Kluwer, Dordrecht, 1996) p. 1.

[3] Hudgins D.M., Sandford S.A., Allamandola L.J. and Tielens A.G.G.M., *ApJS* **86** (1993) 713.

[4] Gerakines P.A., Schutte W.A., Greenberg J.M. and van Dishoeck E.F., *A&A* **296** (1995) 810.

[5] Gerakines P.A., Schutte W.A., Ehrenfreund P. and van Dishoeck E.F., *A&A* **312** (1996) 289.

[6] Kaiser R.I., Jansen P., Petersen K. and Roessler K., *Rev. Sci. Instrum.* **66** (1995) 5226.

[7] Kaiser R.I., Gabrysch A. and Roessler K., *Rev. Sci. Instrum.* **66** (1995) 3058.

[8] Salama F. and Allamandola L.J., *J. Chem. Phys.* **94** (1991) 6964.

[9] Okabe H., *J. Opt. Soc. Am.* **54** (1963) 478.

[10] Strazzulla G. and Baratta G.A., *A&A* **241** (1991) 310.

[11] Moore M.H. and Hudson R.L., *ApJ* **401** (1992) 353.

[12] Hiraoka K., Ohashi N., Kihara Y., *et al.*, *Chem. Phys. Lett.* **229** (1994) 408.

[13] Whittet D.C.B., Schutte W.A., Tielens A.G.G.M., *et al.*, *A&A* **315** (1996) L357.

[14] d'Hendecourt L.B. and Jourdain de Muizon M., *A&A* **223** (1989) L5.

[15] Lacy J.H., Carr J.S., Evans N., *et al.*, *ApJ* **376** (1991) 556.

[16] Schutte W.A., Tielens A.G.G.M., Whittet D.C.B., *et al.*, *A&A* **315** (1996) L333.

[17] Smith R.G., Sellgren K. and Tokunaga A.T., *ApJ* **344** (1989) 413.

[18] Hagen W., Tielens A.G.G.M. and Greenberg J.M., *Chem. Phys.* **56** (1981) 367.

[19] Hagen W., Tielens A.G.G.M. and Greenberg J.M., *A&AS* **51** (1983) 389.

[20] d'Hendecourt L.B. and Allamandola L.J., *A&AS* **64** (1986) 453.

[21] Sandford S.A. and Allamandola L.J., *ApJ* **355** (1990) 357.

[22] Schutte W.A., Tielens A.G.G.M. and Sandford S.A., *ApJ* **382** (1991) 523.
[23] Ehrenfreund P., Breukers R., d'Hendecourt L. and Greenberg J.M., *A&A* **260** (1992) 431.
[24] Palumbo M.E. and Strazulla G., *A&A* **269** (1993) 568.
[25] Palumbo M.E., Tielens A.G.G.M. and Tokunaga A.T., *ApJ* **449** (1995) 674.
[26] Ehrenfreund P., Gerakines P.A., Schutte W.A. and van Dishoeck E.F., *A&A* **312** (1996) 263.
[27] Bernstein M.P., Sandford S.A. and Allamandola L.J., *ApJ* **476** (1997) 932.
[28] Boogert A.C.A., Schutte W.A., Helmich F.P, Tielens A.G.G.M. and Wooden D.H., *A&A* **317** (1997) 929.
[29] Boudin N., Schutte W.A. and Greenberg J.M., *A&A* **331** (1998) 749.
[30] Bohlin R.C., Savage B.D. and Drake J.F., *ApJ* **224** (1978) 132.
[31] Roche P.F. and Aitken D.K., *MNRAS* **208** (1984) 481.
[32] Whittet D.C.B., Boogert A.C.A., Gerakines P.A., *et al.*, *ApJ* **490** (1997) 729.
[33] Whittet D.C.B., Bode M.F., Longmore A.J., *et al.*, *MNRAS* **233** (1988) 321.
[34] Lacy J.H., Baas F., Allamandola L.J., *et al.*, **ApJ 276** (1984) 533.
[35] Grim R.J.A. and Greenberg J.M., *ApJ* **321** (1987) L91.
[36] Tegler S.C., Weintraub D.A., Rettig T.W., *et al.*, *ApJ* **439** (1995) 279.
[37] Keane J. and Schutte W.A. (1998) in preparation.
[38] Schutte W.A. and Greenberg J.M., *A&A* **317** (1997) L43.
[39] Demyk K., Dartois E., d'Hendecourt L.B., *et al.*, *A&A* **339** (1998) 553.
[40] Pendleton Y.J., Tielens A.G.G.M., Tokunaga A.T. and Bernstein M.P., *ApJ* (1998) in press.
[41] Gillett F.C., Jones T.W., Merrill K.M. and Stein W.A., *A&A* **45** (1975) 77.
[42] Whittet D.C.B., Bode M.F., Longmore A.J., Baines D.W.T. and Evans A., *Nature* **303** (1983) 218.
[43] Sofia U.J., Cardelli J.A., Guerin K.P. and Meyer D.M., *ApJ* **482** (1997) L105.
[44] Snow T.D. and Witt A.N., *ApJ* **468** (1996) L65.
[45] Chiar J.E., Gerakines P.A., Whittet D.C.B., *et al.*, *ApJ* **498** (1998) 716.
[46] Boogert, A.C.A., Schutte, W.A., Tielens, A.G.G.M., *et al.*, *A&A* **315** (1996) L377.
[47] Mitchell G.F., Curry C., Maillard J.-P. and Allen M., *ApJ* **341** (1989) 1020.
[48] van Dishoeck E.F. (1998) private communication.
[49] Combes F. and Wiklind T., *A&A* **303** (1995) L61.
[50] Marechal P., Pagani L., Langer W. and Castets A., *A&A* **318** (1997) 252.
[51] van Dishoeck E.F. and Helmich F.P., *A&A* **315** (1996) L177.

[52] Irvine W.M., Goldsmith P.F. and Hjalmarson A., Interstellar Processes, edited by D.J. Hollenbach and H.A. Thronson (Dordrecht, Reidel, 1987) p. 561.
[53] Herbst E. and Leung C.M., *ApJS* **69** (1989) 271.
[54] Tielens A.G.G.M., Tokunaga A.T., Geballe T.R. and Baas F., *ApJ* **381** (1991) 181.
[55] Grim R.J.A., Baas F., Geballe T.R., Greenberg J.M. and Schutte W., *A&A* **243** (1991) 473.
[56] Allamandola L.J., Sandford S.A., Tielens A.G.G.M. and Herbst T.M., *ApJ* **399** (1992) 134.
[57] Schutte W.A., Gerakines P.A., Geballe T.R., van Dishoeck E.F. and Greenberg J.M., *A&A* **309** (1996) 633.
[58] Chiar J.E., Adamson A.J. and Whittet D.C.B., *ApJ* **472** (1996) 665.
[59] de Graauw Th., Whittet D.C.B., Gerakines P.A., *et al.*, *A&A* **315** (1996) L345.
[60] Ehrenfreund P., d'Hendecourt L., Dartois E., *et al.*, *Icarus* **130** (1997) 1.
[61] Whittet D.C.B., Gerakines P.A., Tielens A.G.G.M., *et al.*, *ApJ* **498** (1998) L159.
[62] Skinner C.J., Tielens A.G.G.M., Barlow M.J. and Justtanont K., *ApJ* **399** (1992) L79.
[63] Schutte W.A., edited by P. Ehrenfreund, H. Kochan, C. Krafft and V. Pirronello (Kluwer, Dordrecht, 1998) in press.
[64] Sandford S.A., Allamandola L.J., Tielens A.G.G.M. and Valero L.J., *ApJ* **329** (1988) 498.
[65] Chiar J.E., Adamson A.J., Kerr T.H. and Whittet D.C.B., *ApJ* **426** (1994) 240.
[66] Chiar J.E., Adamson A.J., Kerr T.H. and Whittet D.C.B., *ApJ* **455** (1995) 234.
[67] Teixera T.C., Emerson J.P. and Palumbo M.E., *A&A* **330** (1998) 711.
[68] Elsila J., Allamandola L.J. and Sandford S.A., *ApJ* **479** (1997) 818.
[69] Schutte W.A. and Greenberg J.M., *A&A* **244** (1991) 190.
[70] Tielens A.G.G.M. and Hagen W., *A&A* **114** (1982) 245.
[71] d'Hendecourt L.B., Allamandola L.J. and Greenberg J.M., *A&A* **152** (1985) 130.
[72] Tielens A.G.G.M. and Whittet D.C.B., Molecules in astrophysics: Probes and Processes, edited by E.F. van Dishoeck (Kluwer, Dordrecht, 1997) p. 45.
[73] Tielens A.G.G.M. and Allamandola L.J., Interstellar Processes, edited by D.J. Hollenbach and H.A. Thronson (Dordrecht, Reidel, 1987) p. 397.
[74] VandenBussche B., Ehrenfreund P., *et al.* (1998) in preparation.
[75] Spaans M., *A&A* **307** (1996) 271.
[76] Spaans M., Hogerheijde M.R., Mundy L.G. and van Dishoeck E.F., *ApJ* **455** (1994) L167.
[77] Prasad S.S. and Tarafdar S.P., *ApJ* **267** (1983) 603.
[78] de Boisanger C., Helmich F.P. and van Dishoeck E.F., *A&A* **310** (1996) 315.

[79] Ohishi M., Molecules in astrophysics: Probes and Processes, edited by E.F. van Dishoeck (Kluwer, Dordrecht, 1997) p. 61.
[80] Hagen W., PhD thesis (University of Leiden, Leiden, 1982) p. 1.
[81] Grim R.J.A., Greenberg J.M., de Groot M.S., *et al.*, *A&AS* **78** (1989) 161.
[82] Allamandola L.J., Sandford S.A. and Valero G.J., *Icarus* **76** (1988) 225.
[83] Bernstein M.P., Sandford S.A., Allamandola L.J., Chang S. and Scharberg M.A., *ApJ* **454** (1995) 327.
[84] Schutte W.A., Allamandola L.J. and Sandford S.A., *Icarus* **104** (1993) 118.
[85] Hasegawa T.I. and Herbst E., *MNRAS* **261** (1993) 83.

LECTURE 13

The Physics and Chemistry of Ices in the Interstellar Medium

M.H. Moore

NASA Goddard Space Flight Center, Astrochemistry Branch, Greenbelt, MD 20771, U.S.A.

Abstract. The effects of the interstellar environment on interstellar ices can be extremely complex. We discuss the cosmic ray radiation environment expected in interstellar regions and the interaction of cosmic rays with icy materials. Infrared studies of laboratory ices can provide information on the evolution of ices as a function of different simulated environments. We summarize results from our experiments on proton-irradiated ices. The effects of ion bombardment on the chemical and physical nature of both pure ices and icy mixtures have been examined using a combination of mid- and far-infrared spectroscopy. Effects include, the synthesis of new species, the formation of both more and less volatile species, and the amorphization of ices. Results of recent laboratory studies on ices deposited onto silicate refractory smokes is included.

1. COSMIC RAY ENVIRONMENT AND EFFECTS ON ICES

1.1. Interstellar ices

There is a diverse and complex array of molecules, ions and radicals detected in the interstellar medium. The majority of these species are in the gas phase, but a fraction have been identified in the condensed phase in cold regions of molecular clouds or stellar nebulae [1, 2]. Table I is a list of the currently identified condensed phase molecules.

Table I. — Condensed phase species identified in the interstellar medium towards IRS9, based on the data compiled by Schutte (this volume).

Identified Species	Abundance
H_2O	= 100
CO	15
CO_2	12
CH_3OH	6.3
NH_3	< 6
H_2O_2	< 5
H_2CO_2 (tentative)	3
H_2CO (tentative)	< 3
CH_4	1.6
X-CN	0.5

The relative abundances of these molecules show that water is the dominant observed ice species. There is accumulating evidence from ISO observations that, in some cases, ices are not mixed along the line of sight. Some observations of CH_3OH, CO, CO_2 and OCS species, show distinct ice components existing with different compositions. Observations often indicate that molecular species are mixtures with polar (*e.g.*, H_2O or CH_3OH) and non-polar molecules (*e.g.*, N_2 or O_2 which have no infrared active vibrational modes because they are homo-nuclear species) [1]. The relative fraction contained in a polar and non-polar matrix varies with the environment. These icy mantles are exposed to the radiation environment of the cloud or nebula which includes ultraviolet photons and galactic cosmic rays. Sources of energy, present over the $10^5 - 10^7$ year lifetime of the icy mantle, cause the UV- and radiation-chemical synthesis of complex molecules which eventually are added to the rich source of organics and other complex material in protoplanetary disks around newly formed stars.

1.2. Cosmic ray environment

The galactic cosmic ray particle radiation environment is modeled as a constant cosmic ray flux which, as cosmochemical evidence indicates, has been, on average, roughly constant throughout the age of the solar system [3]. It is estimated that 87% of the interstellar cosmic rays are protons, 12% Helium nuclei, and 1% heavier nuclei [4]. The interstellar electron cosmic ray flux is a factor of 0.1 to 0.01 times that of the proton flux for energies below

100 MeV [5]. Ion-irradiation processing is dominated by protons whose intensity is largest in the low MeV range where the ions also have their largest stopping power (discussed in the next section). Direct measurements of the intensity of galactic cosmic ray protons is not possible in the inner solar system since solar cosmic rays co-exist with galactic cosmic rays whose energies have been modulated by interactions with the solar wind plasma and solar magnetic field. The separation of the solar and galactic cosmic ray components is not straightforward. In general, below 100 MeV the spectrum is dominated by solar cosmic rays. Only for energies above 200 GeV is there no solar modulation and the measured flux of protons follows a power law decrease with increasing energy necessitating a large upward extrapolation to estimate the flux of low energy galactic protons. Direct measurements of the solar wind plasma, magnetic field and radiation environment are in progress at the edge of our solar system (35 – 64 AU) by Pioneer 10, Voyager 1 and Voyager 2. However, the results from these spacecraft show that these distant regions are still affected by the heliosphere. The current prediction is that the edge of the local interstellar medium lies beyond these spacecraft at a minimum distance of $90-180$ AU [6]. One indirect upper limit to the low energy MeV proton flux in the diffuse medium comes from the hydrogen atom ionization rate inferred from observations of the state of ionization of *e.g.* OH, or DCO in the interstellar medium. In a recent review, Kulkarni and Heiles [7] estimate $1-5\times10^{-17}$ ionizations s^{-1} which result from a calculated cosmic ray flux of 0.6 to 3 $cm^{-2}\ s^{-1}$ using an appropriate ionization cross-section. Other upper limits range from 6 $cm^{-2}\ s^{-1}$ [8] to 9 $cm^{-2}\ s^{-1}$ [9]. The value of 10 $cm^{-2}\ s^{-1}$ is used by Strazzulla and Johnson [10] for calculations of energy deposited in materials exposed to the interstellar cosmic ray environment and this value is adopted in this paper. For isotropically distributed, mono-energetic (1 MeV) protons, the energy flux is 1×10^{7} eV $cm^{-2}\ s^{-1}$.

1.3. Interaction of cosmic rays with solids

MeV ions moving through a solid are slowed down by collisions and the energy which they lose is transferred to the lattice in the form of lattice excitations, electronic excitation and ionizations. The stopping power (energy deposited, MeV $cm^2\ g^{-1}$) depends on the ion charge, mass, and energy of the ion as well as the mass and atomic number of the target. Figure 1 shows the stopping power of low MeV protons as a function of incident energy [11]. The stopping power has a maximum value near 0.1 MeV; at higher energies the stopping power decreases. For a density of 1 g cm^{-3}, the stopping power for a 1 MeV proton is about 300 MeV cm^{-1}. Also shown is the range of protons; a 1 MeV proton in 1 g cm^{-3} ice travels 23 μm. For higher energies the range increases.

During interactions, a proton's energy is lost *via* discrete and diverse processes. Loss per collision is a small fraction of the total energy of the proton and a large number of interactions are required to stop the particle. The average primary excitation energy produced by the incident proton is of

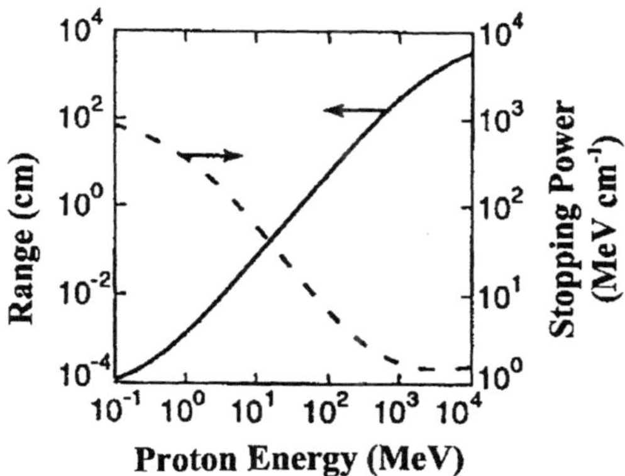

Fig. 1. — Stopping Power and Range of Protons (from Johnson [11]).

the order of $30 - 60$ eV mol^{-1}, a value larger than the few eV mol^{-1} typically required to break an OH, CH or NH bond. Energetic secondary electrons are also produced which on average cause an additional ionization nearby. Atoms, molecules or molecular ions may be sputtered near the surface of the solid. In the bulk of a solid the proton's path is nearly straight. If the solid target thickness is greater than the proton's range, the proton is implanted and most of its energy is deposited near the end of its track leading to severe local alterations. MeV protons pass through interstellar sub-micron sized icy coated grains depositing only a fraction of their energy (*e.g.*, a 1 MeV proton deposits an energy of the order of 1 keV in a 0.01 μm ice layer). Broken bonds, radical formation, electronic excitations, molecular motion and subsequent reactions from diffusing species cause chemical and physical changes even at low temperatures. These processes involve all molecules present in the track regardless of their chemical nature. Most new species formed by the processing are stable (except for hydrogen) at cloud temperatures near 10 K, and the results of these processes accumulate over long time periods.

1.4. Cosmic ray and UV photon interactions with solids

Ultraviolet photons differ from MeV protons in that they produce a single excitation/ionization event and have a large absorption cross section in water since they are absorbed in a column density of $10^{16} - 10^{17}$ mol cm^{-2}, much less than that for $1 - 10$ MeV protons *i.e.*, $10^{20} - 10^{21}$ mol cm^{-2}. UV photons are absorbed within a few 100 Å in an ice and initiate photochemical reactions with a single molecule. Reactions can be limited to a specific type of molecule present in the medium, or to a specific bond, creating a single, well-defined excited state leading to different products and to a distinctly different distribution of reactive species. Higher energy photons, X-rays and gamma-rays, generate ionizing electrons due to the photoelectric and Compton effects and therefore act in some ways differently from UV photons since they can produce multiple events in close proximity. X-ray irradiation of icy grain mantles in interstellar environments [12] or during the T-Tauri phase of star formation may be an additional form of ice processing.

A summary is given in Table II for the predicted UV and ion fluxes for different interstellar cloud regions. Cold diffuse clouds have temperatures of $\approx$ 100 K, densities of $\approx$ 100 molecules cm^{-3}, and lifetimes of $\approx 10^7$ years. More than 99% of the carbon is ionized but all hydrogen and helium are neutral. Following the discussion by Jenniskens *et al.* [13], the UV flux is 9.6×10^8 eV cm^{-2} s^{-1}; about half of this is absorbed in a 200 Å ice layer. The absorbed cosmic ray flux for 1 MeV protons is calculated assuming a 300 MeV cm^{-1} stopping power and assuming that the protons deposit their energy in both the entrance and exit ice layers. The dose is the absorbed energy multiplied by the time spent in the environment and divided by the molecular density. Cold dense regions have temperatures of approximately 10 – 20 K and densities of approximately 10^3 molecules cm^{-3}. Here, neutral molecules dominate and dust shields most of the dense cloud interior. The UV photons inside a dense cloud arise from cosmic ray induced fluorescence of molecular hydrogen. Jenniskens *et al.* [13] assumed that a fraction of 0.12 is absorbed in the ice layer. For radiation processing comparison, the remaining cosmic ray proton flux has been diminished by a factor of 10. Table II shows that in diffuse regions, UV photons deposit more energy in the icy mantles than protons, but in dense regions their energy contributions for processing are similar.

2. SIMULATIONS IN THE LABORATORY

2.1. Some ion irradiation experiments relevant to cosmic ices

The results of radiolysis studies (*e.g.*, protons, helium ions, electrons and discharges) on a variety of ices have been reported. A sample of experiments is listed here to point out some of the earlier experiments along with more recent results. A variety of techniques are used to analyze the results, *e.g.*, IR spectroscopy, mass spectroscopy and gas chromatography-mass spectrometry (GC-MS). Studies of pure molecular solids include: H_2O [14–17]; CH_4 [17–25]; NH_3 [17]; CO [26]; CO_2 [27]; CH_3OH [28,29]; C_2H_2 [24,30]; SO_2 [31]. Studies of binary mixtures with water relevant to interstellar, cometary, and ices on the surface of the outer satellite and planetary surfaces include: H_2O+CH_4 [32,33]; H_2O+CO [34]; N_2+CO and O_2+CO [35], H_2O+CO_2 [21,27,34,36]; H_2O+CH_3OH [28,29,37]; H_2O+C_2H_2 [32]; H_2O+C_2H_6 [33,38]. More complex icy mixtures experiments have also been performed [32,33,39–42]. Experiments on so-called "dirty ices" where refractory solids (minerals and silicates) are mixed with ices are very limited [43,44].

Table II. — Predicted UV and 1 MeV proton fluxes in diffuse and dense interstellar cloud regions and accumulated energy deposited in a 200 Å ice coated grain. Information based on Jenniskens *et al.* [13], Strazzulla and Johnson [10], Wdowiak [45], and Duley and Williams [46].

Time in region (years)	UV Irradiation			ION Irradiation		
	Flux $eV/cm^2/s$	energy absorbed $eV/cm^2/s$	Energy dose eV/mol.	Flux $eV/cm^2/s$	Energy absorbed $eV/cm^2/s$	Energy dose eV/mol.
Cold diffuse clouds	9.6×10^8	5×10^8 (200 Å mantle)		1×10^7	1.2×10^4 (200 Å mantle)	
(10^5)			1×10^4			< 1
(10^7)			1×10^6			3×10^1
(10^9)			1×10^8			3×10^3
Cold dense clouds	1.4×10^4	1.7×10^3 (200 Å mantle)		1×10^6	1.2×10^3 (200 Å mantle)	
(10^5)			< 1			< 1
(10^7)			4			3
(10^9)			4×10^2			3×10^2

2.2. Laboratory set-up for proton irradiation experiments

In our laboratory experiments, thin ice films are formed by gas phase deposition on an aluminum mirror cooled by a closed-cycle cryostat to $T <$ 20 K. Surrounding the tail section of the cryostat is a six-sided vacuum chamber designed for a variety of *in situ* measurements. The chamber pressure is in the 6×10^{-10} bar range. The ice-coated mirror can be rotated to face different chamber ports depending on the type of measurement, *e.g.*, IR spectroscopy, mass spectroscopy, or proton irradiation. A more detailed discussion of this set-up is given in Moore and Hudson [16]. Spectra of ices are recorded after the infrared beam passes through the ice film, reflects from the substrate mirror and makes a second pass through the ice. With this configuration the substrate mirror can be electrically isolated from the cryostat for use as a Faraday cup to measure the ion beam current.

Mid-IR spectra from 4000 – 400 cm^{-1} (2.5 – 25 μm) or far-IR spectra from 500 – 100 cm^{-1} (20 – 100 μm) of the ice films are recorded before and after irradiation. Typically a resolution of 4 cm^{-1} with 60 scans is used. Each spectrum is ratioed with the spectrum of the uncoated mirror. The analysis of the spectral data provides information on the identification and number of new species formed during irradiation. The identification requires comparison with reference spectra recorded for candidate species diluted in water ice. Quantitative information on the formation or loss of a species is calculated from integrated band areas when the appropriate absorption coefficient for that molecule is known. Absorption coefficients can be calculated knowing the ice thickness (measured using laser interference techniques), index of refraction, and ice density. Some experiments include a layer of refractory smokes covered with a layer of ice. The smoke layer (*e.g.*, a silicate smoke) is added to the mirror before it is attached to the cryostat head. Smoke production, performed in a separate grain condensation chamber, is obtained either by evaporation of SiO solid [47] or by combustion of SiH_4 with O_2 followed by vapor phase nucleation and growth in an H_2 atmosphere [48]. Typical smoke thicknesses range from < 0.1 mm to 0.5 mm.

Ion beam parameters are chosen to approximate the galactic cosmic ray environment. We use 1 MeV protons which is typical of the dominant ions. This energy, easily generated in small Van de Graaff accelerators, is sufficient to penetrate microns of ice in a region where the stopping power of the proton is large, and does not result in nuclear reactions. These protons are not implanted in ice films < 23 μm thick. The laboratory flux of protons can be 10^{10} times higher than in the interstellar environment. Although we cannot study reactions that are rate dependent, we estimate that the probability of overlapping events is 1 in 10^{12} protons. Therefore, interactions within a proton track anneal before another proton comes into the same track. The proton fluence (ions cm^{-2}) can be varied to give a calculated dose from 0.1 to 100 eV mol^{-1}, a range similar to that estimated for the cold cloud environments.

3. MID-IR SPECTRA OF PROTON IRRADIATED ICES

The mid-infrared is the "fingerprint" region for the vibrational spectra of molecules and is typically observed between 4000 and 400 cm^{-1} (2.5 – 25 μm). Although gas phase spectra contain a number of sharp lines due to the vibrational-rotational transitions of molecules, the rotational motions are quenched in the condensed phase and only the vibrational transitions between the constituent atoms within the molecule are measured. Each molecule's functional group (*e.g.*, OH group) has a stretching and bending frequency specific to that group. Using these frequency "fingerprints", identification of unknown absorptions can be attempted. Figure 2a is the mid-IR spectrum of H_2O+CH_4

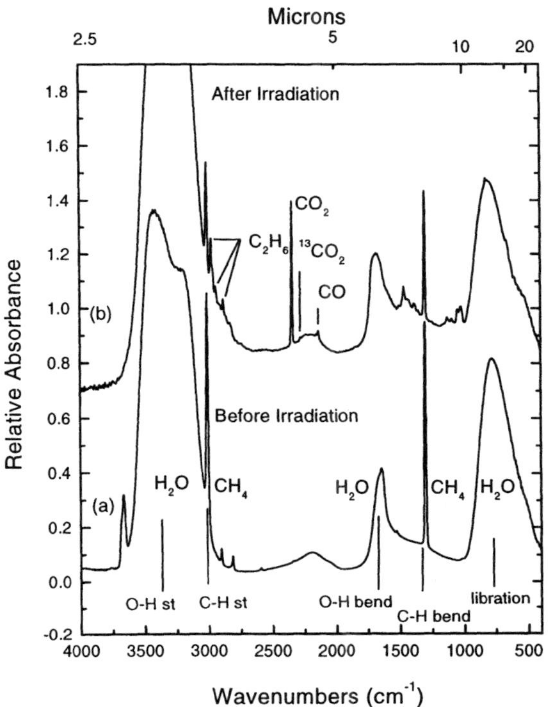

Fig. 2. — Comparison of the mid-infrared spectrum of a (4:1) H_2O:CH_4 ice before and after irradiation at $T < 20$ K.

before irradiation (labeled are: the stretching, bending and librational modes of H_2O (see Sect. 4), and the stretching and bending modes of CH_4).

3.1. H-atom addition reactions

Results from the experiment shown in Figure 2 were part of a group of mid-IR investigations to examine the synthesis of C_2H_6 in simple icy mixtures relevant to interstellar ices. This study was stimulated by observations of comets C/1996 B2 Hyakutake and C/1995 O1 Hale-Bopp in which significant amounts of C_2H_6, CH_4 and C_2H_2 were detected. Mumma *et al.* [49] pointed out that the abundances were consistent with the production of C_2H_6 in interstellar icy grain mantles which could occur from H-atom addition reactions to acetylene condensed from the gas phase or to solid phase UV or radiation processing of interstellar ices. In our experiments, C_2H_6 (also CO_2 and CO) is easily identified

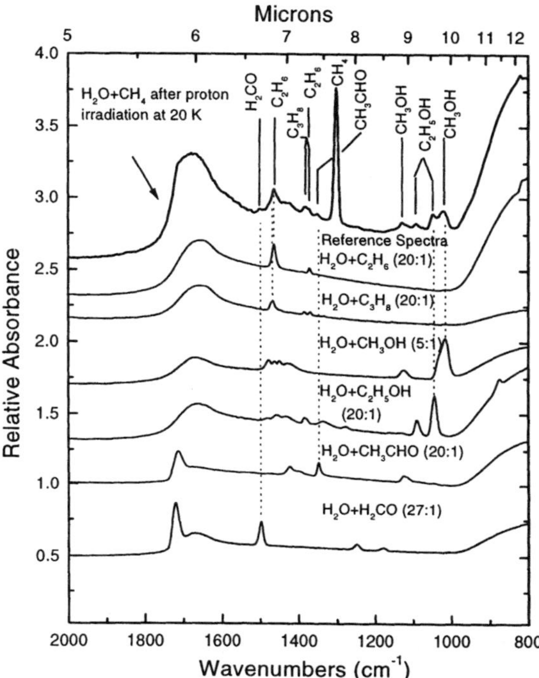

Fig. 3. — Identification of the synthesized species in irradiated H_2O+CH_4 between 5 and 12 μm by comparison with reference spectra.

after irradiation. Other synthesized molecules are identified after expanding the 5 to 12 μm region and comparing it with reference spectra (Fig. 3).

A probable formation mechanism for C_2H_6 is the dimerization of two methyl radicals. Methanol formation is an example of a competitive reaction where a CH_3 radical combines with an OH. The ratios of C_2H_6:CH_4 in these experiments were smaller than those observed in the comets. A higher C_2H_6:CH_4 ratio is formed after the irradiation of H_2O + C_2H_2 ices shown in Figure 4. As in the previous experiment, CO, CO_2 and C_2H_6 are easily identified after irradiation. A detailed study of the spectrum in the 5 to 12 μm region (not shown) reveals that the same products are formed as in the H_2O + CH_4 experiment with the addition of C_2H_4, ethylene. H atom addition reactions to acetylene, proposed by Mumma *et al.* [49], are consistent with these experiments. H atoms come from the proton irradiation of H_2O and add sequentially to C_2H_2 in steps producing C_2H_4 as the first stable molecule and

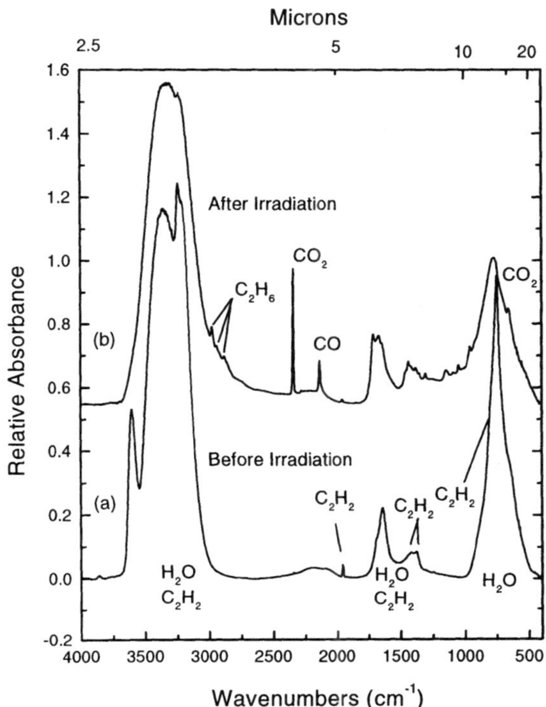

Fig. 4. — Comparison of the mid-infrared spectrum of a (5:1) H_2O:C_2H_2 ice before and after irradiation at $T < 20$ K.

then C_2H_6. Analysis of the column density of C_2H_2 as a function of proton dose shows its continuous decrease. The intermediate product, C_2H_4, first increases and then decreases as it is used to form C_2H_6 which continuously increases [32, 50].

4. FAR-IR SPECTRA OF PROTON IRRADIATED ICES

The far-infrared spectrum is typically observed between 400 and 50 cm^{-1} (25 – 200 μm) although spectra to 16.7 cm^{-1} (600 μm) can be performed with evacuated spectrometers. At these longer wavelengths, the absorptions in solids are due to the intermolecular vibrations (lattice vibrations). The atoms which form each molecular group also undergo intra-molecular vibrations (those observed in the mid-IR). The coupling between these intra-molecular

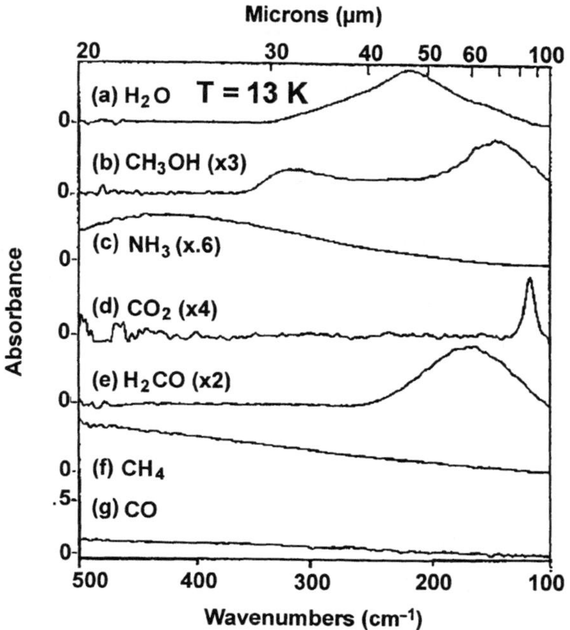

Fig. 5. — Far-IR Spectra of Amorphous Phase Ices, $T < 20$ K.

vibrations and the lattice periodicity will determine how many lattice modes may arise. Higher frequency modes are referred to as the optical branch (detected optically) whereas lower frequency modes are referred to as the acoustic branch (usually not optically detected except in some combination bands or in Brillouin spectra). Lattice vibrations are separated into translational modes and rotational modes. Translational modes involve the motion of the molecule's center of mass and can be longitudinal or transverse motions. Some transverse modes are infrared active; longitudinal modes may be observed in Raman spectra. Rotational, or librational modes, involve a hindered-rotation of the entire molecular group about their centers of mass. The libration mode for H_2O is near 760 cm^{-1} (13.2 μm), see Figure 2.

4.1. Amorphous and crystalline ice phase spectra

Gas phase spectra of molecules in the far-infrared contain a number of sharp lines due to the rotational transitions of the molecule; but when condensed the molecule's orientation becomes relatively fixed in the amorphous solid. The resulting infrared spectrum of water ice (Fig. 5a) is typical of amorphous materials showing a featureless broad absorption. This spectrum results when slowly condensed water vapor is collected on a surface cooled to $T < 20$ K, but

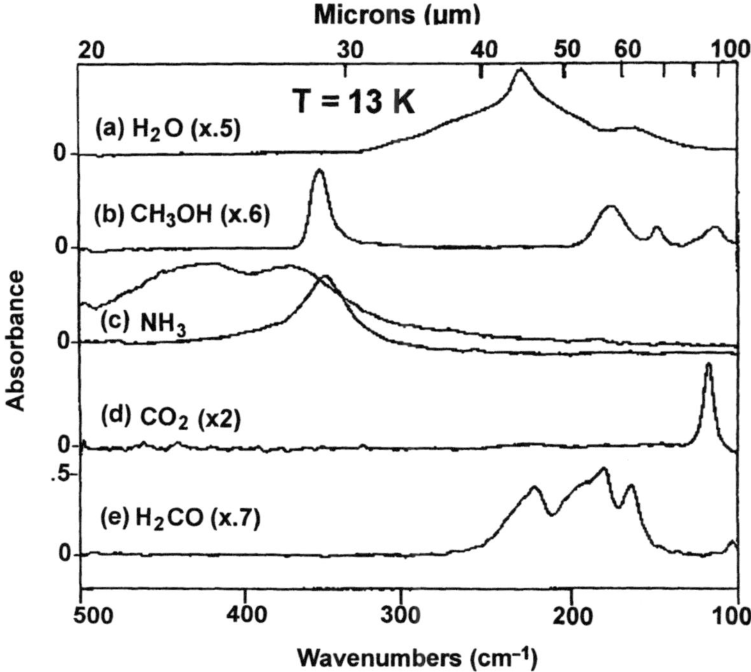

Fig. 6. — Far-IR Spectra of Crystalline Phase Ices, $T < 20$ K.

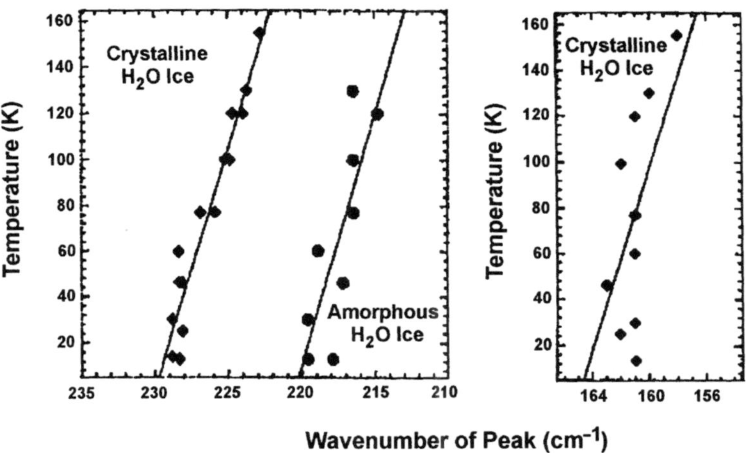

Fig. 7. — Peak positions of crystalline and amorphous ice as a function of temperature. A linear least-squares fit to the data points has been drawn for each maximum.

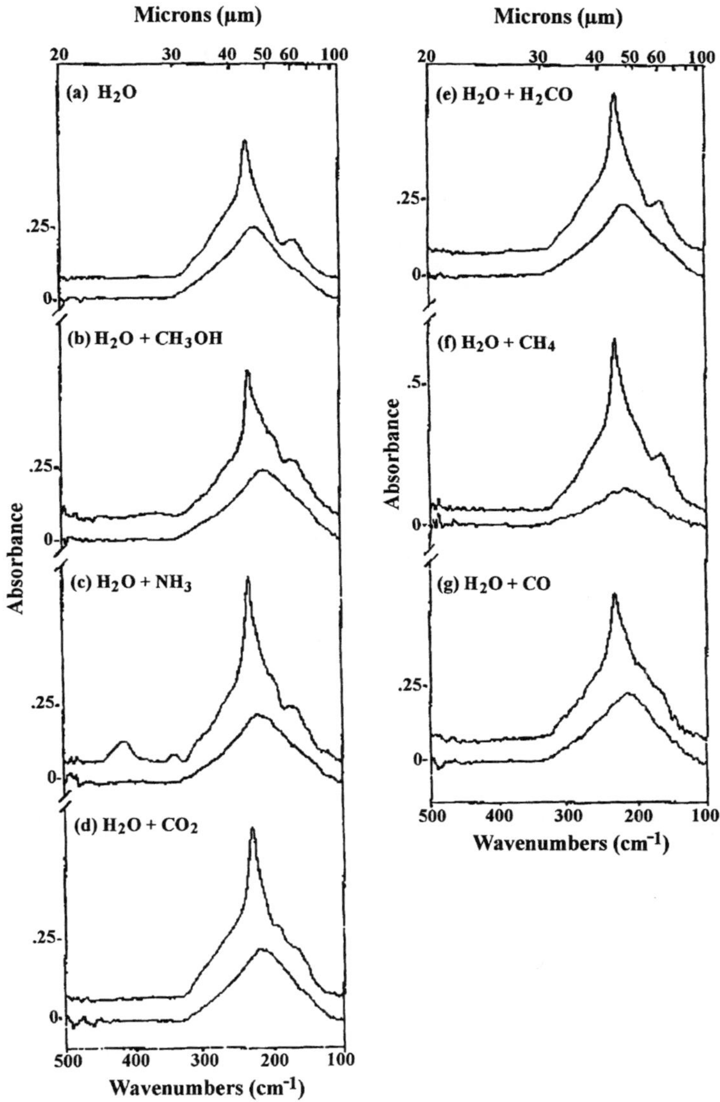

Fig. 8. — Far-IR spectra of pure amorphous and crystalline phase H_2O ice is compared with the far-IR spectra (b–g) of amorphous and crystalline phase ices of H_2O combined with CH_3OH, NH_3, CO_2, H_2CO, CH_4, and CO ($\sim$100:12).

similar results are obtained to temperatures near 77 K (Moore and Hudson [16] and references therein). Amorphous ice is not stable against thermal cycling. It converts irreversibly to a polycrystalline phase during warming to $T > 120$ K. Figure 6a shows the crystalline phase spectrum after re-cooling to 13 K.

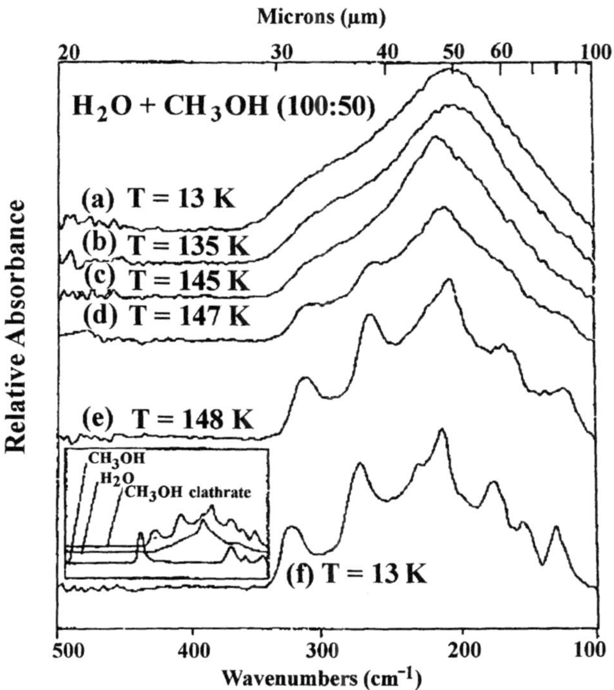

Fig. 9. — Evolution of the far-IR spectrum of an H_2O+CH_3OH mixture during warming showing the development of multi-line features associated with the CH_3OH clathrate hydrate; the insert spectra demonstrate that there are many differences between the spectrum of the CH_3OH clathrate at 13 K and the spectrum of either pure crystalline CH_3OH or H_2O.

The absorptions near 45 and 62 μm are lattice vibrations of the optical branch and result from the coupled translational and hindered rotation modes.

We have recorded the spectral features of several pure ices in both the amorphous (Fig. 5) and crystalline (Fig. 6) phases. Each ice has a unique spectrum which can be compared with far-IR ISO observations to search for likely candidates.

The dramatic difference between amorphous and crystalline phase H_2O was examined in some detail. The amorphous phase peak shifts to smaller wavenumbers as the temperature increases. After conversion to crystalline phase at 160 K, the major peak near 222 cm^{-1} (45 μm) shifts to larger wavenumbers reversibly with cooling (Fig. 7). This type of information can often be used to deduce ice grain temperatures.

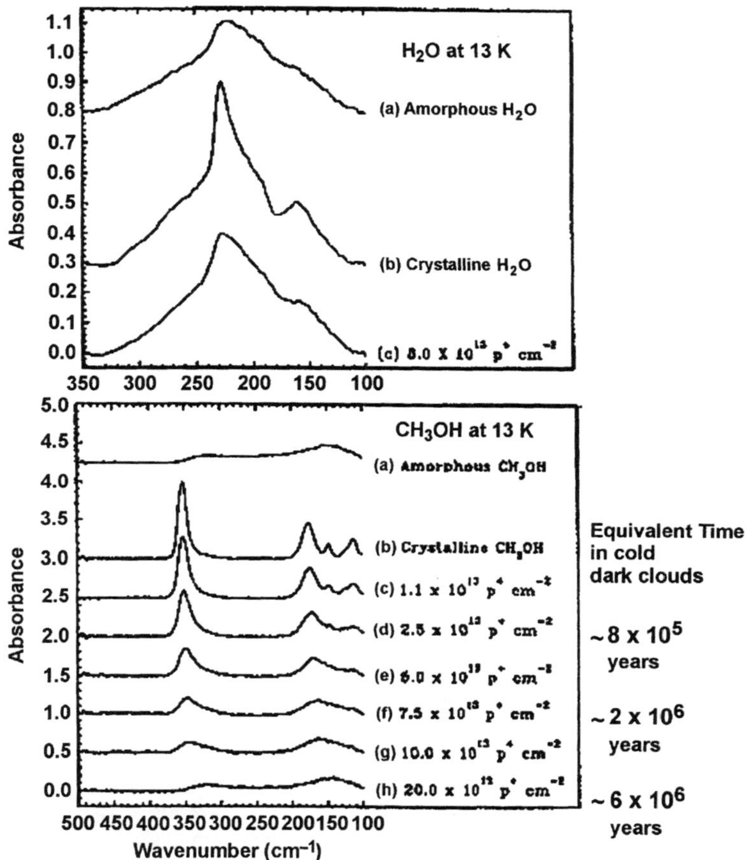

Fig. 10. — Far-IR spectra of amorphous, crystalline, and irradiated H_2O (top) and CH_3OH (bottom) at 13 K. A dose of 20×10^{15} p^+ cm^{-2} corresponds to $\sim 6 \times 10^6$ years exposure to cosmic ray bombardement in a cold dark cloud.

4.2. Spectra of H_2O-dominated icy mixtures

Spectra of H_2O-dominated icy mixtures with 12% of an added molecule are shown in Figure 8 for the amorphous and crystalline phases. Pure amorphous H_2O peaks at 220 cm^{-1} (45.4 μm) whereas the mixtures have peaks shifted to smaller wavenumbers by as much as 9 cm^{-1}. Pure crystalline phase H_2O has its largest absorption at 230 cm^{-1} (43.5 μm) and the mixtures have the same absorption feature within 2 cm^{-1} often along with some extra weak features. Since there are small differences between the crystalline peak positions for pure and mixed water ices and since the fit with the amorphous phase involves a weaker broad band, it may be difficult to use these features to determine the composition of low abundance components in water ice mixtures.

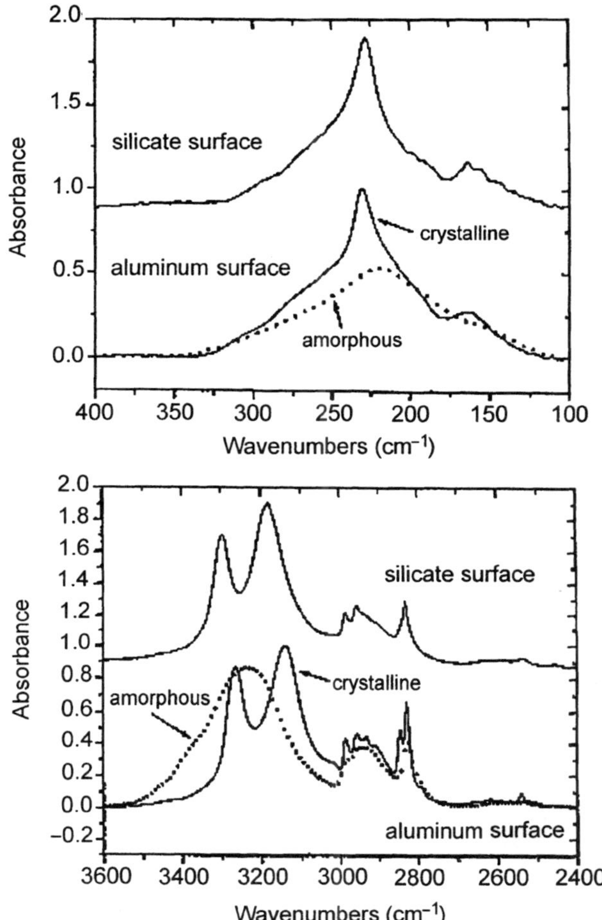

Fig. 11. — IR spectra of H_2O (top) and CH_3OH (bottom) deposited directly onto a silicate smoke compared with spectra of H_2O and CH_3OH in the amorphous and crystalline phase on an aluminum substrate.

4.3. Spectrum of the methanol clathrate

In our study of H_2O mixtures we observed that when the concentration of CH_3OH is near 50%, the broad absorption feature of the amorphous phase ice has a shoulder peaked near 300 cm^{-1} (33.3 μm) at $T < 20$ K. This mixture evolves into a five peaked spectrum when the temperature is increased to 148 K and these sharpen when cooled to $T < 13$ K (see Fig. 9). This multi-line spectrum is associated with the CH_3OH clathrate hydrate [37, 51]. The characteristic multi-line spectrum of the methanol clathrate is weakened when impurities such as NH_3 and CO are added to the ice mixture. When irradiated, the characteristic multi-line spectrum weakens as the result of amorphization until the spectrum is indistinguishable from the broad underlying absorption. Re-warming does not reform the clathrate since radiation products act as impurities. Detection of such a multi-line spectrum in an interstellar source would mean that the icy mixtures have experienced little radiation amorphization and that there is a relatively low concentration of impurities ($< 10\%$) compared to H_2O and CH_3OH.

4.4. Radiation amorphization of crystalline phase ice

Far-IR spectroscopy was used to investigate the crystalline-to-amorphous phase changes in irradiated H_2O, CH_3OH, CO_2, H_2CO [52]. Figure 10 shows the amorphization of crystalline phase H_2O and CH_3OH by proton irradiation. In general, the rate of amorphization increases with decreasing temperature. These results are in qualitative agreement with experiments involving UV irradiation, e^- irradiation, and He^+ irradiation of crystalline H_2O ices.

Amorphous phase ice can result from the irradiation of crystalline phase ice for doses within the range estimated for cold cloud environments. Therefore the identification of the phase of an interstellar ice is not sufficient information for determining its formation temperature or thermal history.

5. MID- and FAR-IR SPECTRA OF ICE ON SILICATE SMOKES

Infrared spectra of H_2O, CH_3OH, and several other molecules condensed at $T <$ 20 K on amorphous silicate smokes reveal that predominantly crystalline phase ice forms directly on deposit. This is called the "Low Temperature Crystalline" effect, LTC. Spectra of these molecules condensed at $T < 20$ K on an aluminum mirror substrate are typical of the amorphous phase. Figure 11 shows two examples of this observation for H_2O and CH_3OH. We chose to study the molecule's spectral band which has the biggest change between the amorphous to crystalline phase. Smokes with different compositions (*e.g.*, Si_xO_y, Mg-silica, Fe-silica, Fe_xO_y) exhibit the LTC effect.

Silicate smokes are oxygen-deficient, very porous, and have a surface area above 100 $cm^2\ g^{-1}$. Various mechanisms which might explain the LTC result are under study. Smokes are a disordered network structure with numerous imperfections and dangling bonds. The idea that these smokes may contain active surface sites (perhaps oxygen vacancy defect sites) which facilitate the amorphous-to-crystalline phase transition during condensation is supported by the observation that the effect can be destroyed by baking. Recent measurements also show, however, that baking changes the surface area of the smokes. Currently surface area studies are compared with IR studies to learn about the connection of these measurements to the understanding of a mechanism.

ACKNOWLEDGMENTS

The author wishes to acknowledge the contributions of R. Hudson, R. Ferrante, R. Khanna, J. Nuth and B. Donn. This work is supported by NASA's NRA 344-02-57 and NRA 344-33-01.

REFERENCES

[1] Ehrenfreund P., d'Hendecourt L., Dartois E., *et al.*, *Icarus* **130** (1997) 1.
[2] Whittet D.C.B., Schutte W.A., Tielens A.G.G.M., *et al.*, *A&A* **315** (1996) L357.
[3] Meyer P., Ramaty P. and Webber W.R., *Phys. Today* **27** (1974) 23.
[4] Johnson R.E., Energetic Charged-Particle Interactions with Atmospheres and Surfaces, edited by R.E. Johnson (Springer-Verlag, New York, 1990) p. 2.
[5] Goldstein M.L., Ramaty R. and Fisk L.A., *Phys. Rev. Lett.* **24** (1970) 1193.
[6] Cooper J.F., Christian E.R. and Johnson R.E., *Adv. Space Res.* **21** (1998) 1611.
[7] Kulkarni S.R. and Heiles C., Interstellar Processes, edited by D.J. Hollenbach and H.A. Thronson (Reidel, Dordrecht, 1987) p. 87.
[8] Meszaros P., *Nature* **248** (1974) 35.
[9] de Jong T. and Kamijo F., *A&A* **25** (1973) 363.
[10] Strazzulla G. and Johnson R.E., Comets In The Post-Halley Era, edited by R.L. Newburn, M. Neugebauer and J. Rahe (Kluwer, Dordrecht, 1991) p. 243.
[11] Johnson R.E., *J.G.R.* **96** (1993) 17553.
[12] Sternberg A., Yan M. and Dalgarno A., Molecules In Astrophysics: Probes and Processes, edited by E.F. van Dishoeck (Kluwer, Dordrecht, 1997) p. 141.
[13] Jenniskens P., Baratta G.A., Kouchi A., *et al.*, *A&A* **273** (1993) 583.
[14] Baratta G.A., Leto G., Spinella F., Strazzulla G. and Foti G., *A&A* **252** (1991) 421.
[15] Strazzulla G., Baratta G.A., Leto G. and Foti G., *Europhys. Lett.* **18** (1992) 517.
[16] Moore M.H. and Hudson R.L., *ApJ* **401** (1992) 353.
[17] Pirronello V. and Averna D., *A&A* **196** (1988) 201.
[18] Davis D.R. and Libby W.F., *Science* **144** (1964) 991.
[19] Lanzerotti L.J., Brown W.L. and Johnson R.E., Ices In The Solar System, edited by J. Klinger, D. Benest, A. Dollfus and R. Smoluchowski (Reidel, Dordrecht, 1985) p. 317.
[20] deVries A.E., Pedrys R., Haring A. and Saris F.W., *Nature* **311** (1984) 39.
[21] Benit J., Bibring J.P. and Rocard F., *Nucl. Instr. Meth. Phys. Res.* **B32** (1988) 349.

[22] Foti G., Calcagno L., Sheng K.L. and Strazzulla D., *Nature* **310** (1984) 126.
[23] Calcagno L., Foti G. and Strazzulla G., *Icarus* **63** (1985) 31.
[24] Kaiser R.I., Eich G., Gabrysch A. and Roessler K., *ApJ* **484** (1997) 487.
[25] Strazzulla G. and Baratta G.A., *A&A* **266** (1992) 434.
[26] Palumbo M.E. and Strazzulla G., *A&A* **259** (1992) L12.
[27] Brucato J R., Palumbo M.E. and Strazzulla G., *Icarus* **125** (1997) 135.
[28] Baratta G.A., Castorina A.C., Leto G., Palumbo M.E., Spinella F. and Strazzulla G., *Planet. Space Sci.* **42** (1994) 759.
[29] Moore M.H., Ferrante R.F. and Nuth J.A. III, *Planet. Space Sci.* **44** (1996) 927.
[30] Floyd G.R., Prince R.H. and Duley W.W., *J. Roy. Astron. Soc. Can.* **67** (1973) 299.
[31] Moore M.H., *Icarus* **59** (1984) 114.
[32] Hudson R.L. and Moore M.H., *Icarus* **126** (1997) 233.
[33] Thompson W.R., Murray B.G.J.P.T., Khare B.N. and Sagan C., *J.G.R.* **92** (1987) 14933.
[34] Moore M.H., Khanna R. and Donn B., *J.G.R.-Planets* **96** (1991) 17541.
[35] Strazzulla G., Brucato J.R., Palumbo M.E. and Satorre M.A., *A&A* **321** (1997) 321.
[36] Pirronello V., Brown W.L., Lanzerotti L.J., Marcantonio K.J. and Simmons E.H., *ApJ* **262** (1982) 636.
[37] Hudson R.L. and Moore M.H., *ApJ* **404** (1993) L29.
[38] Khare G.N. Thompson W.R., Cheng L., *et al.*, *Icarus* **103** (1993) 290.
[39] Kobayashi K., Kasamatsu T., Kaneko T., *et al.*, *Adv. Space Res.* **16** (1995) 21.
[40] Moore M.H., Donn B., Khanna R. and A'Hearn M.F., *Icarus* **54** (1983) 388.
[41] Oro J., *Nature* **197** (1963) 971.
[42] Berger R., *Proc. Natl. Acad. Sci.* **47** (1961) 1434.
[43] Zhao N.S., Ph.D. Thesis (University Leiden, Leiden, 1990) p. 1.
[44] Moore M.H., Ferrante R.F., Hudson R.L. Nuth J.A. III and Donn B., *ApJ* **428** (1994) L81.
[45] Wdowiak T.J., Solid-State Astrophysics, edited by E. Bussoletti and G. Strazzulla (Elsevier Science Pub., New York, 1991) p. 281.
[46] Interstellar Chemistry, edited by W.W. Duley and D.A. Williams (Academic Press, London, 1984) p. 7.
[47] Nuth J.A. III and Donn B., *ApJ* **257** (1982) L103.
[48] Nelson R., Thiemens M., Nuth J.A. III and Donn B., *Proc. Lunar Planet. Sci. Conf.* **19** (1989) 559.
[49] Mumma M.J., DiSanti M.A., Dello Russo N., *et al.*, *Science* **272** (1996) 1310.
[50] Moore M.H. and Hudson R.L., *Icarus* **135** (1998) 518.
[51] Moore M.H. and Hudson R.L., *A&AS* **103** (1994) 45.
[52] Hudson R.L. and Moore M.H., *Radiat. Phys. Chem.* **45** (1995) 779.

LECTURE 14

Crystalline Silicates in Circumstellar Shells

L.B.F.M. Waters[1,2], F.J. Molster[1] and C. Waelkens[3]

[1] *Astronomical Institute Anton Pannekoek, University of Amsterdam, Kruislaan 403, 1098 SJ Amsterdam, The Netherlands*
[2] *SRON Laboratory for Space Research Groningen, P.O. Box 800, 9700 AV Groningen, The Netherlands*
[3] *Instituut voor Sterrenkunde, Katholic University Leuven, Celestijnenlaan 200B, 3001 Heverlee, Belgium*

Abstract. The mid- and far-infrared spectra of many oxygen-rich circumstellar dust shells, as observed by the Infrared Space Observatory, are very rich in narrow solid state emission features, that can be identified with crystalline silicates. Crystalline silicates are found in the outflows of evolved stars (both massive and low mass), in the disks surrounding young intermediate-mass stars, and in the solar system comet Hale-Bopp. The shape of the emission bands varies significantly between objects. Comparison with laboratory data shows that Mg-rich and Fe-poor olivines and pyroxenes are present, mostly at modest ($5 - 10\%$ by mass) abundance. The different chemical composition of crystalline (Mg-rich) and amorphous (Fe-rich) silicates suggests a different formation history for the grains.

1. INTRODUCTION

Solid state spectroscopy of dust particles in circumstellar shells is a valuable tool to study the nature of the underlying star, its evolutionary status, and the physics and chemistry of dust formation and dust processing. The launch in 1995 by ESA of the Infrared Space Observatory (ISO) [1] has opened the $2 - 200$ μm spectral region for continuous quantitative solid state spectroscopy at spectral resolutions $\lambda/\Delta\lambda$ between 300 and 2500, *i.e.*, very well suited for this type of study. This paper reports on some early results obtained with the Short Wavelength Spectrometer (SWS) [2], in particular, of oxygen-rich circumstellar dust. The ISO-SWS spectra show a remarkably rich structure, which can be attributed to different kinds of crystalline silicates. These silicates are found both in the dust shells surrounding young stars [3], and in those of evolved red giants and supergiants [4]. We will discuss the occurence of crystalline silicates in both types of objects, and suggest some mechanisms for their formation.

2. EVOLVED STARS

Stars with a wide range of initial masses are known to evolve to the red part of the HR diagram at the end of their main-sequence life. Stars with masses less than about 8 $M_\odot$ evolve into red giants, and later into Asymptotic Giant Branch (AGB) stars. More massive objects evolve to the Red Supergiant (RSG) phase. While the internal structure of AGB stars and of RSG is rather different, both types of objects possess very extended, cool atmospheres and show extensive mass loss *via* a slowly expanding wind. For an excellent review about AGB evolution, we refer to the paper by Habing [5].

The infrared spectral region of evolved stars is characterized by thermal emission from solid particles that have condensed in their cool, low velocity outflows. While the dust particles contain only 1% of the mass in the outflow, their emission efficiency is high and so they dominate over the stellar photosphere at long wavelengths. For high mass loss rates (typically $> 10^{-5}$ $M_\odot$/yr), the optical depth of the dust shell becomes so large that the star is obscured at optical and near-IR wavelengths. Such high mass loss rates are important for the evolution of stars on the Asymptotic Giant Branch (AGB), since the rate at which the outer envelope of the star is removed competes with the nuclear burning rate. Models for the winds of AGB stars involve dust formation in a pulsating atmosphere [6]. The dust formation process is therefore important for such models, and quantitative measurements of the type of grains that condense, their chemical composition, and the grain nucleation environment, provide important constraints for these models.

The composition of dust surrounding oxygen-rich AGB stars has been the subject of many studies. These studies all point to a dominant contribution from amorphous (*i.e.*, disordered) silicates, which show the 9.7 μm Si–O stretching mode and the 18 μm O–Si–O bending mode (either in absorption or emission, depending on the optical depth of the dust shell). In addition, emission at 13 μm has been attributed to Al_2O_3 [7], and emission at 11, 13.1 and 19 μm to metal oxides [8]. At wavelengths longward of 20 μm, very little structure was reported before ISO. In the Frosty Leo object, strong emission at 43 and 60 μm was found [9] and attributed to crystalline H_2O ice. Unpublished Kuiper Airborne Observatory spectra of cool AGB stars (see [10] and references therein) showed structure longward of 20 μm. Based on laboratory studies of simple

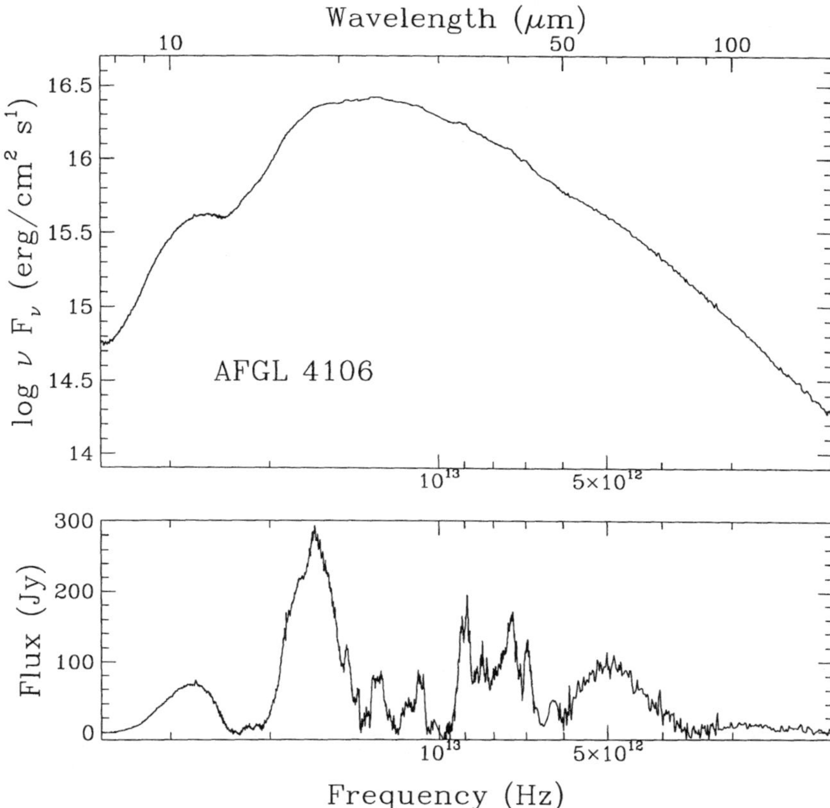

Fig. 1. — Full SWS and LWS spectra of AFGL 4106 (top panel), and of the continuum subtracted spectrum (lower panel). The continuum is dominated by the contribution from amorphous silicates, with weak crystalline silicates superimposed at the longer wavelengths. A bump near 60 μm is also evident, probably due to crystalline H_2O ice. The dotted line in the upper panel represents the adopted continuum.

oxides it was suggested that the $20-25$ μm region contains a contribution from FeO and MgO [11].

The ISO-SWS spectra for the first time convincingly show the presence of many new emission bands longward of 20 μm in the spectra of oxygen-rich AGB and post-AGB stars, and, surprisingly, also in the spectra of some objects that were believed to contain only carbon-rich dust [4,10,12,13]. In Figure 1, we show the ISO-SWS spectrum of AFGL4106 [14], which is believed to be rapidly evolving from low to high temperatures, and which has recently experienced a phase of very high mass loss. Also shown in Figure 1 is a continuum-subtracted spectrum, revealing the richness of the solid state structure at long wavelengths in this object. Table I contains a list of the strongest bands, with proposed identifications. For these identifications we used laboratory measurements of crystalline olivines, $(Mg_xFe_{1-x})_2SiO_4$, and pyroxenes, $(Mg_xFe_{1-x})SiO_3$ [15]. The Mg-rich end members of the olivines and pyroxenes ($x = 1$) are Forsterite and Enstatite respectively. These laboratory data are in good agreement with earlier measurements [16, 17]. The spectrum is dominated by amorphous silicates, with bands near 10 and 18 μm. The 18 μm silicate band has a peculiar shape which suggests a contribution from another component. This could be a simple oxide, or a crystalline silicate band. The long wavelength part is dominated by crystalline silicates, and by crystalline H_2O ice. The wavelength of the band positions suggest that the crystalline silicates are Mg-rich and Fe-poor, in contrast with the amorphous silicates which are believed to be rich in Fe. It is, however, difficult to constrain the Fe/Mg ratio, because it depends on particle shape effects which cause the peak wavelengths of the bands to shift. However, we can exclude Fe/Mg ratios larger than 10%. The Fe/Mg ratio strongly influences the 69 μm olivine peak, which would shift to longer wavelengths, by several μm, for Fe/Mg = 0.05 [16]. Unfortunately, the 69 μm peak is not seen in AFGL 4106, probably due to the intrinsic weakness of the band and the low abundance of crystalline silicates. It is useful to compare the wavelengths of the observed olivine bands in AFGL 4106 to those of sources where the 69 μm feature is seen, and which would indicate a Fe/Mg ratio close to 0. A good example is HD 100546 (see below), which has olivine bands at similar wavelengths as seen in AFGL4106. The 69 μm band in HD 100546 is at the value expected for pure Forsterite. This suggests that AFGL 4106 also has a very low Fe/Mg ratio (much lower than 0.1). A model fit to the ISO data for AFGL 4106 gives an abundance of crystalline silicates of about $3-5\%$ [14].

Not all oxygen-rich AGB stars show crystalline silicates. Only stars with dust shells with colour temperatures less than about 300 K (with some notable exceptions however) show the emission bands [4]. This may be due to the fact that the crystalline grains are not produced at low mass loss rates, *i.e.*, when the optical depth of the dust shell is low. However, the crystalline grains tend to be Fe-poor and therefore quite transparent at near-IR and optical wavelengths, resulting in significantly lower temperatures compared to the amorphous dust. Therefore in warm dust shells the contrast between hot (amorphous) and cold (crystalline) grains may prevent easy detection. This effect would be important if there is no thermal contact between the crystalline and amorphous grains, *i.e.*, if they are two separate populations.

It is not clear how the crystalline grains form, and why their chemical composition differs from that of the amorphous component. The crystalline lattice points to high formation temperatures, above the glass temperature for olivines (about 1050 K). Since the bulk of the dust in oxygen-rich outflows is amorphous, this dust must condense at temperatures below 1000 K, where the low temperature prevents the grains from annealing. The difference in

Table I. — Solid state bands measured in the SWS and LWS spectrum of AFGL4106.

Wavelength μm	F_{band} 10^{-14} Wm^{-2}	$FWHM$ μm	$FWHM/\lambda$	$I_{\text{peak}}/I_{\text{cont}}$	Identification
10.8	$4 \cdot 10^2$	2.5	0.23	1.8	amorph. silicate
13.6	10.	.5	0.037	1.04	cryst. forsterite
14.2	6.5	.4	0.028	1.05	cryst. enstatite (:) instrumental (:)
16.1	40.	.7	0.043	1.06	cryst. forsterite
16.8	15.	.6	0.036	1.03	unidentified
17.6	$5 \cdot 10^2$	3.	0.17	1.2	amorph. silicate
18.1	75.	1.0	0.055	1.07	cryst. enstatite
19.2	50.	1.0	0.052	1.045	cryst. enstatite and cryst. forsterite
20.6	20.	0.48	0.023	1.035	cryst. enstatite (:)
21.5	7.3	0.25	0.012	1.024	cryst. enstatite and cryst. forsterite(:)
22.9	25.	0.5	0.021	1.04	unidentified
23.6	30.	0.7	0.030	1.04	cryst. forsterite
24.1	.46	0.09	0.0037	1.004	unidentified
26.2	16.	1.05	0.040	1.02	cryst. forsterite
27.8	44.	1.15	0.041	1.05	cryst. forsterite and cryst. enstatite
32.8	20.	.6	0.018	1.07	cryst. enstatite
33.6	40.	1.0	0.030	1.08	cryst. forsterite
34.0	.96	.12	0.0035	1.016	cryst. enstatite (:)
35.8	5.5	.54	0.015	1.03	cryst. enstatite
36.5	2.4	.28	0.0077	1.02	cryst. forsterite
40.4	13.	1.1	0.027	1.05	cryst. enstatite
41.1	1.1	.17	0.0041	1.02	unidentified
43.1	17.	1.2	0.028	1.07	cryst. enstatite and cryst. H_2O
47.8	9.	1.9	0.040	1.036	unidentified
61	68.	20.	0.33	1.05	cryst. H_2O
			Spurious features		
11.07	1.98	0.082	0.0074	1.05	instrumental artifact

Note: (:) indicates that the identification is uncertain; amorph. (cryst.) indicates an amorphous (crystalline) state. The 32.8 μm peak was not found in the laboratory data of Jäger *et al.* [15] and we base it on the laboratory measurements of ortho-enstatite by Koike and Shibai [19].

chemical composition may be the result of the lower temperatures that are required to adsorb Fe into the lattice [18]. This absorption process modifies the lattice structure and, because the reaction occurs below the glass temperature, an amorphous, Fe-rich structure may result. A possible scenario for the formation of grains would then be that crystalline, Mg-rich grains form, and as they cool and adsorb Fe, most are converted to amorphous silicates or grow Fe-rich amorphous mantles [20]. Whatever their formation history, it is clear the crystalline grains give insights into the dust formation process (which seems to occur at high temperatures), and therefore into the mass loss mechanism in AGB stars and red supergiants.

3. YOUNG STARS

Spectra, taken with the SWS instrument, of young, intermediate mass pre-main-sequence stars, often referred to as Herbig Ae/Be or HAEBE stars, have revealed an unexpectedly rich structure of crystalline silicates. While most embedded stars show red continua with amorphous silicate absorption or emission, some *isolated, optically bright* HAEBE stars have prominent crystalline olivine and/or pyroxene bands [22]. The best studied case so far is HD 100546 (Fig. 2) [3], for which both SWS and LWS spectra have been published [23]. The spectrum is characterized by prominent emission bands at 10.2, 11.4, 16.5, 19.8, 23.8, 27.9, 33.7 and 69 μm (and some minor ones in addition). The wavelength of these bands is in good agreement with laboratory data of pure forsterite [16], *i.e.*, Fe-poor olivine. Some peaks can be identified with pyroxenes (*e.g.*, an emission band at 40.5 μm), but their emission is less prominent than for the olivines. A ratio of crystalline to amorphous dust of less than 10% is found [3].

The SWS spectrum of HD 100546 is very similar to that of the solar system comet Hale-Bopp [24]. This suggests that the grains in both objects have a similar formation history. In this respect it is important to note that so far no strong evidence for the presence of crystalline silicates in the interstellar medium has been found. This implies that these grains must have been formed, from amorphous grains, during the star and planet formation process. Annealing of amorphous grains requires heating to temperatures above about 1050 K in the case of olivines. The conclusion seems inevitable that the crystalline dust in HD 100546, which now has a temperature of the order of 200 K or less [3], has been much hotter in the past, pointing to mixing of grains (either radially or in the z-direction). By analogy, a similar process may have been at work in the proto-solar cloud, resulting in the crystallisation of the grains in the comet Hale-Bopp.

So far, HD 100546 has remained a rather exceptional case where the strengths of the crystalline silicate bands are concerned. Several other HAEBE stars show weaker crystalline silicates [25–27], but these have not yet been analysed in detail. We show in Figure 2 the full SWS and LWS spectrum of the young

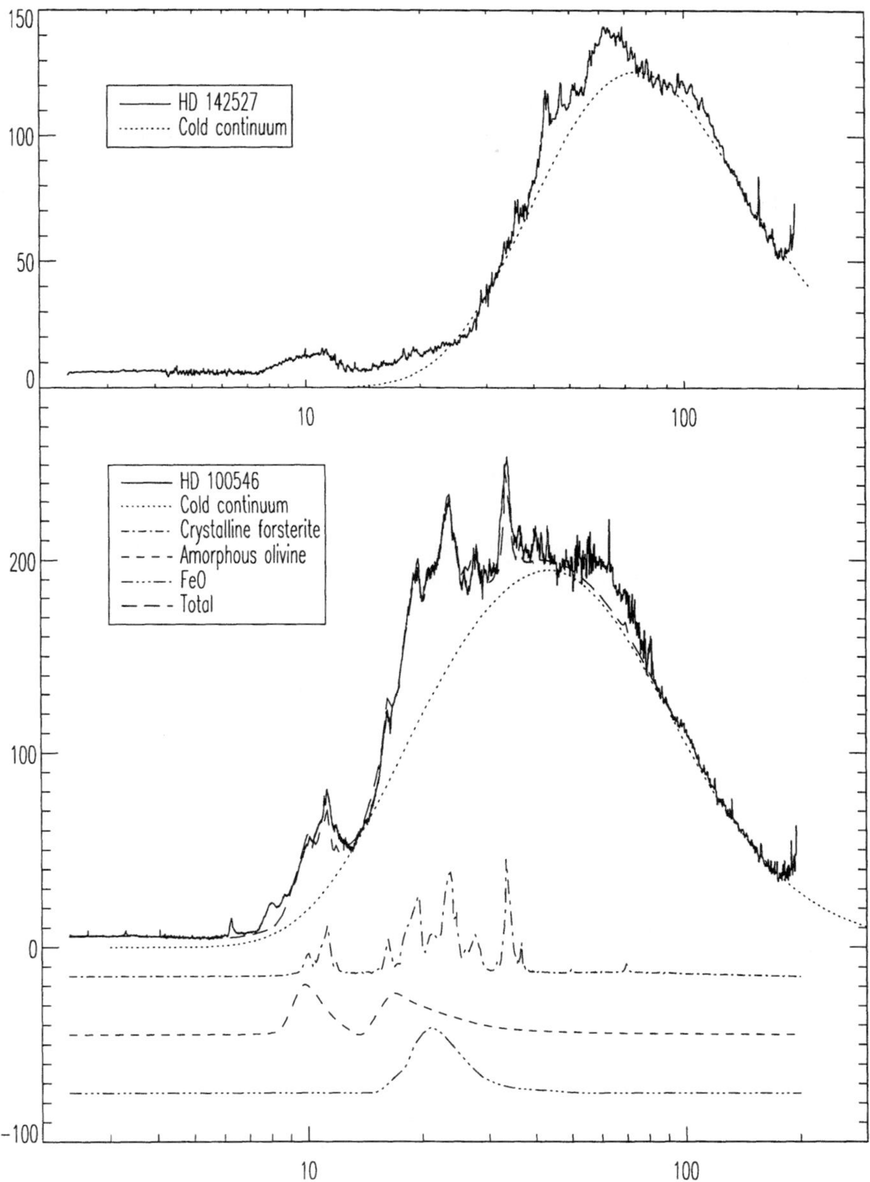

Fig. 2. — Full SWS and LWS spectra of the young Fe star HD 142527 (top) and the Herbig Ae/Be star HD 100546 (bottom). Plotted is the flux (in Jy) *versus* wavelength (in μm). Notice the prominent crystalline silicates in HD 100546, but also in HD 142527. In addition, crystalline H_2O ice can be found at 43 and 60 μm. An unidentified 100 μm feature is seen in HD 142527, and weakly in HD 100546. This feature may be due to hydrous silicates [21]. Finally, FeO is present in HD 100546 ([3]; see also [21]).

Fe star HD 142527 (see also [21]), which shows a cool dust component which is significantly cooler than HD 100546. There is evidence for crystalline silicates in the 35 – 45 μm region, suggesting that these grains are confined to this cool component (the amorphous silicate emission observed at 10 μm is due to somewhat warmer grains). Remarkably strong crystalline H_2O emission bands at 43 and 60 μm are apparent in HD 142527, as well as an unidentified 100 μm feature. These emission bands are also seen in HD 100546, but significantly weaker. The presence of crystalline silicates as well as H_2O ice is reminiscent of the composition of comets.

4. THE 33-34 MICRON COMPLEX

The crystalline silicate bands show a remarkable variety of shapes and strengths. As an example, we show the continuum subtracted 30 – 38 μm spectra of 4 representative cases in Figure 3. These spectra are normalized to 1 at the peak to facilitate intercomparison. It is clear that the 33 – 34 μm emission in fact consists of 2 main components, one at about 32.8 – 33 μm (hereafter called "33" peak) and the other near 33.5 – 33.8 μm (hereafter called "33.8" peak). Note that the peak wavelengths shift by typically 0.1 to 0.2 μm from star to star. The large variation in the ratio 33/33.8 peak strength between stars indicates that (at least) two independent dust components are present. Comparison with laboratory spectra suggest that the 33 peak is due to pyroxenes, while the 33.8 peak is due to olivines.

Closer inspection of the peaks reveals possible sub-structure in both the 33 and 33.8 peaks. This sub-structure is particularly clear in AFGL4106, which has very narrow, sharp emission bands. For instance, the 33 band has sub-structure at about 32.4 μm, while the 33.8 band has a clear sub-peak near 34 μm. Much less structure is seen in OH26.5 and NGC 6302 (Fig. 3), and almost none in HD 100546. Note that laboratory spectra (*e.g.*, of olivine) do not show such sub-structure, but have significantly wider bands than observed in the evolved stars. Only HD 100546 and the comet Hale-Bopp show peak shapes similar to those observed in the laboratory. These two objects contain mainly forsterite and the pyroxene abundance is much lower. Particle shape effects are unlikely to explain this substructure, since they tend to broaden and shift the peaks. Radiative transfer calculations using optical constants [15] and small grains (0.01 – 0.1 μm radius) also do not improve the fit.

The lack of agreement in band width between laboratory and observations is not well understood at present. It is possible that the 33 and 33.8 bands are in fact blends of several individual resonances, whose intrinsic width varies from source to source. If this is the case, the laboratory measurements apparently also have several blended resonances. Alternatively, on top of the broad underlying peak, other dust components contribute. A better understanding of the origin of these resonances is desirable.

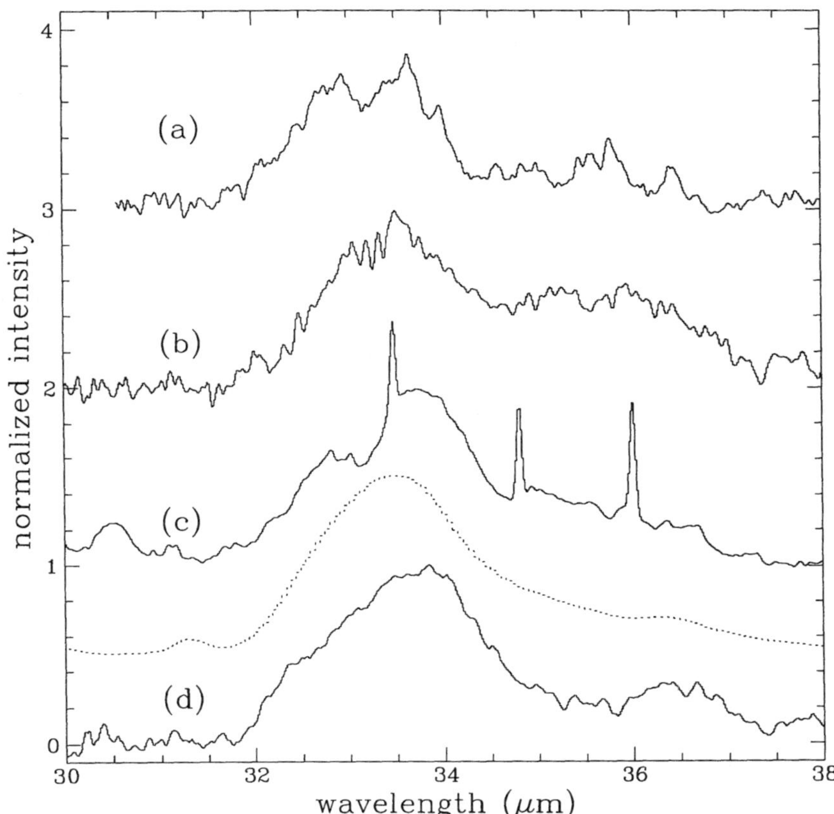

Fig. 3. — Continuum subtracted SWS spectra of some representative oxygen-rich dust envelopes with crystalline silicates. The spectra have been normalized for ease of comparison. Shown are AFGL4106 (a), OH26.5+0.5 (b), NGC 6302 (c), and HD 100546 (d). The dotted line is a laboratory transmission spectrum of forsterite [15].

In most spectra that show the 33 and 33.8 peaks, a conspicuous red emission wing is observed, extending to about 37 μm, referred to as "the plateau" [4]. The shape of the plateau is highly variable from source to source, again suggesting that multiple components contribute. This is particularly clear in the case of OH 26.5, that shows very strong plateau emission, probably due to additional emssion bands near 35.3 and 36 μm. In HD 100546, the plateau is not observed, but a red tail to the 33.8 band is visible as well as an emission feature near 36.5 μm. Both the red tail and the 36.5 μm feature may be ascribed to forsterite. Since HD 100546 contains few pyroxenes, it is tempting to attribute the plateau emission to this component in the other objects.

5. SUMMARY

The very rich spectra of oxygen-rich circumstellar dust shells discussed in this paper offer a new and unexpected view on the composition and formation history of silicates and oxides. This is truly an "ISO revolution", whose full impact has yet to be established. However, it is already apparent that the presence and surprisingly high abundance of crystalline silicates has implications for our understanding of the formation process of grains in the outflows of oxygen-rich stars, of the chemical evolution of stars on the Asymptotic Giant Branch (both single stars and binaries), and on the evolution of dust in proto-planetary disks.

ACKNOWLEDGMENTS

We thank D. Beintema, X. Tielens, T. de Jong, K. Malfait, G. Meeus, A. de Koter, J. Bouwman, I. Yamamura, R. Voors and M. van den Ancker for valuable discussions. LBFMW acknowledges financial support from an NWO *Pionier* grant.

REFERENCES

[1] Kessler M.S., Steinz J.A., Anderegg M.E., *et al.*, *A&A* **315** (1996) L27.
[2] de Graauw Th., Haser L.N., Beintema D.A., *et al.*, *A&A* **315** (1996) L49.
[3] Malfait K., Waelkens C., Waters L.B.F.M., *et al.*, *A&A* **332** (1998) L25.
[4] Waters L.B.F.M., Molster F.J., de Jong T., *et al.*, *A&A* **315** (1996) L361.
[5] Habing H.J., *A&AR* **7** (1996) 97.
[6] Sedlmayr E. and Dominik C., *Sp. Sc. Rev.* **73** (1995) 211.
[7] Vardya M.S., de Jong T. and Willems F.J., *ApJ* **304** (1986) L29.
[8] Goebel J., Volk K., Walker H.J., *et al.*, *A&A* **222** (1989) L5.
[9] Omont A., Forveille T., Moseley S.H., *et al.*, *ApJ* **355** (1990) L27.
[10] Cohen M., Barlow M.J., Sylvester R.J., *et al.*, *ApJ* (1998) in press.
[11] Henning Th., Begemann B., Mutschke H. and Dorschner J., *A&AS* **112** (1995) 143.
[12] Waters L.B.F.M., Beintema D.A., Zijlstra A.A., *et al.*, *A&A* **331** (1998) L61.
[13] Waters L.B.F.M., Waelkens C., van Winckel C., *et al.*, *Nature* **391** (1998) 868.
[14] Molster F.J., Waters L.B.F.M., Trams N.R., *et al.* (1998) in preparation.
[15] Jäger C., Molster F.J., Dorschner J., *et al.*, *A&A* **339** (1998) 904.
[16] Koike C., Shibai H. and Tuchiyama A., *MNRAS* **264** (1993) 654.

[17] Jäger C., Mutschke H., Begemann B., Dorschner J. and Henning Th., *A&A* **292** (1994) 641.
[18] Tielens A.G.G.M., From Mira's to Planetary Nebulae: which path for stellar evolution?, edited by M.O. Mennessier and A. Omont (Éditions Frontières, Gif-sur-Yvette, 1990) p. 186.
[19] Koike C. and Shibai H., ISAS report No. 671 (1998).
[20] Tielens A.G.G.M., Waters L.B.F.M., Molster F.J. and Justannont K., *Astrophys. Space Sci.* **255** (1998) 415.
[21] Malfait K., Waelkens C., Bouwman J., *et al.*, *A&A* (1998) in press.
[22] Waelkens C., Waters L.B.F.M., Malfait K., *et al.*, *A&A* **315** (1996) L245.
[23] Clegg P., Ade P.A.R., Armand C., *et al.*, *A&A* **315** (1996) L38.
[24] Crovisier J., Leech K., Bockelee-Morvan D., *et al.*, *Science* **275** (1997) 1904.
[25] Waters L.B.F.M. and Waelkens C., *ARA&A* **36** (1998) 233.
[26] van den Ancker M.E., Bouwman J., Wesselius P.R. and Waters L.B.F.M., *A&A* (1998) in press.
[27] Meeus G., *et al.* (1998) in preparation.

LECTURE 15

Comparison of Ices and Silicates in Comets and in the Interstellar Medium

P. Ehrenfreund

Leiden Observatory, P.O. Box 9513, 2300 RA Leiden, The Netherlands

Abstract. A large amount of dust grains may have survived unmodified in the outer solar nebula before incorporation into comets. To trace interstellar ices and heterogenous organic matter (formed during the evolution of interstellar dust grains) in comets is one of the future goals of the STARDUST and ROSETTA comet rendezvous missions. Interstellar ices, released as cometary volatiles will probe the origin and history of comets. ISO (Infrared Space Observatory) allowed to study for the first time the complete inventory of interstellar ices over the entire spectrum between 2.5 – 200 μm. The recent ISO observations of interstellar ices reveal that the abundances of solid H_2CO, CO_2, OCS, and CH_4 measured towards embedded protostars are similar to current cometary observations, indicating that interstellar ices may have been incorporated unaltered into comets. The striking similarity of crystalline olivines observed toward young Herbig Ae/Be stars and comet Hale-Bopp will provide further information about the link between interstellar and cometary dust. These measurements also allow to study the connection between interstellar, cometary and meteoritic dust and consequently provide constraints on the formation of the Solar System and the early evolution on Earth.

1. INTRODUCTION

Molecular clouds contain about 5% of the mass of the Galaxy and have a typical life-time of $10^7 - 10^8$ years. We can distinguish diffuse clouds with temperatures of 40 – 100 K and a density of about 100 particles per cm^3 and dense clouds which are very cold (10 – 30 K) and very dense ($10^4 - 10^8$ particles per cm^3). The dense cloud cores are the sites of active star formation. To date, 120 interstellar and circumstellar gas phase molecules are known in the literature with $HC_{11}N$ as the largest gaseous species. About a third of those interstellar molecules are also found in the cometary coma.

Interstellar ices are formed in dense clouds [1, 2]. Silicate particles which are produced in stellar outflows, accrete during their passage through a dense molecular cloud an icy grain mantle. These cold, small dust particles are interspersed with the interstellar gas with an abundance of $\sim 10^{-12}$ per H atom. Icy grains act as an important catalyst in the interstellar medium. Accretion is a very efficient process in cold environments because most of the gaseous species (except H_2, He) stick on grains. Some of the accreted species are mobile and can diffuse throughout the grain mantle. Accreting atoms, molecules and radicals can react on the grain surface and lead to the formation of new molecules. An efficient desorption mechanism from grains returns molecules back to the gas phase [3]. The basis of gas-grain chemistry is that new molecules are formed on the grain surface which are thereafter released in the interstellar gas [4]. Additional processes such as UV radiation, cosmic ray bombardment and temperature variations determine the grain mantle growth and evolution. Such processes become important when molecular clouds evolve from an initial cold quiescent phase to warm dense and active protostellar regions. Energetic protostellar outflows create shocks which can raise the temperature locally to more than 2000 K. Grain processing may lead to molecule desorption from grains but can also lead to total grain destruction. Irradiation processes in the cloud outskirts as well as temperature elevations lead in general to the formation of radicals, complex molecules and even organic refractory material [5]. Astronomical observations indicate the existence of different types of ices in proto-stellar environments and toward field stars. Hydrogen-rich ices (polar ices), dominated by H_2O ice, are formed when H is abundant in the interstellar gas. They contain besides water ice: CO, CO_2, CH_4, NH_3, CH_3OH, and possibly traces of HCOOH and H_2CO. Such polar ices evaporate around 90 K under astrophysical conditions and can therefore survive in higher temperature regions close to the star. Apolar or hydrogen-poor ices are formed far away from the proto-star and are composed of molecules with high volatility (evaporation temperatures of < 20 K) such as CO, O_2 and N_2 [6]. As a consequence of the accretion process, grain mantles may be arranged in "onion" structures in certain environments.

2. ISO OBSERVATIONS

The Infrared Space Observatory ISO allowed for the first time to study the complete inventory of interstellar ices. The ISO-SWS (Short Wavelength Spectrometer) instrument offered a large wavelength coverage and a

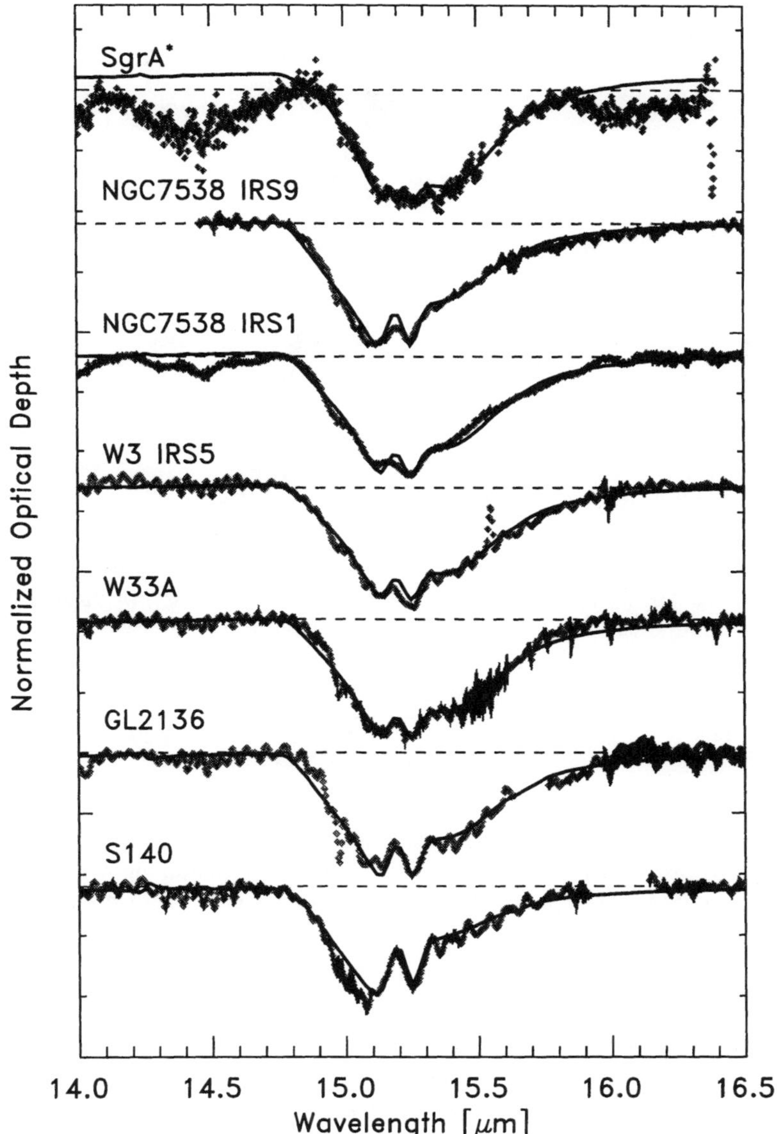

Fig. 1. — Observations of the CO_2 bending mode at 15.2 μm of different interstellar targets toward dense and diffuse clouds. The CO_2 features shows a particular multipeak structure which is indicative for segregated CO_2 [7].

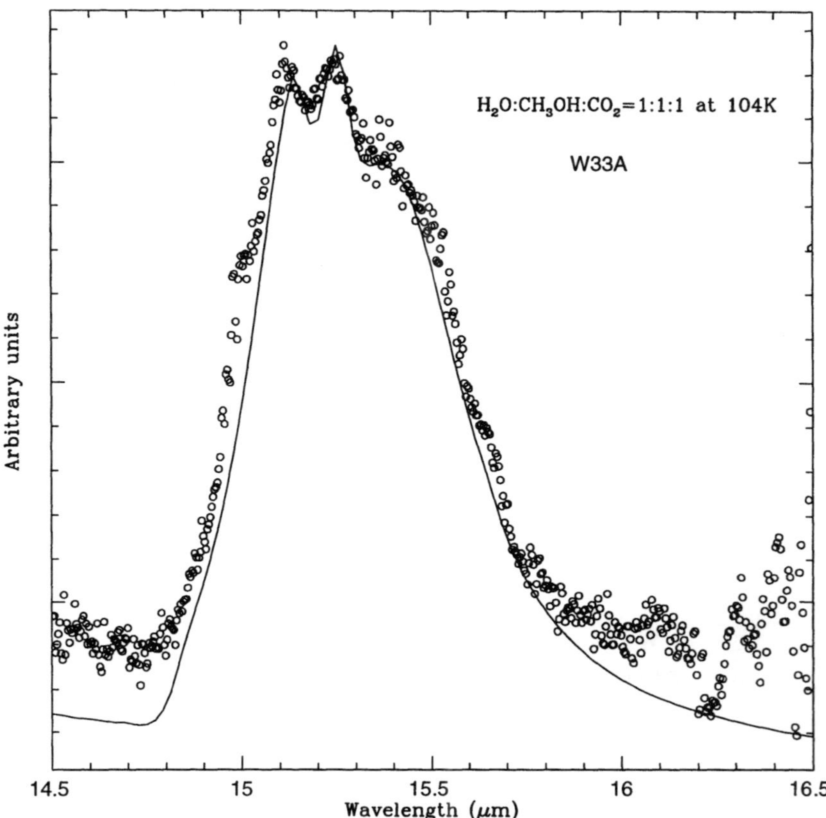

Fig. 2. — A laboratory fit of the CO_2 bending mode observed at 15.2 μm toward the protostar W33A and an annealed ice mixture containing CO_2, CH_3OH and H_2O in equal proportions. The perfect fit suggests extensive ice segregation in the environment of protostars.

resolution well adapted to the solid phase. The numerous observations changed our knowledge of the physical-chemical properties of ices in space. In particular, the discovery of many new ice features has allowed in conjunction with dedicated laboratory experiments, to determine more accurate abundances of major ice components [8]. First results showed the ubiquitous presence of abundant CO_2 ice in space and provided new insights into the gas to solid ratio of many interstellar molecules.

A large amount of laboratory work performed in preparation for ISO has allowed a rapid interpretation of incoming satellite data and provided new insights into interstellar and cometary ice chemistry. ISO has confirmed the ubiquity of solid CO_2 detected by IRAS-LRS [9]. A relative high abundance of CO_2, namely 15 – 20% compared to H_2O ice has been reported [10, 11]. Figure 1 displays ISO observations of the CO_2 bending mode towards many protostellar targets [7]. Laboratory data show that solid CO_2 invokes particular interactions with other polar and apolar molecules, which result in strong spectroscopic diversity [12, 13]. Therefore astrophysical data of ice species can be used not only to determine column densities but also to investigate the grain mantle composition through the analysis of intrinsic profiles. Recent ISO results show segregation of ices according to temperature zones in the vicinity of proto-stars [6].

Figure 2 shows the bending mode of CO_2 toward the massive protostar W33A, which is characterized by a triple-peak structure. This characteristic multi-peak feature is currently observed with ISO toward 20 protostars [7, 14]. This particular mode shows two sharp peaks at 15.15 and 15.29 μm and a broad wing at 15.4 μm. Recent laboratory results identify the CO_2 bending mode feature in dense clouds with molecular complexes formed between CO_2 and CH_3OH ice, which become spectroscopically discernible during the ice segregation process [6, 15]. The comparison of laboratory data and the ISO spectrum of RAFGL 7009S indicates the presence of an ice layer with a composition of CO_2, CH_3OH and H_2O in equal proportions [6]. Comparison of astronomical data toward W33A with laboratory spectra of segregated CO_2 ice also show a perfect fit, indicating extensive ice segregation and thermal processing toward protostars.

Figure 3 displays a schematic drawing of the ice segregation process in the line of-sight toward massive protostars. The region close to the protostar is dominated by energetic processes, such as strong UV irradiation, shocks and temperatures in the hot core, which may exceed 1000 K. Under such conditions ices are usually desorbed from grains or grains are completely destroyed. Assuming a protostar of a luminosity $L \sim 20\,000\ L_{\odot}$ (corresponding to a mass of 10 $M_{\odot}$) it is calculated that, at a distance of 3000 AU (0.015 pc) from the star, the temperature usually drops to 80 K, which is below the sublimation temperature of H_2O [16]. Such models also indicate that volatile ices such as CO, O_2 and N_2 do not occur inside the radius within 30.000 AU from the star, because the temperature never decreases to 20 K in the circumstellar region. In the following, we characterize the evolution of interstellar grains, composed of a silicate core and an ice mantle in the environment of protostars. Polar ices are dominated by H_2O, and contain also CH_3OH, CO, CO_2, some CH_4, NH_3 and and other minor species. Far from the protostar in colder (below 20 K) and denser regions or in "clumps", where CO is abundant in the interstellar gas, apolar ice mantles, dominated by CO, N_2 and some O_2 may accrete as an additional grain mantle layer. The temperature rise in the vicinity of protostars is responsible for the evolution of interstellar ice mantles. All pure and trapped ices sublimate at specific temperatures [4]. Above $\sim$ 50 K the major ice species H_2O, CH_3OH and CO_2 dominate the interstellar ice spectrum and show in comparison with laboratory data that ice layers rearrange. Above

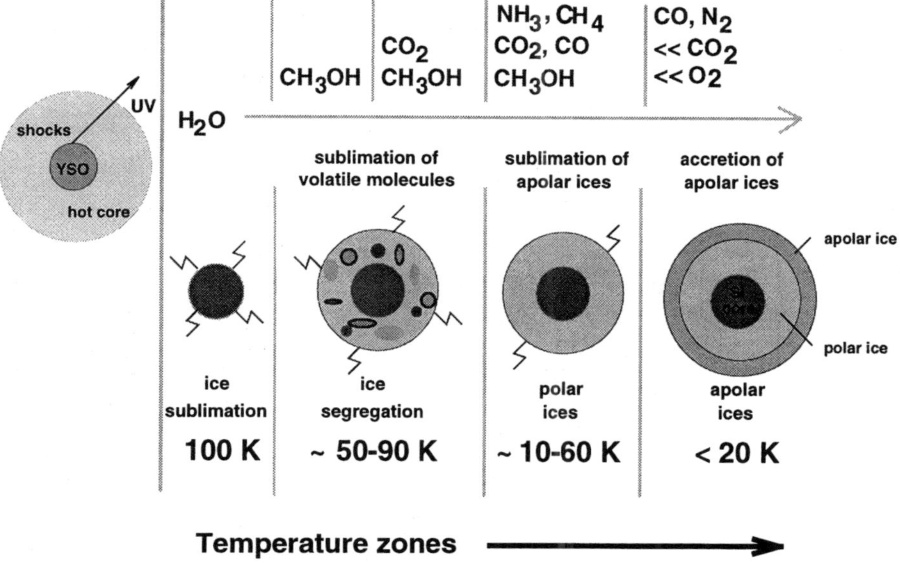

Fig. 3. — Different ice types are present in proto-stellar environments. Apolar ices containing volatile species reside in cold environments far from the proto-stellar source. Temperature zones above 50 K show due to thermal processing only polar and segregated ice layers. At 100 K all ices are sublimated.

80 K, CH_3OH seperates from H_2O and beyond $\sim 90-100$ K all major ice species sublimate.

CH_4, a newly discovered interstellar ice species is an important molecule to trace the origin of comets. CH_4 bands, which in previous ground-based observations were heavily contaminated by telluric features, could be confirmed with ISO and allowed the measurement of reliable abundances [11,17,18]. The abundance of CH_4 ice relative to water ice is only $\sim 2\%$ and even less CH_4 is found in the gas phase [18, 19]. These observations indicate that the gaseous CH_4 may be evaporated from the grain surface in hot cores ([18] and contribution by Dartois in this volume).

OCS is currently the only sulphur bearing molecule observed in interstellar ices [20]. The estimated abundance of OCS toward RAFGL 7009S is 0.18% relative to water ice. IR measurements showed OCS abundances in comet Hyakutake and Hale-Bopp of $2-5\times10^{-3}$ relative to water [21].

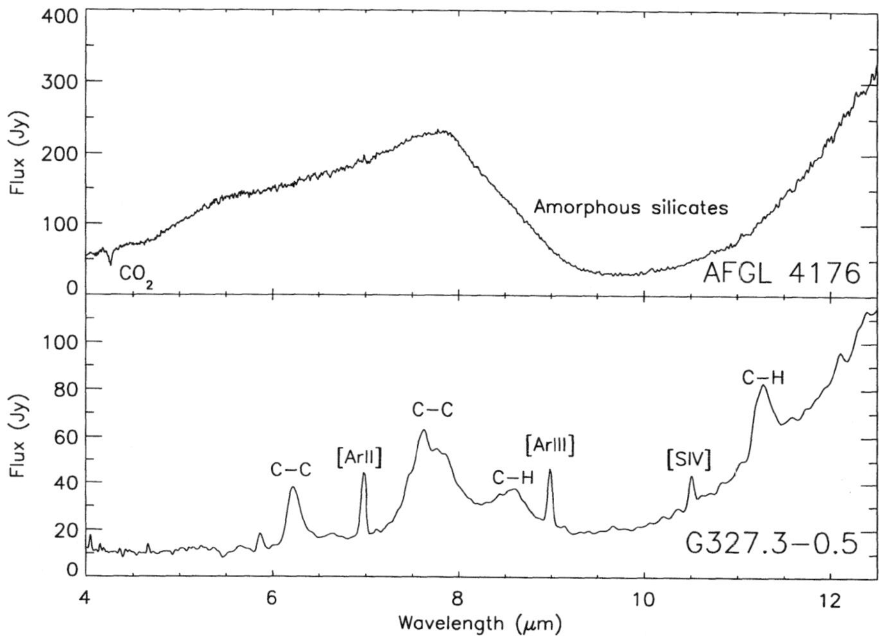

Fig. 4. — Comparison of southern star-forming regions AFGL 4176, which displays a large silicate band, as well as ices and gaseous CO, H_2O and CO_2 and G327.3-0.5 which samples an ultracompact HII region in the line-of-sight where PAHs and atomic emission lines dominate the spectrum.

3. EVOLUTION OF DENSE INTERSTELLAR CLOUDS

According to their evolutionary state, young objects show interstellar ice features such as H_2O, CO, CO_2, CH_4 (and other minor species), as well as interstellar gas features, such as H_2O, CO and CO_2 [22]. The more evolved objects are dominated by strong atomic emission lines and polycyclic aromatic hydrocarbons (PAHs). Figure 4 shows the comparison between two bright southern star-forming regions. AFGL 4176 is characterized as a young embedded proto-star which has a large fraction of high temperature gas and dust along the line of sight. Towards AFGL 4176, H_2O ice and 15% CO_2 ice (relative to water ice) as well as gaseous species such as CO, CO_2 and H_2O have been observed [23, 24]. The source G327.3-0.5, on the other hand, has already developed an ultracompact HII region.

H_2O ice and gaseous H_2O as well as strong atomic emission lines and PAH features are present in this object, but no volatile ices such as CO or CO_2 have been observed. Recent ISO spectra have shown the strong spectral diversity among star-forming regions. The combination of results on interstellar ices and their gaseous counterparts, solid state features, atomic emission lines and PAHs allows a reconstruction of the line of sight conditions and facilitates a determination of the evolutionary stages of these objects.

4. INTERSTELLAR SILICATES

ISO has revealed for the first time the presence of crystalline silicates, such as olivines and pyroxenes in AGB stars, Planetary Nebulae, as well as young stellar objects [25,26]. In the interstellar medium silicates are only observed in amorphous form. Note that it could be very difficult to detect a 5% crystalline silicate contribution against most sources such as the Galactic Center that are used to characterize the interstellar silicate grain population. Crystalline silicates can be converted into amorphous silicates due to the cosmic ray irradiation in the interstellar medium [27] but the timescales for such processes have not yet been investigated.

In the Solar System – in interplanetary dust particles and comets, crystalline silicates are observed. Amorphous silicates as well small crystalline olivine particles have been identified in the spectrum of comet Halley by comparison with spectra of interplanetary dust particles and laboratory silicates [28]. Several comets have been extensively observed between $7.8-13\ \mu$m [29]. A broad $10\ \mu$m emission feature due to amorphous silicate is observed in short as well as long-period comets [29,31]. The strength of the feature is quite variable and stronger in dust-rich comets. Sharp peaks (*e.g.*, at $11.3\ \mu$m) superimposed on the broad emission, indicative for the presence of crystalline silicates are observed in many comets and even in dynamically new comets, such as Mueller 1993a [30]. Crystalline olivine $(Mg, Fe)_2\ SiO_4$ has additional bands at 16.5, 19.8, 24.0, 27.6 and $33.9\ \mu$m and crystalline pyroxene (Mg, Fe, Ca) SiO_3 displays bands at 15.6, 26.5, 37.5, and $49\ \mu$m which could be searched for in IR spectra of comet Hale-Bopp with ISO. A detailed comparison with laboratory spectra reveals that the emission features observed in Hale-Bopp match those of Mg-rich olivines (fosterite) rather precisely ([32], see also Waters this volume).

ISO revealed a striking identification of fosterite around young Herbig Ae/Be stars which matches well the spectrum of comet Hale-Bopp, indicating a link between disks around young stars and comets [33]. In the cold environment of the Oort cloud the density and temperature do not in general allow the formation of crystalline silicates. Annealed silicates may have been formed in the hot and very turbulent environment of the nebula [34] and thereafter incorporated into comets. Radiation damaged silicate grains, which were previously crystalline may be re-annealed on a shorter time scale and at a lower temperatures. However, laboratory studies indicate that the temperature conditions in the cometary coma do not seem sufficent to allow this process. It has to be noted that the fraction of crystalline silicates in the circumstellar environments accounts only for a few percent. To study the relation between circumstellar

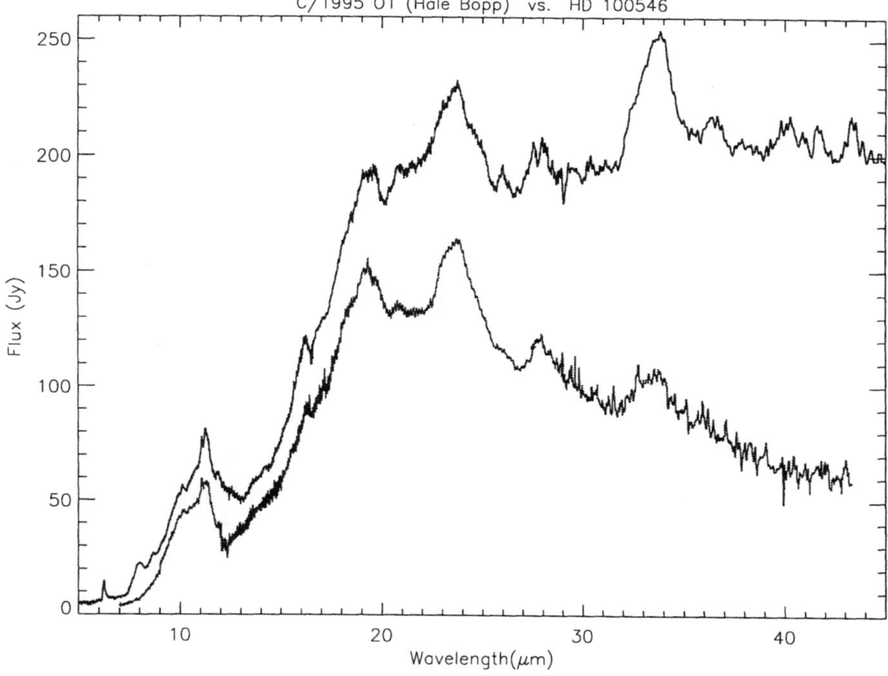

Fig. 5. — Comparison between the ISO-SWS spectrum of HD 100546 (top trace) with the spectrum from comet C/1995 O1 (Hale-Bopp) (bottom trace) (taken from [33]).

and cometary silicates should however provide important constraints for the formation of the Solar System. Figure 5 shows a comparison of the IR spectrum of HD 100546 and Hale-Bopp. The nearly perfect match of the crystalline fosterite features of HD 100546 with the spectrum of Hale-Bopp indicate the presence of a massive cloud of comets surrounding this star [33].

5. DUST CONNECTION

The life cycle of interstellar dust is bound to the evolution of stars in the Galaxy. The formation of complex molecules occurs already during the evolutionary cycle of interstellar icy grains which are originally formed in dark clouds. Figure 6 shows the grain mantle growth and evolution during the cycle of interstellar grains in the dense and diffuse medium. The icy grain mantles may be processed, heated and return complex molecules to the interstellar gas. Photoprocessing in the outskirts of dense clouds as well as cosmic ray irradiation and thermal processing modify the grain surface and form new molecules and radicals. When simulating these interstellar processes in the laboratory, it

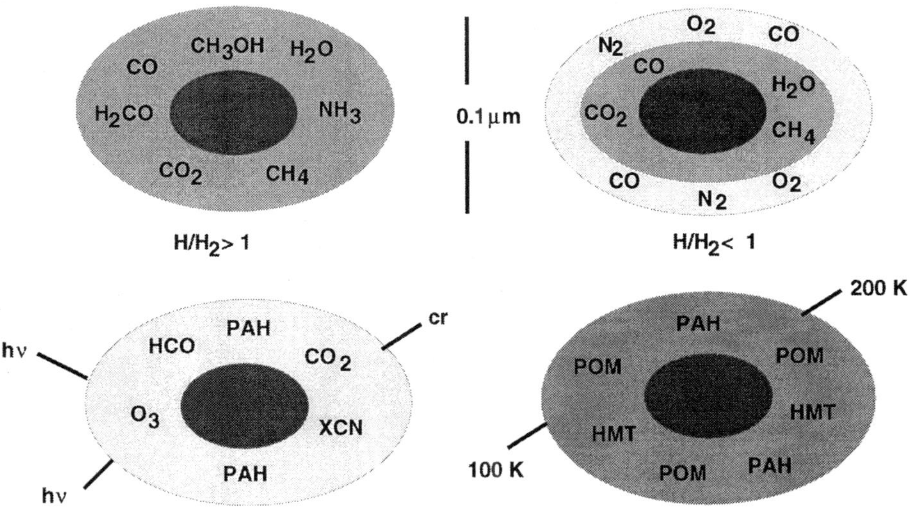

Fig. 6. — Polar and apolar ices are formed in different regions in the protostellar environment. Energetic processing and heating convert simple ice species during the cycle of interstellar grains into complex organic molecules, such as POM (polyoxymethylene) and HMT (hexamethyltetramine, $C_6H_{12}N_4$).

has been shown that complex molecules are formed. Photoprocessing of H_2O, CH_3OH, CO and NH_3 ices may result in the formation of radicals and irradiation products such as HCO and H_2CO. Gradual heating of such mixtures in laboratory experiments revealed the presence of complex molecules such as alcohols, ketones, polyoxymethylene (POM), and HMT ($C_6H_{12}N_4$). The latter molecule may react with H_3O^+ to form amino acids [5, 35].

6. DISCUSSION

Comets are likely the least evolved bodies in the Solar System and comet nuclei may be a low density aggregate of interstellar dust [36]. The observed low NH_3 and CH_4 abundance, the CO/CH_4 ratio and the D-enrichment (relative to Solar System values) of H_2O are key indicators for the interstellar origin of cometary volatiles. Observations of H_2CO and CH_3OH in comets show higher abundances than expected for solar nebula processes [37]. In the scenario of low mass star formation an embedded protostar with an accretion disk and strong outflows is formed after a gravitational collapse of the molecular cloud in a time-scale of $10^4 - 10^5$ years. This embedded protostar evolves into a T Tauri star and enters after $\sim 10^6$ years the main sequence during which time it may form a planetary system. The formation of complex molecules occurs already during the evolutionary cycle of interstellar icy grains which are originally formed in dark clouds.

An important constraint for the origin and evolution of comets is to search for pre-solar ices and to obtain evidence to what extent interstellar ices have been preserved in comets. In the solar nebula, interstellar dust volatiles may have evaporated and undergone subsequent chemistry before becoming incorporated in comets. Recent ISO observations, the advances in high resolution instrument technology and a number of interesting cometary targets in the last years has however strongly improved the knowledge on cometary volatiles [38]. The interstellar nature of ices in some comets is also supported by such recent cometary observations but still many open questions remain.

The recent detection of CH_4 and C_2H_6 in Hyakutake was reported with respective abundances of 0.4 and 0.7% relative to water ice and could be indicative for an interstellar origin [39]. Interstellar CH_4 shows abundances of $1 - 4\%$ relative to water ice in the solid form and very low values in the gas phase [18]. C_2H_6 is not observed in the interstellar gas and only a very low upper limit could be derived from recent ISO observations for C_2H_6 ice [40]. The measured abundance of HNC relative to HCN is very similar to that observed in quiescent interstellar molecular clouds and quite different from the equilibrium ratio expected in the outermost solar nebula, where comets are thought to form [41]. However, recent studies of the HNC/HCN abundance ratios as well as HCO^+ observations in the coma of Hale-Bopp suggest an important role for superthermal H atoms and ion-molecules reactions in the cometary coma [42]. Future observations will reveal the role of interstellar processes in the formation of these molecules. C_2H_2 has been observed in comet Hyakutake with an abundance consistent with observations of the interstellar gas [43]. C_2H_2 is however not yet detected on interstellar grains. A very recent detection of ammonia in comet Hale-Bopp was reported from radio observations [44]. The derived values of $1 - 1.8\%$ NH_3 relative to water is consistent with the previously reported limit of a few per cent of NH_3 in interstellar ices [1]. New ground-based observations of solid NH_3 indicate abundances of $\sim 10\%$ relative to water ice [45] which would not be compatible with cometary measurements. ISO observations have revealed large abundances of solid CO_2 throughout the interstellar medium as well as the outgassing of abundant CO_2 from comet Hale-Bopp at different heliocentric distances [32]. The abundance of CH_3OH has been a debated subject for several years [46,47]. Recent ground-based observations of CH_3OH bands in the NIR (near-infrared) toward RAFGL 7009S seem to confirm high methanol abundances toward some massive protostars up to 30% relative to water ice [48]. Therefore the abundance of CH_3OH in comets is rather compatible with the low abundances measured toward low-mass protostars. Finally the infrared inactive molecules O_2 and N_2 have been recently searched with ISO in dense cold clouds. In consistency with recent interstellar gas measurements of O_2, the abundance of O_2 ice is very low and an upper limit of $\sim$ 5% of the total oxygen budget can be attributed to solid O_2 [49]. The reactivity and unstability against photo-dissociation is likely responsible for the low O_2 abundance in solid and gaseous form in the interstellar medium. Large abundances of atomic oxygen in the interstellar gas have recently been observed with ISO towards SgrB2 [50], which indicate that OI may be the dominant oxygen-bearing species in the interstellar medium. In contrast, N_2 is supposed to be abundant in the interstellar gas and on grains in low temperature environments. Searches for solid N_2 have however been not successful up to now.

To summarize, the comparison of interstellar and cometary ices using recent ISO data have revealed important similarities [20]. The abundance of CH_4, H_2CO and OCS are fairly similar in the cometary coma and in the interstellar medium. The strong outgassing of CO_2 observed for comet Hale-Bopp is consistent with the large amounts of solid CO_2 found ubiquitously in the interstellar medium. The large discrepancies in the CH_3OH ice abundance in different interstellar environments will have to be further investigated in order to relate them to cometary methanol abundances. The high NH_3 abundance in interstellar ices, recently observed in one object, has yet to be confirmed by future observation of additional NH_3 bands in the near-infrared. An important fact is that cometary volatiles are rather comparable to interstellar solid state abundances and not to the interstellar gas component.

More indications for the origin of comets from interstellar dust can be constrained from the refractory component. The $3 - 20$ μm thermal emission spectrum of comet Hale-Bopp at heliocentric distance of 1.15 AU was derived with a model of comet dust consisting of porous aggregates of interstellar dust and a perfect fit could be obtained [51]. The striking similarity of forsterite, the most common silicate in the Solar System, observed around the young Herbig Ae/Be star HD 100546 and comet Hale-Bopp provides further information about the link between circumstellar and cometary dust [33]. The presence of crystalline silicates in YSOs implies that crystallization may occur during star and planet formation. The link between processes in dark embedded clouds and comets seems to be evident and future studies on the connection between interstellar, cometary and meteoritic dust will provide constraints on the formation of the Solar System and the early evolution on Earth.

ACKNOWLEDGMENTS

The author wishes to thank the Austrian Academy of Sciences for funding. The author wants to thank the ISO-SWS team and especially her close collaborators A. Boogert, P. Gerakines, W. Schutte and E. Dartois. Special thanks goes to K. Malfait for the courtesy to publish a spectrum of his recent work.

REFERENCES

[1] Whittet D.C.B., Dust and chemistry in astronomy, edited by T.J. Millar and D.A.Williams (IOP Publ. Ltd., Bristol, 1993) p. 1.
[2] Schutte W.A., The Cosmic Dust Connection, edited by M. Greenberg (Kluwer Academic Publisher, Dordrecht, 1996) p. 1.
[3] Schmitt B., Molecules and Grains in Space, edited by I. Nenner (AIP Press, New York, 1994) p. 735.
[4] Tielens A.G.G.M. and Whittet D.C.B., Molecules in Astrophysics: Probes and Processes, edited by E. van Dishoeck (Kluwer Academic Publisher, Dordrecht, 1997) p. 45.
[5] Allamandola J.M., Bernstein M.P. and Sandford S.A., Astronomical and Biochemical Origins and the search for Life in the Universe, edited by C.B. Cosmovici, S. Bowyer and D. Werthimer (Editrice Compositori, Bologna, 1997) p. 23.
[6] Ehrenfreund P., Dartois E., Demyk K. and d'Hendecourt L., *A&A* **339** (1998) L17.
[7] Gerakines P.A., Whittet D.C.B., Ehrenfreund P., *et al., ApJ* (1999) in press.
[8] Ehrenfreund P., Boogert A.C.A., Gerakines P.A., *et al.*, First ISO Workshop on Analytical Spectroscopy, edited by A.M. Heras, K. Leech, N.R. Trams and M. Perry (ESA Pub., Noordwijk, 1997) p. 3.
[9] d'Hendecourt L.B. and Jourdain de Muizon M., *A&A* **223** (1989) L5.
[10] de Graauw Th., Whittet D.C.B., Gerakines P.A., *et al.*, *A&A* **315** (1996) L345.
[11] d'Hendecourt L., Jourdain de Muizon M., Dartois E., *et al.*, *A&A* **315** (1996) L365.
[12] Sandford S.A. and Allamandola L.J., *ApJ* **355** (1990) 357.
[13] Ehrenfreund P., Boogert A.C.A., Gerakines P.A., Tielens A.G.G.M. and van Dishoeck E.F., *A&A* **328** (1997) 649.
[14] Boogert A., Ehrenfreund P., Gerakines P.A., *et al.*, *A&A* (1998) in press.
[15] Dartois E., PhD thesis, Université Paris 11, Orsay (1998).
[16] van der Tak F., *et al., ApJ* (1998) in press.
[17] Boogert A.C.A., Schutte W.A., Tielens A.G.G.M., *et al.*, *A&A* **315** (1996) L377.
[18] Boogert A.C.A., Helmich F.P., van Dishoeck E.F., *et al.*, *A&A* **336** (1998) L352.
[19] Dartois E., d'Hendecourt L., Boulanger F., *et al.*, *A&A* **331** (1998) 651.
[20] Ehrenfreund P., d'Hendecourt L., Dartois E., *et al.*, *Icarus* **130** (1997) 1.
[21] Dello Russo N., DiSanti M.A., Mumma M.J., Magee-Sauer K. and Rettig T.W., *Icarus* (1998) in press.
[22] Ehrenfreund P., van Dishoeck E.F., Burgdorf M., *et al.*, *Astrophys. Space Sci.* **255** (1998) 83.
[23] van Dishoeck E.F. and Helmich F.P., *A&A* **315** (1996) L177.
[24] van Dishoeck E.F., Helmich F.P., de Graauw Th., *et al.*, *A&A* **315** (1996) L349.

[25] Waters L.B.F.M., Molster F.J., de Jong T., *et al.*, *A&A* **315** (1996) L361.
[26] Waelkens C., Waters L.B.F.M., Malfait K., *et al.*, *A&A* **315** (1996) L245.
[27] Nuth J.A., *Earth Moon Planets* **47** (1989) 33.
[28] Campins H. and Ryan E.V., *ApJ* **341** (1989) 1059.
[29] Hanner M.S., Lynch D.K. and Russel R.W., *ApJ* **425** (1994) 274.
[30] Hanner M.S., Hackwell J.A., Russel R.W. and Lynch D.K., *Icarus* **112** (1994) 490.
[31] Hanner M.S., Lynch D.K., Russel R.W., *et al.*, *Icarus* **124** (1996) 344.
[32] Crovisier J., Leech K., Bockelée–Morvan D., *et al.*, *Science* **275** (1997) 1904.
[33] Malfait K., Waelkens C., Waters L.B.F.M., *et al.*, *A&A* **332** (1998) L25.
[34] Hallenbeck S., Nuth J. and Daukantas P., *Icarus* **131** (1998) 198.
[35] Bernstein M.P., Sandford S.A., Allamandola L.J., Chang S. and Scharberg M.A., *ApJ* **454** (1995) 327.
[36] Greenberg J.M., Comets, edited by L.L. Wilkening (University of Arizona Press, Tucson, 1982) p. 131.
[37] Mumma M.J., Weissman P.R. and Stern S.A., Protostars and Planets III, edited by E.H. Levy and J.I. Lunine (University of Arizona Press, Tucson, 1993) p. 1177.
[38] Mumma M.J., From Stardust to Planetesimals, ASP Conf. 122, edited by Y.J. Pendleton and A.G.G.M. Tielens (ASP, San Francisco, 1997) p. 369.
[39] Mumma M.J., Di Santi M.A., Dello Russo N., *et al.*, *Science* **272** (1996) 131.
[40] Boudin N., Schutte W.A. and Greenberg J.M., *A&A* **331** (1998) 749.
[41] Irvine W.M., *Nature* **383** (1996) 418.
[42] Irvine W.M., Dickens J.E., Lovell A.J., *et al.*, Chemistry and Physics of Molecules and Grains in Space, Faraday Discussion No. 109 (The Royal Society of Chemistry, London, 1998) p. 475.
[43] Brooke T.Y., Tokunaga A.T., Weaver H.A., *et al.*, *Nature* **383** (1996) 606.
[44] Bird M.K., Huchtmeier W.K., Gensheimer P., *et al.*, *A&A* **325** (1997) L5.
[45] Lacy J., Faraji H., Sandford S. and Allamandola L., *ApJ* **501** (1998) L105.
[46] Allamandola L.J., Sandford S.A., Tielens A.G.G.M. and Herbst T.M., *ApJ* **399** (1992) 134.
[47] Skinner C.J., Tielens A.G.G.M., Barlow M.J. and Justtanont K., *ApJ* **399** (1992) L79.
[48] Dartois E., Schutte W., Geballe T., *et al., A&A* (1999) inpress.
[49] Vandenbussche B., Ehrenfreund P., *et al., A&A* (1999) submitted.
[50] Baluteau J.–P., Cox P., Cernicharo J., *et al.*, *A&A* **322** (1997) L33.
[51] Li A. and Greenberg J.M., *ApJ* **498** (1998) L83.

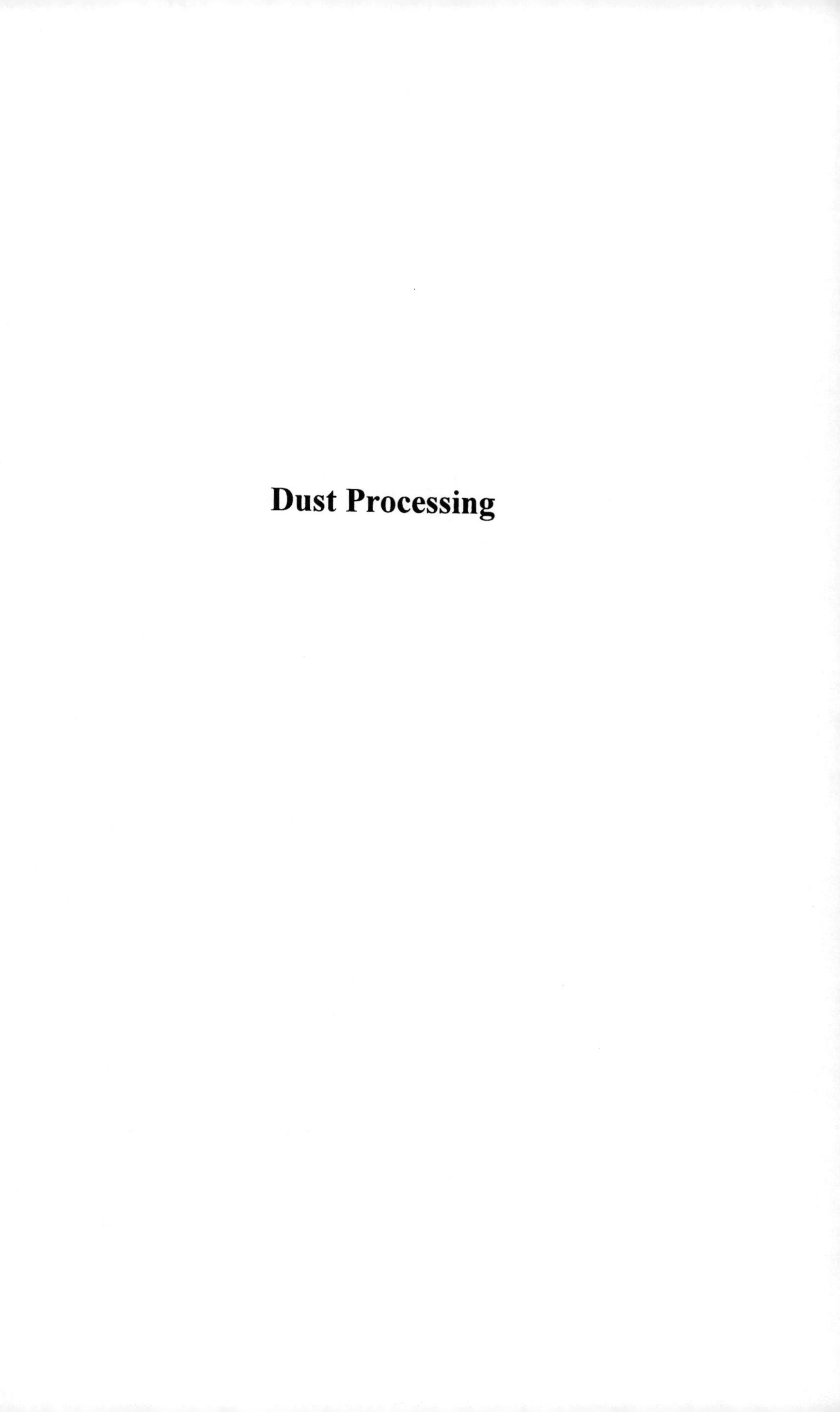

Dust Processing

LECTURE 16

Grain Formation and Evolution in the Interstellar Medium

Th. Henning

Astrophysical Institute and University Observatory (AIU), Schillergäßchen 3, 07745 Jena, Germany

Abstract. Depending on the environment in which they are located in the interstellar medium, different populations of cosmic grains evolve. The lifecycle of the grains from their formation in the outflows of evolved stars to their processing by shocks, UV light, and ion irradiation in the diffuse interstellar medium until their incorporation into dense molecular cloud cores and their growth and consumption by star formation is reviewed.

1. INTRODUCTION

Cosmic dust grains play an important role in the evolution of our and other galaxies. In dusty regions of the interstellar medium (ISM), the solid particles strongly influence the thermal, dynamical, ionisation, and chemical states of matter. Because of their large optical cross sections, dust grains are effective absorbers of radiation energy and momentum. When present, dust is the dominant opacity source exceeding the contributions from atoms and molecules by several orders of magnitude. The collisional coupling between grains and gas in dense envelopes around evolved stars, in molecular cloud cores, and protoplanetary disks has serious consequences for the temperature and the dynamical evolution of the gas component. The instability of molecular cloud cores and their fragmentation into smaller clumps, the upper and lower mass limits for stars, thermal convection in protoplanetary disks, and mass loss from evolved stars are directly linked to the presence of dust.

Grains effectively shield molecules in the regions around red giants from the dissociation by the interstellar UV radiation field and play a similar role in molecular clouds. The formation of molecular hydrogen takes place on the surfaces of grains. Chemical surface reactions lead to the production of mantles of volatile material ("molecular ices") around refractory grain cores, especially in the molecular cloud cores. The evaporation of such mantles in regions of massive star formation may trigger a completely different gas-phase chemistry from that of quiescent molecular cloud cores.

The previous examples demonstrate that dust grains are indeed active players in galactic evolution. On the other hand, the participation of dust in the different interstellar processes modifies the particles and leads to the formation of a complicated multi-component system. Constraints on the nature of dust grains come from the analysis of the abundance patterns, the study of extinction and polarisation curves, infrared spectroscopy, observations of thermal dust continuum radiation, the analysis of primitive meteorites, *in situ* studies of interplanetary dust particles (IDPs) and comets, laboratory astrophysics, and theoretical models of the evolution of the ISM.

This review mainly deals with the evolution of cosmic grains. Section 2 outlines the lifecycle of the dust particles. In the subsequent sections, I discuss grain formation in the mass outflows from evolved stars (Sect. 3), grain modification in the diffuse ISM (Sect. 4), and grain and mantle growth in molecular clouds (Sect. 5). A consideration of the evolution from micron-sized grains to planetesimals in protoplanetary disks is beyond the scope of this review because of space limitations (see, *e.g.*, [1–3]). For other aspects of the nature of cosmic dust, the interested reader is referred to other reviews by the author [4–6], Mathis [7, 8], and Sandford [9] and books by Whittet [10], Millar and Williams [11], and Greenberg [12].

2. FROM AGB STARS TO THE INTERSTELLAR MEDIUM

Mass loss from evolved stars is a very important process as it injects newly formed dust grains into the ISM. The main sources of this "stardust" are luminous red giants which have reached the Asymptotic Giant Branch (AGB). AGB stars are intermediate-mass stars ($M_* \leq 6\,M_\odot$). Their dense and compact cores consist mainly of carbon and oxygen and a degenerate electron gas surrounded by a very extended hydrogen region of low density (see [13] for a review). These objects have relatively high mass loss rates (10^{-7} to $10^{-4}\ M_\odot\,\mathrm{yr}^{-1}$) for a period of about 10^5 yr. The total dust production rate from these stars in the solar neighbourhood ranges between 5.3×10^{-3} and $6.5\times10^{-3}\ M_\odot\,\mathrm{pc}^{-2}\,\mathrm{Gyr}^{-1}$ [14, 15]. Other stardust sources of minor importance are novae, supergiants, and the cool Wolf-Rayet (WC 8 and WC 9) stars.

The dust production rates in the ejecta of Type II supernovae might compete with the AGB rates. However, neither the efficiency of dust formation nor the dust composition in these ejecta is presently known. Assuming efficient dust production in Type II supernovae, the maximum dust production rate is of the order of 3×10^{-2} $M_\odot\,\mathrm{pc}^{-2}\,\mathrm{Gyr}^{-1}$. Most of the mass of the ISM is contained in its denser phases. The total gas mass in molecular clouds (particle densities $n \geq 300\,\mathrm{cm}^{-3}$, temperature $T = 10$ K) and diffuse HI regions ($n = 50$ cm^{-3}, $T = 80$ K) amounts to approximately 5×10^9 $M_\odot$. With the canonical value of the gas-to-dust mass ratio of 100, the total dust mass of the ISM can be replenished by stardust injection on a timescale of 3 Gyr [14].

Strong supernova shock waves are the dominant mechanism for grain destruction and modification of the grain size distribution in the ISM [16, 17]. While contributing only a minor fraction to the total mass of the ISM, the warm medium and the tenuous hot coronal gas fill most of the total volume of the ISM. The hot intercloud medium is characterised by gas densities that are far too low (a few 10^{-3} cm^{-3}) to be of any importance for shock destruction of grains. Therefore, destruction will predominantly occur in the warm neutral/ionised medium ($n \simeq 0.25 - 0.5$ cm^{-3}, $T \simeq 8000$ K). Recent calculations of the lifetime of grains resulted in values of $4 - 6 \times 10^8$ yr, which is much shorter than the stardust injection timescale [16, 17]. This led to the conclusion that some dust must be formed in the ISM. We will discuss this problem in more detail in Section 3. However, we note that these destruction rates may have been overestimated because a homogeneous distribution of supernova events was assumed, in contrast to the general belief that the formation of massive stars and, accordingly, Type II supernovae occur in a cluster mode. This means that Type II supernovae should be concentrated both in space and time. We note, however, that in the calculations [16, 17] the effective interval between supernovae was enhanced by a factor of five, compared to the typical Galactic value, in order to take into account this clustering effect.

The chemical and physical structure of grains in the diffuse ISM may be strongly modified by extended periods of UV and ion irradiation. The mean flux density of the visual/UV radiation field is about 10^{10} photons m^{-2} s^{-1} nm^{-1}. Cosmic rays are mainly composed of relativistic protons and electrons. The differential energy spectrum of the various cosmic ray species pervading interstellar clouds at energies higher than 1 GeV nucleon^{-1} can be well represented by a power law that goes as $E^{-2.6}$ where E is the kinetic energy per nucleon [18]. At lower energies, this spectrum is practically unknown. The bombardment of materials with MeV ions in the laboratory may well simulate the expected effects of cosmic ray irradiation in interstellar clouds. On the basis of such experiments, we expect that the grains present in the ISM will have a more amorphous structure than the stardust. The modification of carbonaceous material leads to the rather invariant interstellar UV bump at 217 nm [19], see also Section 4. Furthermore, the UV irradiation or cosmic-ray bombardment of ices may lead to the formation of organic refractory mantles or even complete carbonization [20]. The 3.4 μm absorption feature seen in the spectra of infrared sources probing the diffuse ISM does not occur in stardust spectra (see contribution by Pendleton in this volume). This feature is generally attributed to C–H stretching modes of CH_2 and CH_3 groups in aliphatic hydrocarbons.

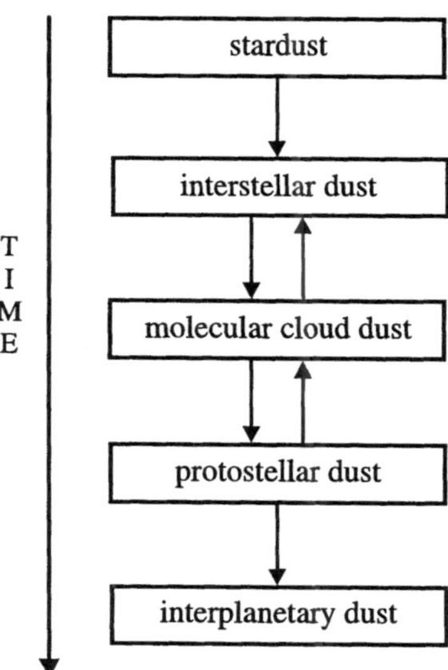

Fig. 1. — Dust populations.

The lack of features related to carbonyl (C=O) vibrations clearly indicates that the 3.4 μm material does not contain significant amounts of oxygen and may have been produced by prolonged periods of irradiation or ion bombardment. However, direct observational evidence for photoprocessing of ices is scarce.

The observation of molecular ice signatures in the infrared spectra of background stars behind molecular clouds and of young stellar objects with circumstellar envelopes/disks has revealed a further source of galactic dust. The adsorption of gas phase species and surface chemistry results in the formation of ice mantles around refractory grain cores. Infrared spectroscopy led to the detection of solid H_2O, CO, CO_2, and CH_3OH with other minor ice components possibly present ([21–23] see also the contributions by Dartois and Schutte in this volume). The exact composition of the ices and the ratio between solid and gas phase components vary from source to source and depends on a variety of factors, such as density, temperature, and UV radiation fields [24]. The dust density determines the H/H_2 ratio which acts as a key parameter for the ice composition.

Grain coagulation modifies the size distribution of grains in dense molecular cloud cores and produces micron-sized composite and fluffy grain agglomerates [25–27]. In protoplanetary disks, dramatic changes in the grain properties due to coagulation, chemical destruction, sublimation and condensation are expected to occur. Further processing takes place in the case of interplanetary dust grains where thermal annealing and further chemical processing are important.

The different dust populations are summarized in Figure 1. There is clear evidence for a further subdivision of these dust populations according to the chemical state of the matter (oxygen-rich *versus* carbon-rich stardust) and the physical conditions (ionizing radiation, temperature, and densities). We expect matter to exchange between the molecular cloud phase and the diffuse atomic gas components by the destruction of molecular clouds during massive star formation, and during the formation of these clouds by a phase transition from the more dilute gas phases of the ISM. The diffuse HI regions are converted to the warm neutral/ionized gas by photoionization. The cycling time between the warm and cold phases of the ISM is not well determined, but is of the order of 3×10^6 yr [16].

3. STARDUST FORMATION

The nature of dust grains produced in the outflows of AGB stars depends on the physical conditions in the extended stellar atmospheres (*e.g.* pulsations, formation of clumps, chromospheric UV radiation) and the molecular composition of the outflows (see Tab. I, [28–30]). A key parameter for dust formation is the elemental C/O ratio because the atoms of the less abundant element are mostly tied up in stable CO molecules and do not participate in grain formation. Therefore, M-type giants ("oxygen-rich" stars) produce refractory dust components which are different from the dust condensates of C-type stars ("carbon-rich" stars). The small transition group of S-type stars is characterized by a C/O ratio close to 1.

Dust formation in oxygen-rich ejecta is based on the excess oxygen and the elements Al, Fe, Si, and Mg. Therefore, we expect the formation of silicates and other metal oxides. This expectation is confirmed by spectroscopic observations of circumstellar envelopes around M-type giants (see [4, 5, 9] and the contribution by Sylvester in this volume). The formation of silicates should start with the formation of SiO_x clusters. The role of the metal cations during grain formation is poorly understood. Laboratory experiments with H_2-SiO, FeO-SiO, and H_2-Mg-SiO systems suggest the formation of amorphous unequilibrated solids with non-stoichiometric elemental ratios (see [31] and references therein). The detection of a crystalline silicate component in the spectra of evolved stars by ISO was completely unexpected and awaits a detailed explanation [32], see the contribution by Waters in this volume (for experimental data see [33]). Also surprising was the detection of carbon dust in some of these environments (see the contribution by Sylvester in this volume). A better theoretical understanding of grain formation in oxygen-rich atmospheres is hampered by a lack of relevant kinetic reaction rates.

An interesting possibility is the formation of small silicon grains. The dangling bonds could be connected to oxygen atoms in a SiO_2 mantle or to hydrogen atoms. Silicon nanoparticles were recently proposed as a component of interstellar dust and as a possible carrier of the so-called extended red emission [34, 35].

Table I. — Dust formation in AGB stars.

Characteristic features	M-type giants	C-type giants
Dominant molecular species C/O ratio Precursor molecules for grains Other important species for grain formation	H_2 < 1 SiO H, metal cations (Fe, Mg, Al), MgS, MgO	H_2 > 1 C_2H_2, C_2H H, Si, SiS, Mg, MgS, Fe
LTE-derived molecular composition	CO, O, H_2O, OH, O_2, N_2, SiO, SiS, Mg, MgO MgOH, Fe, $Fe(OH)_2$	CO, C_2H_2, N_2, C_2H, HCN, Si, CS, SiS, C_3H
Other observed molecules	N, SO, SO_2, H_2S, NH_3, simple organic molecules (*e.g.* HCN)	Cyanopolyynes, carbon-chain molecules, acetylenic compounds, simple organic molecules (*e.g.* HCN), silicon-containing molecules (*e.g.* SiC, SiN, SiC_2), NH_3
Key objects	IRC+60 150, IRC+10 011	IRC+10 216

In the envelopes of C-type giants, disordered carbon grains and silicon carbides should form ([19, 36] also see the contribution by Sylvester in this volume). Another component for which some spectroscopic evidence exists (features at 21 and 30 μm) is sulfide grains [37, 38]. The first molecular steps towards grain formation in hydrocarbon-containing gaseous systems are better understood than in oxygen-rich environments because of the wealth of data available from combustion and aerosol science [39, 40]. The formation of carbonaceous grains should start with acetylene, the most abundant molecule after H_2 and CO. The next decisive steps are [36, 41, 42]:

1. Formation of the first aromatic ring, benzene (C_6H_6).
2. Growth to polyaromatic hydrocarbons (PAHs) by repeated radical formation due to hydrogen abstraction and addition of hydrocarbons.
3. Condensation of primary nanoparticles.

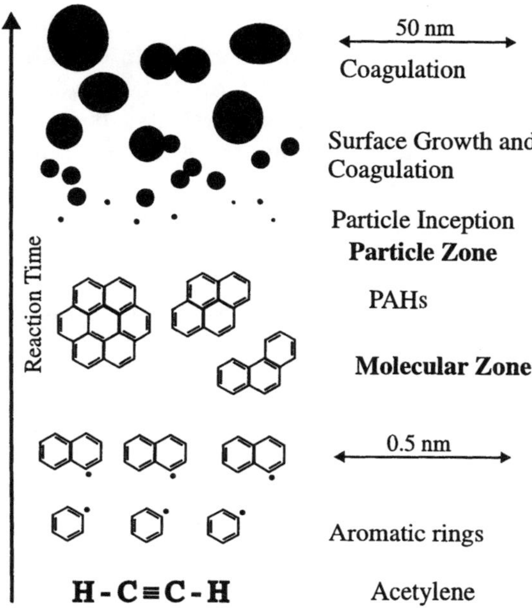

Fig. 2. — Schematic picture of carbon grain formation. Adapted from Bockhorn [40].

4. Growth of the primary particles through coagulation, deposition of carbon *via* acetylene surface reactions, and surface condensation of gas-phase species.

The different steps during grain formation are schematically summarized in Figure 2. The formation of PAHs can only occur in a small temperature window between 900 and 1100 K. For temperatures larger than 1100 K, both the hydrogen abstraction reaction and the addition of acetylene cannot occur because of the reversibility of these reactions. At temperatures below 900 K, the endothermic hydrogen abstraction reaction cannot proceed and no aromatic radicals can be formed. The chemical processes during the condensation of the primary carbonaceous particles are still poorly understood. Here we should also stress that the question of whether the PAHs are really intermediate species during soot formation or merely by-products is not satisfactorily answered by experimental studies. Laboratory investigations demonstrate that the internal structure and the spectroscopic properties of carbon-black particles are sensitive to the physical and chemical conditions in the cooling gas (see, *e.g.*, [43]). Small amounts of oxygen-bearing molecules may also be present in C-rich circumstellar envelopes and could influence the dust nucleation process [44]. Although most of the nitrogen should be bound to N_2, the role of this element in grain formation has to be investigated in more detail.

Silicon carbide grains are industrially produced by high-temperature reactions between solid SiO_2 and carbon materials. In the laboratory, silicon carbide grains can also be produced via the formation of hydrocarbon radicals and reactions with various Si-containing components. It is this process that we expect to occur in circumstellar envelopes as well [45]. Here we should add that there is no spectroscopic evidence for the presence of SiC grains in the diffuse ISM. Possible explanations for this fact include the formation of mantles around the SiC grains, a low efficiency for SiC formation, or gray opacities due to very large SiC particles.

From an astrochemical point of view, the formation of pure carbon grains in the winds of hydrogen-deficient but carbon-rich Wolf-Rayet (WC) stars is a particularly interesting case [45]. From laboratory experiments it is known that simple linear carbon chain molecules are the molecular precursors of carbon material in hydrogen-poor atmospheres. At higher densities, fullerenes form via monocyclic rings and may later initiate the formation of carbon dust particles. Since the formation of fullerenes is not efficient at low gas pressures in the laboratory, it is not expected to occur in the low-density WC winds. In this case, carbon grains are expected to form directly from simple linear and cyclic carbon molecules.

The interpretation of astronomical infrared spectra needs special care because the simple separation scheme based on the C/O ratio may not be sufficient for their explanation. Complicating factors include the binarity of stars, the presence of ionizing stellar UV radiation, surface chemistry, and transition phases during stellar evolution. We mention, as an example, carbon stars showing silicate bands in emission in their infrared spectra ([46–50] and also the contribution by Sylvester in this volume). In these cases, the oxygen-rich dust is probably the remnant from earlier mass loss phases. In the meantime, the photosphere of the M-type star was enriched with carbon as the result of carbon dredge-up by repeated thermal pulses and the AGB star turns into a carbon star.

In this section, we have mainly discussed the chemical aspects of stardust formation. However, we should note that the hydrodynamical and radiation field conditions in the extended atmospheres and the circumstellar envelopes are of equal importance. Examples illustrating the importance of these factors for grain formation are the formation of dense clumps in the outflows and time-variable mass loss processes.

4. GRAIN MODIFICATION IN THE INTERSTELLAR MEDIUM

Interstellar dust grains undergo destructive and disruptive processing due to supernova shock waves in the warm ISM phase. Empirical evidence of such processing comes from the presence of a population of very small grains (non-thermal excess infrared emission) and the analysis of elemental depletion patterns (see the review by [51]). The depletion of certain elements is larger in the dense phases of the ISM than in the more dilute and warmer diffuse halo gas, suggesting the return of solid material to the gas in the diffuse ISM.

Detailed numerical modelling demonstrated that thermal and non-thermal sputtering is the dominant destructive process for interstellar grains and operates at high shock velocities ($\geq$ 50 km s^{-1}) [16,17]. Thermal sputtering by gas-phase species impinging with thermal velocities on the grains is important at very high shock velocities ($\geq$ 150 km s^{-1}). For shock velocities between 50 and 150 km s^{-1}, non-thermal sputtering (due to the motion of the grains through the gas) dominates the grain destruction.

Grain-grain collisions are the dominant process for grain shattering and the resulting re-distribution of grain mass. Differential velocities between grains are produced by the dependence of the post-shock betatron acceleration, and the gyration radius of the charged grains around the magnetic field lines, on the grain size and density. Grain-grain collisions with minimum collision velocities of the order of 1 km s^{-1} predominantly lead to grain disruption and a modification of the grain size distribution [17]. Starting with an MRN power law for the grain size distribution [52] ($n(a)\mathrm{d}a \propto a^{-3.5}\ \mathrm{d}a$), calculations [17] indicate a redistribution of the dust mass from the larger grains ($a \geq 0.1\ \mu$m) into smaller grains ($a < 50$ nm). Figure 3 shows the initial MRN distribution and the final grain size distribution for different shock velocities [17]. Grain-grain collisions are less important in high-velocity shocks (200 km s^{-1}) because the grains are not accelerated by the betatron effect until late on in the shock. Larger grains are lost from the grain size distribution on timescales of the order of 10^8 yr.

The observed extinction laws in the ISM, the depletion of dust-forming elements (*e.g.*, C, O, Si, Mg, Fe), and the strength of the interstellar 9.7 and 18 μm silicate absorption features imply that a large fraction of the heavy elements is locked up in the dust and that submicron sized silicate grains are present. Together with the discrepancy between the stardust injection timescale and the lifetime of the grains in the ISM, this implies that either the models overestimate the destruction and disruption efficiency of grains in supernova shocks or that efficient grain growth must occur in the dense phases of the ISM (see Sect. 4). Factors which could lead to an overestimate of the destruction/disruption rates are the assumed homogeneous distribution of supernova events, an overestimate of the cycling frequency between the warm and cold phases of the ISM, and a better survival of porous grains or grains with very refractory mantles. For a detailed model of the dust evolution in the Galaxy and a discussion of the different key parameters, we refer to the paper by Dwek [15].

Until now, we have mainly discussed the modification of the grain size distribution. The intense interstellar UV radiation or cosmic rays are factors which may change the physical and chemical structure of grains. In this review, we will not discuss the ion- and photoprocessing of ices (see, *e.g.*, [53,54] and the contributions of Schutte and Moore in this volume) but will concentrate on structural changes in the *refractory* grains by UV irradiation and particle bombardment. In contrast to the many studies of ice photoprocessing, relatively few data are available for radiation and particle modification of

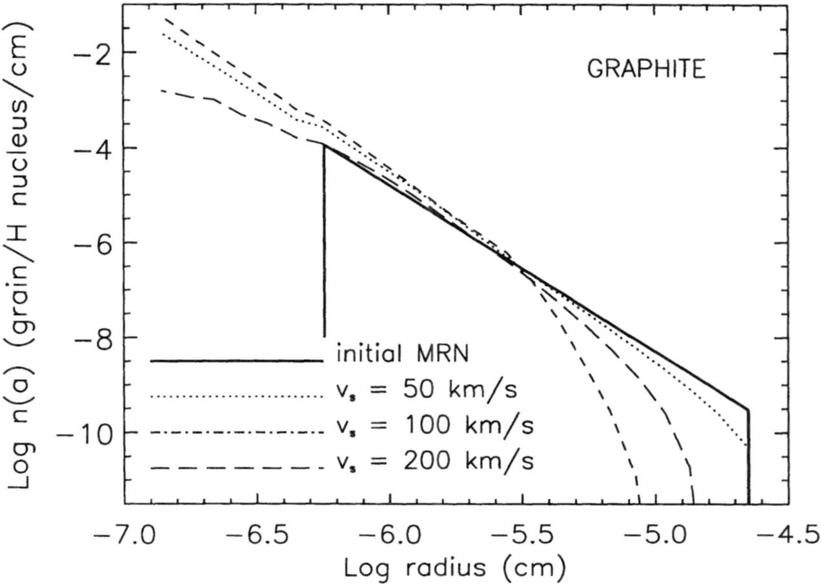

Fig. 3. — Grain destruction in shocks of different velocities. After Jones *et al.* [17].

astrophysically relevant refractory materials. The spectral behaviour of hydrogenated carbonaceous grains after UV irradiation has been investigated [55], see also the contribution by Colangeli in this volume. It was found that the $\pi - \pi^*$ resonance was activated and remained stable at 215 nm, and that the intensity of the UV bump increased as a function of the dose deposited in the samples.

In the case of ion irradiation two regimes can be distinguished, which depend on the energy deposition in the material. These are the electronic ("ineleastic") and nuclear ("elastic") contributions, respectively. In the case of the electronic interaction, the ions interact with the electron cloud of the material and cause excitation, ionization, and charge transfer. The nuclear contribution to the energy deposition comes from the interaction of the impinging ions with the nuclei and the shell electrons of the material. The nuclear stopping power (energy lost per unit path length) dominates at low ion energies, whereas the electronic contribution becomes important at higher energies.

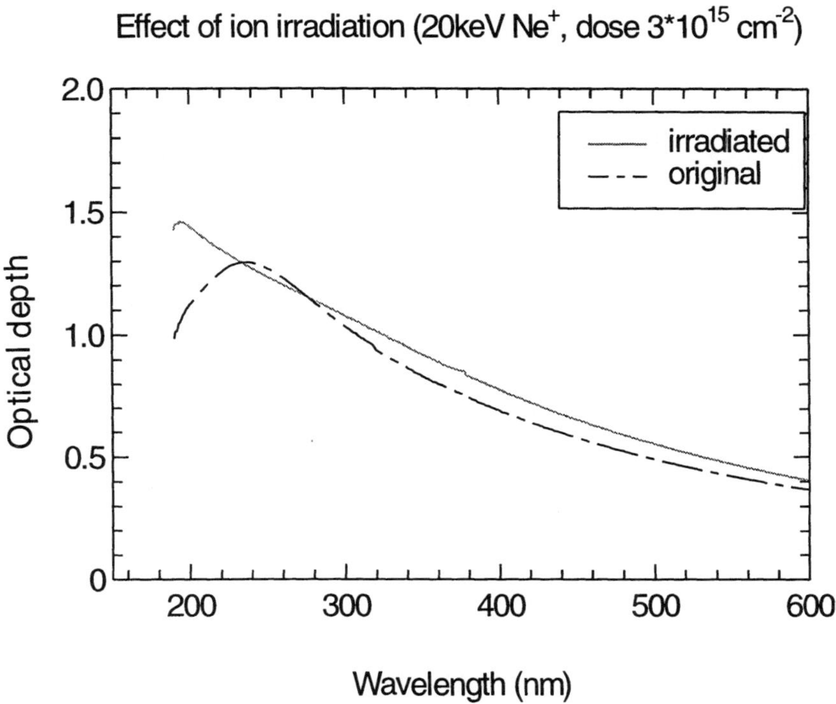

Fig. 4. — Spectral changes of carbonaceous material by ion bombardment.

Amorphous silicates were produced by heavy ion irradiation of natural olivine with a relatively high dose of 1.5 MeV neon ions [56], producing structureless silicate features. A study of the UV spectral changes after irradiation with 3 keV He^+ ions led to band shifts from 203 to 215 nm for a hydrogenated amorphous carbon sample and a change from 240 to 218 nm in the case of samples produced in an argon atmosphere ([57], see also the contribution by Colangeli in this volume). These changes were interpreted as due to the competing effects of amorphization and graphitization leading to similar microstructures for the different samples at the highest ion fluence considered (10^{17} ions cm^{-2}). In an earlier study [58] it was shown that the atomic structure of hydrogenated carbon materials reaches a stable configuration after keV-ion bombardment without any memory of the initial sample structure. In Figure 4 the results of ion irradiation of carbonaceous dust, produced in an argon atmosphere, with keV Ne^+ ions are shown. The spectrum of the irradiated sample shows a shift of the UV peak to shorter wavelengths and a less pronounced UV bump compared to the original material caused by a more amorphous structure. An interesting result was obtained by the irradiation of graphitic carbonaceous material with Ne^+ ions of 3 MeV energy [59]. Spherical carbon onions are produced and the transformation of their cores to cubic diamond crystals is observed under continued irradiation. The optical characterization of this material would be of great astrophysical interest.

Here we should not forget that the process of grain destruction during star formation is also very important. The rates for carbon and silicate destruction are about 8×10^{-3} and 30×10^{-3} $M_\odot$ pc^{-2} Gyr^{-1}, respectively [15]. These rates are about an order of magnitude lower than the destruction rates by supernova shock fronts.

5. GRAIN GROWTH IN MOLECULAR CLOUDS

In dense molecular clouds, the accretion of gas phase species onto grains and the coagulation of grains leads to grain growth. The timescale for accretion is shorter than that for coagulation. For typical star-forming molecular clouds ($n = 10^4$ cm^{-3}, $T = 40$ K), the accretion timescale for atoms and molecules (heavier than H_2) is of the order of 10^5 yr [53], which is much shorter than the expected lifetime of a cloud of about 10^7 yr. Therefore, mantle desorption must be important during the lifetime of the cloud. Accretion results in an increase of the grain mass. However, the size of the larger grains cannot increase significantly by accretion because only relatively thin mantles can be formed. A quantitative estimate of the increase of grain mass is difficult because the accretion process competes with a variety of mantle destruction processes [53], for which the rates are not well known. A rough estimate gives that a mantle of the thickness of about 20 nm is accreted during the lifetime of the molecular cloud [20] (for a discussion see [60]). An estimate of the maximum increase in dust mass by mantle accretion based on cosmic abundance considerations leads to a factor of about 1.5 relative to the mass of the refractory material [25].

Coagulation causes a re-distribution of the grain sizes and may reform the larger grains present in the ISM. This would imply that such grains should be inhomogeneous and a certain fraction of their volume should be made of vacuum (*i.e.*, they are "fluffy" or "porous" grains) [25, 26, 61]. This prediction can be checked by detailed light scattering calculations for such particle structures and the comparison of the results with observed extinction and polarization curves, taking into account the abundance constraints (see, *e.g.*, [62, 63]). A detailed modelling of the coagulation process in dense molecular cloud cores and the resulting opacities can be found in [25, 26]. The calculations showed a rapid growth of the smallest grains, whereas the upper grain size limit is only slightly shifted. The timescale for coagulation was about 10^5 yr in a cold and dense molecular cloud core ($n = 10^6$ cm^{-3}, $T \leq 20$ K). The coagulation of grains will not only modify the grain size distribution, but will also change the total surface area of grains with some consequences for molecular abundances [64,65].

Empirical evidence for the accretion of gas phase species and the formation of ice mantles comes from infrared spectroscopy and the detection of molecular ices. This topic has been covered in recent reviews [21, 53] and other contributions in this volume.

The observation of flatter extinction curves (less extinction in the ultraviolet relative to the visual), a lower extinction per unit mass, and a different polarization behaviour in the outer regions of molecular clouds are generally taken as evidence for the presence of larger grains due to coagulation. Quantitative investigations of extinction curves [66,67] led to the conclusion that larger grains grow by collisions with smaller grains (significant reduction of small grains) and that the formation of micron-sized grains by coagulation is possible.

6. CONCLUSIONS

Small solid particles are a ubiquitous component of the Universe, undergoing a wide range of evolutionary processes. The complex interplay between gas, dust, and stars drives galactic evolution. Without a global characterization of the physical conditions in the ISM provided by sophisticated observations, a better understanding of the underlying microphysical processes guided by dedicated laboratory experiments, and the large-scale numerical modelling of the combined effects of radiation transfer, hydrodynamical evolution and chemistry, real progress in the investigation of cosmic dust will not be possible.

The following questions await answers:

1. How does the transition from simple molecules to very large molecules/very small grains and vice versa proceed?
2. What is the relative importance of stardust formation compared to dust formation in molecular clouds?
3. What is the efficiency of dust formation in Type II supernovae?
4. What are the destruction rates for porous grains?
5. What is the exchange rate between the different phases of the ISM?
6. Can carbon-bearing molecular ices be completely transformed to carbonaceous material by UV irradiation or ion bombardment?
7. What is the role of surface chemistry?
8. How do silicates form from gas phase species and how do they evolve?
9. How does the formation of different carbon allotropes depend on the physical and chemical conditions of the dust-forming regions?
10. What role do other dust species play (*e.g.*, metallic iron, metal oxides, carbides, sulfides, nanodiamonds)?

Infrared spectroscopy and imaging can provide the answers to some of these questions. With the complete analysis of the ISO data, we should find out what the spatial distribution of PAHs is and how this distribution correlates with the distribution of AGB stars and molecular clouds. We will also learn more about the chemical and crystal structure of silicates. The role of photoprocessing of ices and the abundance of molecular species in the gas phase relative to the solid phase will be clarified. Furthermore, the nature of other dust components such as iron, metal oxides, sulfides, silicon, carbides, and the various allotropes of carbon materials will be better understood with the new spectroscopic data.

With the imaging capabilities of ESO's Very Large Telescope Interferometer (VLTI) at infrared wavelengths, a detailed mapping of protoplanetary disks and circumstellar envelopes will be possible. In a similar way, millimetre interferometry will contribute to our knowledge of the distribution of cold dust and its relation to the gas in dense molecular regions. Together with the UV spectroscopic data from the HST, far-infrared observations with SOFIA and FIRST will lead to a complete inventory of the elements carbon and oxygen in the Galaxy.

ACKNOWLEDGMENTS

The author wishes to thank H. Mutschke, M. Schnaiter, W. Wesch, and W. Witthuhn for the collaboration leading to the ion irradiation experiments. I also thank A. Jones for providing Figure 3 and J. Gürtler for critical reading the paper. The experimental research of the author is supported by a grant from the Max Planck Society.

REFERENCES

[1] From stardust to planetesimals, ASP Conf. Series 122, edited by Y.J. Pendleton and A.G.G.M. Tielens (ASP, San Francisco, 1997) p. 1.

[2] Henning Th., The cosmic dust connection, edited by J.M. Greenberg (Kluwer, Dordrecht, 1996) p. 399.

[3] Beckwith S., Henning Th. and Nakagawa Y., Protostars and Planets IV, edited by A.P. Boss and V. Mannings (Arizona University Press, Tucson, 1998) in press.

[4] Dorschner J. and Henning Th., *A&AR* **6** (1995) 271.

[5] Henning Th., Molecules in astrophysics – Probes and Processes, edited by E.F. van Dishoeck (Kluwer, Dordrecht, 1997) p. 343.

[6] Henning Th., *Chem. Soc. Rev.* **27** (1998) 315.

[7] Mathis J.S., *ARA&A* **28** (1990) 37.

[8] Mathis J.S., *Rep. Prog. Phys.* **56** (1993) 605.

[9] Sandford S.A., *Meteor. Planet. Sci.* **31** (1996) 449.

[10] Dust in the galactic environment, edited by D.C.B. Whittet (IOP Publishing Ltd., London, 1992) p. 1.

[11] Dust and chemistry in astronomy, edited by T.J. Millar and D.A. Williams (IOP Publishing, Bristol and Philadelphia, 1993) p. 1.
[12] The cosmic dust connection, edited by J.M. Greenberg (Kluwer, Dordrecht, 1996) p. 1.
[13] Habing H.J., *A&AR* **7** (1996) 97.
[14] Jones A.P. and Tielens A.G.G.M., The cold universe, edited by Th. Montmerle, Ch.J. Lada, I.F. Mirabel and J. Tran Than Van (Éditions Frontières, Gif-sur-Yvette, 1994) p. 35.
[15] Dwek E., *ApJ* **501** (1998) 643.
[16] Jones A.P., Tielens A.G.G.M., Hollenbach D.J. and McKee C.F., *ApJ* **433** (1994) 797.
[17] Jones A.P., Tielens A.G.G.M. and Hollenbach D.J., *ApJ* **469** (1996) 740.
[18] High energy astrophysics, Vol. 1, Particles, photons, and their detection, 2nd edition, edited by M.S. Longair (Cambridge Univ. Press, Cambridge, 1992) p. 1.
[19] Henning Th. and Schnaiter M., Laboratory astrophysics and space research, edited by P. Ehrenfreund, H. Kochan, C. Krafft and V. Pirronello (Kluwer, Dordrecht, 1999) p. 249.
[20] Jenniskens P., Baratta G.A., Kouchi A., *et al.*, *A&A* **273** (1993) 583.
[21] Whittet D.C.B., The cosmic dust connection, edited by J.M. Greenberg (Kluwer, Dordrecht, 1996) p. 133.
[22] Whittet D.C.B., Schutte W.A., Tielens A.G.G.M., *et al.*, *A&A* **315** (1996) L357.
[23] Gürtler J., Henning Th., Kömpe C., *et al.*, *A&A* **315** (1996) L189.
[24] van Dishoeck E.F., Helmich F.P., Schutte W.A., *et al.*, Star formation with the Infrared Space Observatory, ASP Conference Series 132, edited by J. Yun and R. Liseau (ASP, San Francisco, 1998) p. 54.
[25] Ossenkopf V., *A&A* **280** (1993) 617.
[26] Ossenkopf V. and Henning Th., *A&A* **291** (1994) 943.
[27] Weidenschilling S.J. and Ruzmaikina T.V., *ApJ* **430** (1994) 713.
[28] Olofsson H., Molecules in the stellar environment, edited by U.G. Jørgensen (Springer-Verlag, Berlin, 1994) p. 133.
[29] Glassgold A.E., *ARA&A* **34** (1996) 241.
[30] Gail H.-P. and Sedlmayr E., Physical processes in interstellar clouds, edited by G.E. Morfill and M. Scholer (Reidel, Dordrecht, 1987) p. 275.
[31] Nuth III J.A., The cosmic dust connection, edited by J.M. Greenberg (Kluwer, Dordrecht, 1996) p. 205.
[32] Waters L.B.F.M., Molster F.J., de Jong T., *et al.*, *A&A* **315** (1996) L361.
[33] Jäger C., Molster F.J., Dorschner J., *et al.*, *A&A* **339** (1998) 904.
[34] Ledoux G., Ehbrecht M., Guillois O., *et al.*, *A&A* **333** (1998) L39.
[35] Witt A.N., Gordon K.D. and Furton D.G., *ApJ* **501** (1998) L111.
[36] Cherchneff I., The molecular astrophysics of stars and galaxies, edited by T.W. Hartquist and D.A. Williams (Oxford University Press, Oxford, 1998) p. 265.
[37] Begemann B., Dorschner J., Henning Th., Mutschke H. and Thamm E., *ApJ* **423** (1994) L71.

[38] Begemann B., Dorschner J., Henning Th. and Mutschke H., *ApJ* **464** (1996) L195.
[39] Carbon black, 2nd edition, edited by J.-B. Donnet, R.C. Bansol and M.-J. Wang (Marcel Dekker, New York, 1993) p. 1.
[40] Soot formation in combustion, edited by H. Bockhorn (Springer-Verlag, Berlin, 1994) p. 1.
[41] Frenklach M. and Feigelson E.D., *ApJ* **341** (1989) 372.
[42] Cherchneff I., Barker J.R. and Tielens A.G.G.M., *ApJ* **401** (1992) 269.
[43] Jäger C., Henning Th., Spillecke O. and Schlögl R. (1998) in preparation.
[44] Willacy K. and Cherchneff I., *A&A* **330** (1998) 676.
[45] Cherchneff I., Dust and molecules in evolved stars, edited by I. Cherchneff and T.J. Millar (Kluwer, Dordrecht, 1998) p. 333.
[46] Little-Marenin I.R., *ApJ* **307** (1986) L15.
[47] Willems F.J. and de Jong T., *ApJ* **309** (1986) L39.
[48] Chan S.J. and Kwok S., *ApJ* **383** (1991) 837.
[49] Kwok S. and Chan S.J., *AJ* **106** (1993) 2140.
[50] Waters L.B.F.M., Waelkens C., van Winckel C., *et al.*, *Nature* **391** (1998) 868.
[51] Savage B.D. and Sembach K.R., *ARA&A* **34** (1996) 279.
[52] Mathis J.S., Rumpl W. and Nordsieck K.H., *ApJ* **217** (1977) 425.
[53] Schutte W.A., The cosmic dust connection, edited by J.M. Greenberg (Kluwer, Dordrecht, 1996) p. 1.
[54] Pirronello V., Dust and chemistry in astronomy, edited by T.J. Millar and D.A. Williams (IOP Publishing, Bristol and Philadelphia, 1993) p. 297.
[55] Mennella V., Colangeli L., Palumbo P., *et al.*, *ApJ* **464** (1996) L191.
[56] Krätschmer W. and Huffman D.R., *Astrophys. Space Sci.* **61** (1979) 195.
[57] Mennella V., Baratta G.A., Colangeli L., *et al.*, *ApJ* **481** (1997) 545.
[58] Compagnini G., Calcagno L. and Foti G., *Phys. Rev. Lett.* **69** (1992) 454.
[59] Wesolowski P., Lyutovich Y., Banhart F., Carstanjen H.D. and Kronmüller H., *Appl. Phys. Lett.* **71** (1997) 1948.
[60] Preibisch Th., Ossenkopf V., Yorke H.W. and Henning Th., *A&A* **279** (1993) 577.
[61] Mathis J.S. and Whiffen G., *ApJ* **341** (1989) 808.
[62] Henning Th. and Stognienko R., *A&A* **280** (1993) 609.
[63] Mathis J.S., Cohen D., Finley J.P. and Krautter J., *ApJ* **449** (1995) 320.
[64] Cardelli J.A., *ApJ* **335** (1988) 177.
[65] Henning Th. and Sablotny R.M., *Adv. Space Res.* **16** (1995) 17.
[66] Kim S.-H., Martin P.G. and Hendry P.D., *ApJ* **422** (1994) 164.
[67] Kim S.-H. and Martin P.G., *ApJ* **462** (1996) 296.

LECTURE 17

Emission from Carbon-Rich Stellar Envelopes*

R.J. Sylvester

Department of Physics and Astronomy, University College London,
London WC1E 6BT, U.K.

Abstract. Recent observations and research on carbon-rich circumstellar outflows are presented. Silicon carbide, formed in these outflows, is a well-studied carbon star dust component, and can be investigated in the laboratory in the form of pre-solar SiC grains recovered from meteorites. The infrared spectrum of one of the rare carbon stars showing SiC in absorption is presented. ISO SWS and LWS spectra show that many C-rich sources show emission from silicate dust, an oxygen-rich condensate. Conversely, a number of oxygen-rich M supergiants show emission in the UIR bands, due to hydrocarbon materials.

1. INTRODUCTION

Just as interstellar dust was first observed as "dark nebulae" caused by obscuration of background stars, circumstellar dust was first detected in absorption. The carbon star R Coronae Borealis (R CrB) was long known to exhibit variations in brightness of up to 9 magnitudes, with no discernible periodicity. The variations take the form of transient events where the star becomes fainter, before returning to its usual magnitude. During these events, the optical spectrum of the star does not change, suggesting that the variation is not caused by changes in the star's photosphere, but is due to some

* Presented on behalf of members of the ISO LWS Post Main Sequence Specialist Astronomy Group: J.-P. Baluteau, M.J. Barlow, J. Cernicharo, M. Cohen, R.J. Cohen, P. Cox, R.J. Emery, E. Gonzales-Alfonso, C. Gry, T. Lim, X.-W. Liu, A. Omont, D. Péquinot, N.-Q. Rieu, B. Swinyard, R.J. Sylvester, and Truong-Bach.

obscuring material entering the line of sight to the star. The lack of periodicity ruled out an Algol-type behaviour, where a fainter binary companion eclipses the brighter star. It was proposed [1, 2] that the events were due to graphitic grains condensing in the atmosphere of the star.

Grains absorb radiation and will undergo an increase in temperature due to the deposition of the photon energy within the lattice of the solid material. Absorbing grains will therefore emit radiation, with a spectral energy distribution (SED) which depends on the grain temperature. Solid materials evaporate at temperatures around 1500 K, so the SED of the emission from dust grains (if they are treated as blackbodies) must peak at wavelengths longer than about 2 μm. Most of the energy radiated by grains will therefore lie in the infrared spectral region.

With the advent of sensitive IR instruments, observers were able to detect excess IR emission (*e.g.* [3, 4]) from a number of stars, including R CrB [5]. The stars which showed excess IR emission were cool giants and supergiants, such as Betelgeuse, Mira and μ Cephei.

It was shown on the basis of molecular equilibrium calculations [6, 7], that O-rich stars would indeed be expected to produce silicate dust, while C-rich stars would produce carbon and silicon carbide dust. This is because the highly stable CO molecule forms in preference to any other molecule, until all of the less abundant of carbon and oxygen is bound up into CO, leaving only the more abundant species to form dust.

The main types of carbon-rich circumstellar dust are amorphous or graphitic carbon, silicon carbide, and the hydrocarbon carriers of the so-called unidentified infrared (UIR) bands (see the contributions by Salama and Moutou in this volume for detailed discussions on the nature of the UIR emitters).

An early exception to the rule that O-rich and C-rich dust cannot form together around a single star came with the observations of Nova Cen 1986, where silicate emission and the UIR bands, ascribed to carbonaceous media, were detected contemporaneously in mid-IR spectra [8, 9]. This could come about if there were inhomogeneities [9] in the outflow, with abundance gradients in the gaseous outflow leading to C-rich and O-rich dust condensing in spatially distinct regions, such as shells or clumps.

Due to their high mass-loss rates,typically $10^{-10} - 10^{-6}$ $M_\odot$ yr^{-1} [10], cool evolved stars are important sources of dust and heavy elements for the interstellar medium.

In some situations, circumstellar dust is formed in brief episodes rather than continuously over an extended period of time. This provides opportunities for the study of dust formation in "real time". Examples of this include the R CrB stars which, as already mentioned, show large photometric variations due to obscuration by newly-formed dust grains. R CrB stars are carbon-rich and hydrogen-poor, and the dust they form is mostly pure carbon, in an amorphous or glassy form [11]. R CrB dust is thought to form in clouds rather close to the stars, where the gas is normally too hot to allow condensation into grains. It is thought that expansion of part of the circumstellar envelope after the passage of a pulsation-driven shock causes the gas to cool below its equilibrium temperature, allowing grain nucleation to occur close to the star [12].

Another example is provided by novae, which often show evidence of dust formation, in particular rapid visual fading, due to obscuration of the star by newly-formed dust, which is accompanied by a simultaneous IR brightening, due to thermal re-emission of the absorbed starlight. Most dust-forming novae produce amorphous carbon, which is featureless in the mid-IR. Silicates, silicon carbide and hydrocarbons can also be formed by novae, occasionally all by the same one, as in the case of nova NQ Vul [13].

2. SiC EMISSION AND ABSORPTION IN CARBON STARS

Silicon carbide is one of the best-studied circumstellar dust materials. It has long been known that equilibrium condensation models [6,7] predict that SiC should form in the atmospheres of stars when oxygen is less abundant than carbon. Early ground-based mid-infrared spectroscopy [14,15] showed that many carbon stars display an emission feature peaking at roughly $11.0 - 11.5\ \mu$m, which is due to the Si–C bond stretching mode in SiC grains.

Laboratory studies of SiC grains which formed around other stars are possible, because they are brought to Earth as inclusions in meteorites. Measurements of the isotope ratios of these grains, and of the noble gas atoms trapped in their crystal structure, can be compared with the predictions of stellar nucleosynthesis theory to demonstrate the presolar origin of the grains, and even to determine what type of stars produced them. It is found that while a small minority (~1%) of meteoritic grains originate in supernovae, most are formed in the outflows of carbon-rich AGB stars ("carbon stars"); see recent reviews on meteorites and SiC [16–18].

Silicon carbide occurs in about 70 different structural forms, which can be grouped into two main types: α-SiC, which has a hexagonal or rhombohedric structure, and β-SiC which has a cubic structure. At high temperatures, β-SiC can be transformed into α-SiC, but the reverse process is thermodynamically unlikely to occur. The two types of SiC have slightly different infrared spectra: α-SiC has an emission feature peaking at 11.4 μm, while the β-SiC feature peaks at 11.0 μm [19]. Observations of SiC emission from carbon stars are best fitted with the optical constants of α-SiC [18,20].

Despite being so prevalent in circumstellar emission, SiC absorption is not detected in the interstellar medium. By contrast, silicate dust, which is produced by oxygen-rich stars (where it is observable in emission) can be detected in absorption in sightlines with sufficient column density. It is estimated [21] that less than 5% of interstellar silicon is bound up in the form of SiC. This may be due to some dust destruction process or mechanism in the ISM which operates more effectively on SiC grains than on other materials, or due to low relative abundance and/or formation efficiency for SiC grains with respect to silicates.

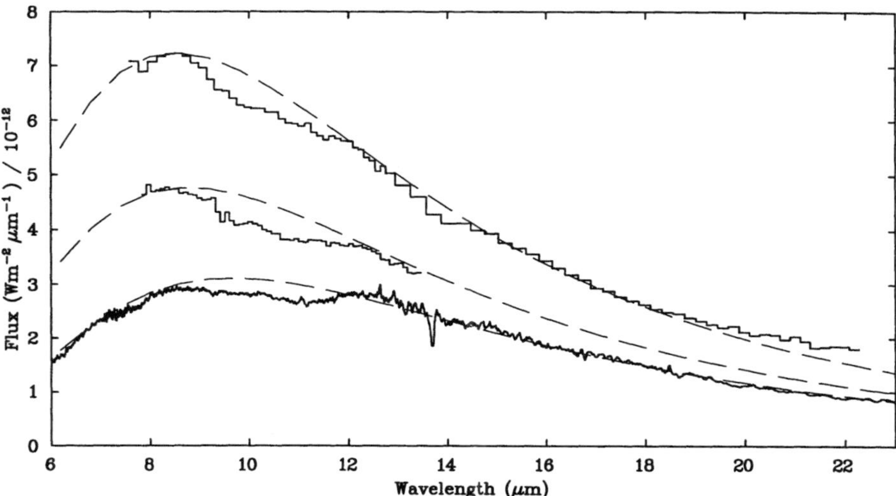

Fig. 1. — Spectra of the carbon star AFGL 5625. Spectra were obtained with the IRAS LRS (upper line), CGS3 at UKIRT (middle) and ISO SWS (lower) instruments. Dashed lines are blackbody fits to the spectra which have very similar temperatures (see text).

Clearly, some SiC grains do survive a sojourn in the ISM, and reach the Earth as the presolar SiC grains in meteorites. Crystallographic studies of these grains reveal that all of the presolar SiC grains are of the β-SiC type [22]. This appears inconsistent with the assignment of circumstellar (*i.e.* newly-formed) SiC to the α form, especially since α-SiC is not thought to transform into β. However, Speck *et al.* [18] note that (i) the techniques used to characterise SiC grains in astrophysical environments (IR spectroscopy) are different to those used on meteoritic grains (Raman spectroscopy and ion microprobe methods), and (ii) the shape of the SiC IR emission feature is very sensitive to particle shape and impurities. These suggestions could be the key to resolving the apparent contradiction.

As mentioned above, SiC has never been seen in absorption in the ISM. Until recently, circumstellar SiC was only seen in absorption towards one source, AFGL 3068 [23]. A ground-based survey of carbon stars [18] using the CGS3 instrument at UKIRT has revealed three more such sources.

An ISO SWS spectrum of one of these sources, AFGL 5625, was obtained in May 1997. This spectrum, kindly made available by Dr. P.F. Roche, shows a broad absorption feature. It was noted [18] that the CGS3 spectrum could be fitted with a combination of SiC and silicate absorption — this appears to

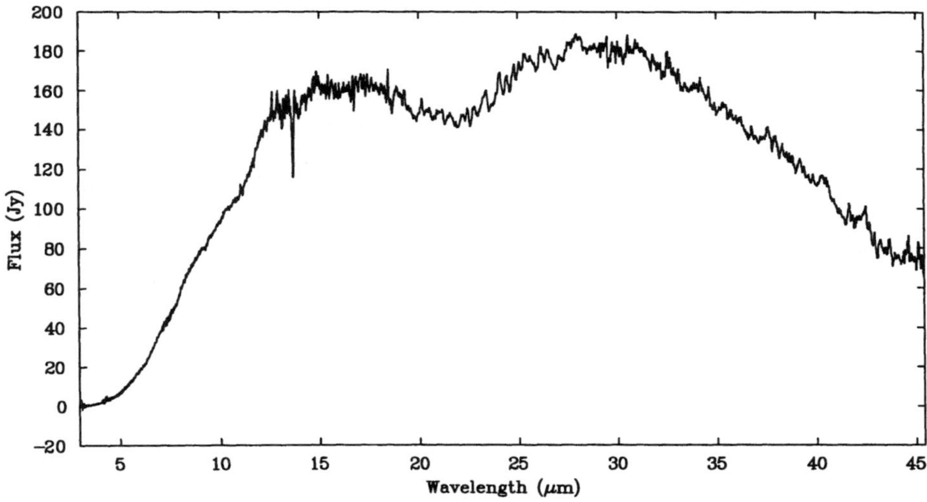

Fig. 2. — The complete SWS spectrum of AFGL 5625 plotted in F_ν (Jy). The strong emission peak at 30 μm may be due to MgS, which could condense in C-rich environments. The absorption line at ~13.7 μm is real, and is due to C_2H_2.

be true also for the SWS spectrum, because the absorption is strong in the wavelength region $\sim 9.5 - 12.5$ μm.

A comparison of the CGS3 and SWS spectra of AFGL 5625 (Fig. 1), indicates that both show similar spectral shapes, but that there is significantly more flux in the CGS3 spectrum (obtained in 1993) than in the SWS spectrum. The IRAS LRS spectrum obtained in 1983 shows an even higher flux level, confirming the variability of the dust emission from this source in the mid-IR. The data presented in Figure 1 should not be interpreted as a single monotonic decline, however, since another CGS3 spectrum taken in 1997 (unpublished) shows a flux level intermediate between the 1983 and 1993 values. Fitting blackbodies to the observed spectra in Figure 1 gives very similar temperatures for all three epochs: 340 K for the LRS, 330 K for the 1993 CGS3 data, and 300 K for the SWS data.

Plotting the full SWS spectrum of AFGL 5625 in F_ν (Jy), rather than F_λ, emphasises the long-wavelength portions of the spectrum, as shown in Figure 2. In this plot, the spectrum is shown to be double-peaked in the infrared, due to a broad emission feature at 30 μm. We interpret this as being due to magnesium sulphide, MgS, which is present in the spectra of a number of carbon-rich proto-planetary nebulae [24–26].

3. ISO SPECTRA OF C-RICH SOURCES

The ISO Long Wavelength Spectrometer (LWS) Post Main-Sequence Guaranteed Time programme contains observations of many mass-losing evolved sources. Here we present results for some well-known carbon-rich objects.

The spectra in the LWS wavelength region (43 – 198 μm) are often dominated by the circumstellar dust emission in the Rayleigh-Jeans limit ($F_\lambda \propto \lambda^4$). Another important effect is the variation of grain emissivity with wavelength: for wavelengths greater than a critical wavelength, λ_c, grain emissivity varies as $\lambda^{-\beta}$. β is usually in the range $1-2$, with the higher values being associated with crystalline grains. Below the critical wavelength, the emissivity is roughly constant. The critical wavelength where the grain emissivity "turns over" is related to the grain size: $\lambda_c \approx 2\pi a$, where a is the grain radius.

A straightforward linear plot of F_λ *versus* wavelength would cover too much dynamic range in flux to be informative. In Figure 3, we have therefore multiplied the spectra by λ^4 to suppress the Rayleigh-Jeans (R-J) dependence. The four sources plotted are all evolved carbon-rich objects. IRC+10216 is the brightest carbon star in the infrared, while AFGL 618, AFGL 2688 (the "Cygnus Egg" nebula) and HD 44179 (the "Red Rectangle") are all C-rich post-AGB stars, evolving from the AGB to the planetary nebula phase. The central star of AFGL 618 has a temperature of around 32 000 K [27], while those of AFGL 2688 and HD 44179 are cooler, at 6500 K [28] and 7500 K [29] respectively.

It is interesting to compare the four spectra in Figure 3. One of the most noticeable properties of the spectrum of AFGL 618 is the set of strong emission lines increasing in both contrast and wavelength spacing toward long wavelengths. These lines are the pure-rotational sequence of CO: more than 20 CO lines are visible in the AFGL 618 spectrum. Fitting a modified blackbody, $B_\lambda(T) \times \lambda^{-\beta}$ to the spectrum, we obtain the best agreement with a temperature $T = 60 \pm 10$ K, and an emissivity index $\beta = 1.9 \pm 0.2$, suggesting that the maximum grain size is less than a few tens of μm. This simple function does not give a perfect fit to the continuum: broad "excess" emission remains around 60 and 110 μm, where the gradient of the spectrum changes most rapidly.

AFGL 2688 also shows the CO lines [28], but at somewhat lower contrast than does AFGL 618. Some broad features, which have yet to be identified [30] are discernible in between the CO lines. The best fitting modified blackbodies have $T = 85 \pm 10$ K and $\beta = 1.1 \pm 0.1$, slightly warmer than for AFGL 618. The smaller value of β is reflected in the more gentle decline of the spectrum longwards of $\sim$110 μm suggesting that the dust is amorphous, or that the grains are cooler. There is a steepening of the spectrum at $\sim$180 μm which could be due to the largest grains reaching the turnover wavelength. Alternatively, it could be indicative of emission from the coolest grains (in a temperature distribution) reaching its peak. A broad emission feature is clearly visible at 110 μm, presumably related to the inflection at the same wavelength in the GL 618 spectrum.

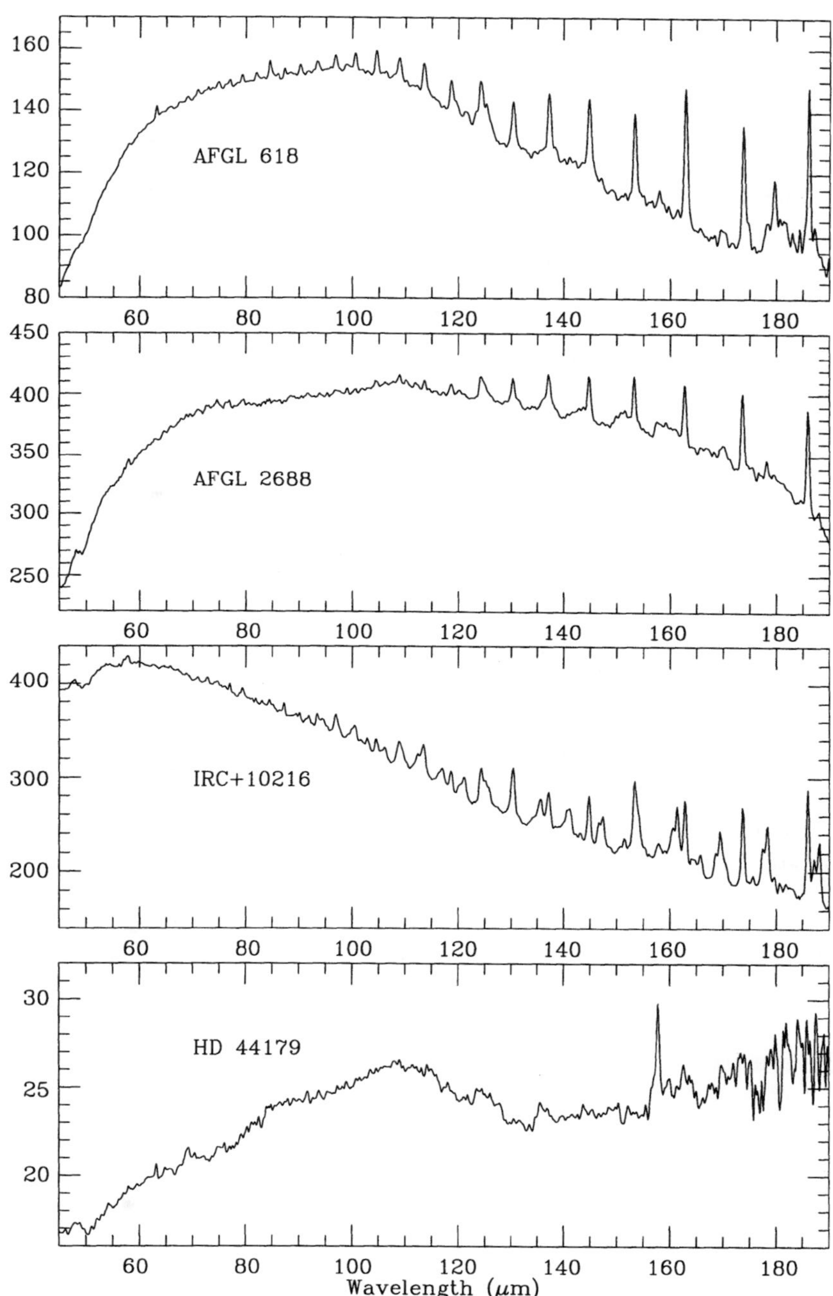

Fig. 3. — LWS spectra of C-rich sources with the Rayleigh-Jeans λ^{-4} wavelength dependence removed. Units are 10^{-15} Wm^{-2} μm^{-1} $\times$ $(\lambda/100\ \mu m)^4$.

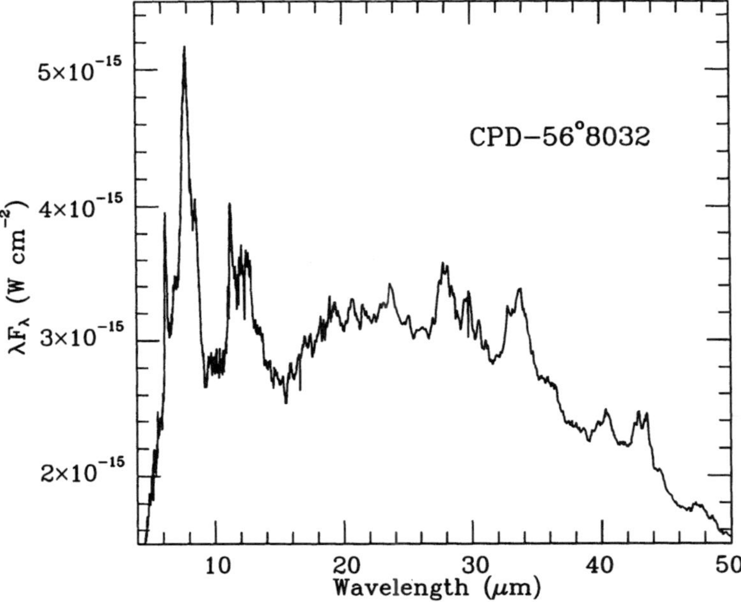

Fig. 4. — Composite SWS and LWS spectra of CPD-56°8032. Crystalline silicates and the UIR emission bands can clearly be seen.

The spectrum of IRC+10216 is richer in molecular lines than the previous two objects. The CO lines are again present, but a more complex set of lines dominates the spectrum. These are due to to rotational transitions of HCN molecules in various vibrational states [31]. The continuum has the shortest-wavelength maximum (in $\lambda^4 F_\lambda$) of any of the spectra in Figure 3. This is consistent with the modified blackbody fitting of Cernicharo *et al.* [31], which gives a warm temperature, 90 ± 10 K, combined with a steep emissivity index, 1.7 ± 0.1.

The spectrum of HD44179, the Red Rectangle, is very different from those of the other sources. No CO lines are visible, and the continuum falls more slowly than a R-J spectrum, causing a rising slope in Figure 3. This implies that the dust temperature is sufficiently cool to be outside the R-J domain, and that the emitting grains are of a size at least comparable with the wavelength. A very strong emission feature is seen, again peaking at around 110 μm, and possessing definite substructure. The SWS spectrum of this source is noteworthy [29] because it shows emission from oxygen-rich crystalline silicates, so the features seen here may well be due to O-rich materials.

In Figure 4, we present a portion of the combined SWS and LWS spectrum of the planetary nebula CPD-56°8032 (henceforth referred to as CPD), which is discussed in detail in [32]. CPD consists of a hydrogen-poor central star of spectral type WC10 [33] surrounded by a dusty nebula, which is very bright in the infrared. CPD is very carbon rich, showing strong UIR-band emission [34], and having a very high nebular C/O ratio of 13 for the gas phase [33]. The ISO spectrum (Fig. 4) confirms the presence of the strong UIR bands at 6.2, 7.7, 8.6 and 11.3 μm. However, at longer wavelengths, we observe features due to crystalline silicates at 24, 28, 30, 33, 36 and 41 μm, and crystalline H_2O ice at 43 μm. Both silicates and water ice are oxygen-rich condensates, which are not expected to form in a C-rich outflow.

Other recent ISO data [35] have revealed the presence of crystalline silicates in two more planetary nebulae with central stars of type WC: together with the CPD data, this suggests that the presence of O-rich dust around these highly C-rich objects is the rule, rather than the exception. The fact that silicate emission is only seen at long wavelengths implies that the O-rich dust is cool (temperatures of around $65-90$ K were derived for the silicate dust from fitting the CPD spectrum). This in turn suggests that the O-rich material lies further from the central star than does the C-rich dust, which could be explained in terms of a recent transition from an O-rich to a C-rich outflow following a late helium shell flash, or by a C-rich expanding nebula encountering an orbiting cloud (similar to the Kuiper belt) of O-rich comets [32, 35], see also the contribution by Waters in this volume.

In Figure 5, we present ISO spectra of the five sources discussed in this section, plus two others: MWC 922, which is a carbon-rich pre-planetary nebula, and He2-113 which is another planetary nebula with a WC-type central star. He2-113 shows silicate emission in its SWS spectrum [35].

The spectra have all been divided by a continuum, derived from a modified blackbody, or a simple spline fit, to enhance the contrast in any dust features that may be present.

The most spectacular dust features belong to MWC 922 and CPD. Both sources show strong, broad, emission features around 65 μm, which are due to clinopyroxene, a crystalline silicate. In the CPD spectrum, this feature is blended with 60 μm emission from crystalline water ice; it is less clear whether there is an ice contribution to the emission around 60 μm.

A strong 44 μm feature, showing some substructure, is seen in most of the spectra. It is strongest in MWC 922, and is probably due to crystalline ice. MWC 922, CPD and the Red Rectangle all show a distinct emission feature at 69 μm, which is due to forsterite, another crystalline silicate. A feature at 47 μm is seen in most of the spectra: this is presently unidentified. However, given that so many features of O-rich material are present in these spectra, the 47 μm emission could well be another crystalline silicate feature.

All of the sources in Figure 5 passed through a phase of oxygen-rich mass loss before they evolved into carbon-rich objects, hence the detection of O-rich material in the cool, outer parts of their circumstellar environments is not inconsistent with their present-day mass loss, as probed by shorter wavelength spectroscopy, which shows the contribution from carbon-rich material.

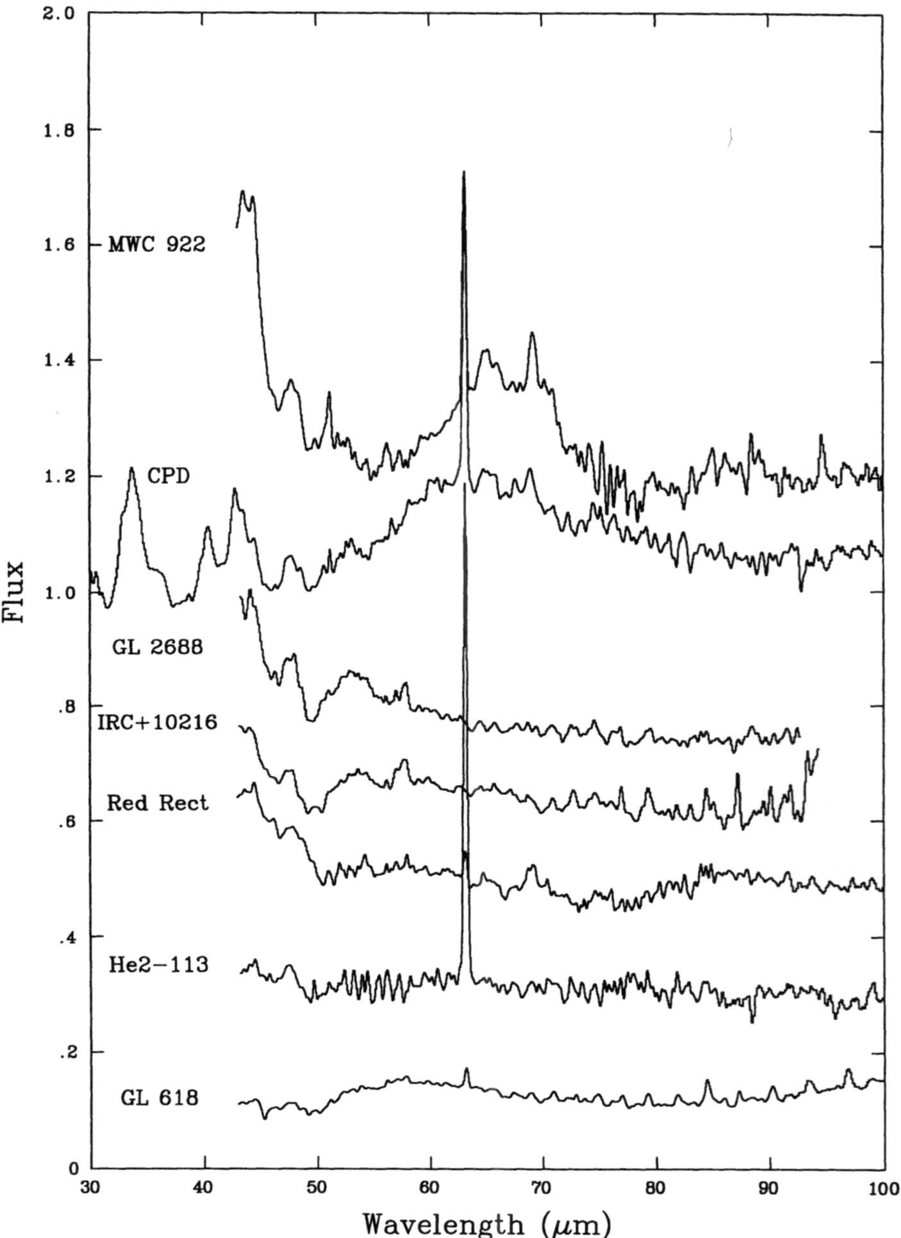

Fig. 5. — ISO LWS spectra of seven C-rich sources after division by a featureless continuum. The spectra have been shifted along the ordinate for clarity.

4. UIR BAND EMISSION FROM OXYGEN-RICH SUPERGIANTS

M supergiants are massive evolved stars with high luminosities ($L > 5\times10^4\ L_\odot$) and low surface temperatures ($T_{\rm eff} \sim 3000$ K). They undergo significant mass loss, which can be evident from blue-shifted optical absorption lines, sub-mm CO emission, and excess infrared emission due to circumstellar dust. Unlike the other stars discussed in this contribution, M supergiants are oxygen-rich. In consequence, the dust that forms around them should be composed only of oxygen-rich materials, *e.g.*, silicates and oxides.

The LRS spectra of two oxygen-rich M supergiants, MZ Cas and AD Per, were found [36] to show an emission peak at $\sim$11.5 μm, which was attributed to SiC. More recently, ground-based mid-IR spectra of both these stars, and of 14 other M supergiants in the h and χ Per association (see Fig. 6, [37]) demonstrated that MZ Cas, AD Per and five other sources showed emission features at 11.3 and 8.6 μm: the well-known UIR bands, attributed to carbonaceous species such as polycyclic aromatic hydrocarbons (PAH). These bands were superposed on broad silicate features, which peaked at around 10.5 μm, while many of the the stars without UIR bands showed strong silicate features peaking closer to 9.7 μm, a peak wavelength more usual for circumstellar silicates.

A possible explanation for the presence of the UIR bands was given in terms of non-equilibrium chemistry models [38], where UV photons from a warm chromosphere dissociate CO molecules, giving rise to a dynamic population of free C atoms. Carbonaceous PAH-like species can then form from these atoms. The UV radiation from the chromosphere also provides the energy to excite UIR-band emission.

Crystalline silicates also give a narrow emission feature near 11.3 μm, which would give a simpler explanation for the spectra of the M supergiants, if it were the only narrow emission band superposed on the normal (amorphous) silicate feature. However, as shown in Figure 7, the SWS spectrum of the supergiant AD Per shows the UIR bands at 6.2, 7.7 and 8.6 μm, as well as the 11.3 μm band. This demonstrates conclusively that emitting hydrocarbons *are* present around some oxygen-rich M supergiants. Measurement of the flux ratios between the individual bands should give information on the size and degree of hydrogenation of the carriers of the UIR bands in these sources.

A new ground-based survey of some 60 Galactic M supergiants [39] showed that the occurrence of UIR-band emission was an order of magnitude lower for field stars than for the stars h and χ Per. Emission in the UIR bands was unambiguously detected in only two new stars, BD+35 4077 and IRC+40427. The h and χ Per M supergiants are thus significantly different from their counterparts in the rest of the Galaxy, but the cause of the difference remains as yet unexplained.

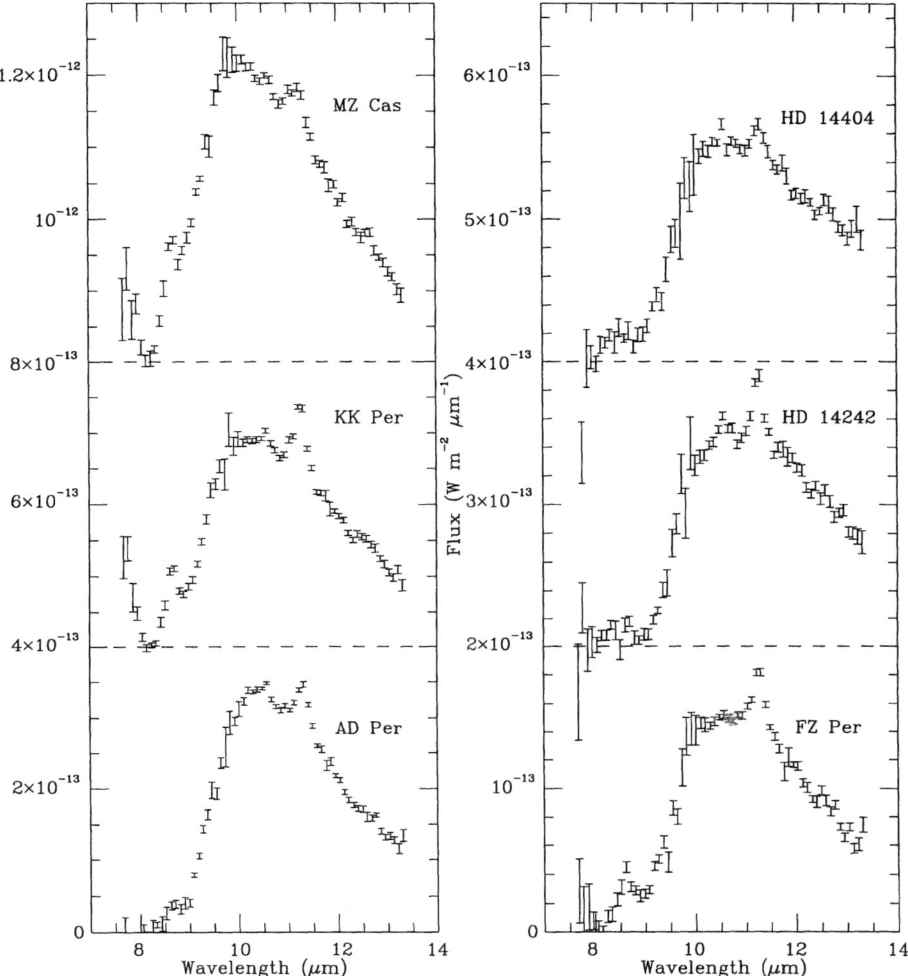

Fig. 6. — Ground-based CGS3 spectra of M supergiants taken from [37]. Blackbodies representing the stellar photospheres have been subtracted from the spectra, which have then been shifted in flux for clarity. The 11.3 μm UIR band is clearly visible in all six sources. The 8.6 μm band can be seen best in KK Per and FZ Per, where the contrast is highest.

ACKNOWLEDGMENTS

I wish to thank Dr. P.F. Roche for making available the SWS spectrum of AFGL 5625 and Angela Speck for useful discussions. This work is based on observations with ISO, an ESA project with instruments funded by ESA Member States (especially the PI countries: France, Germany, the Netherlands and the United Kingdom) with the participation of ISAS and

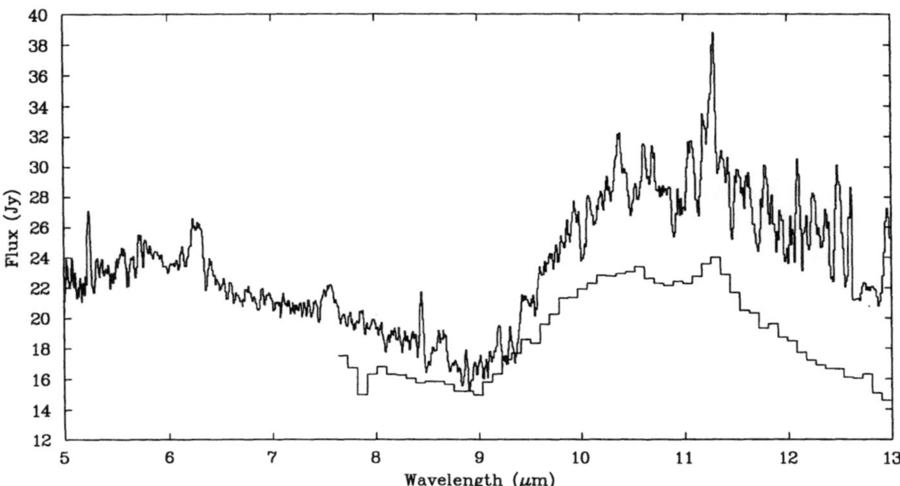

Fig. 7. — The SWS spectrum of AD Per. Narrow features ($FWHM \approx 0.1$ μm) at 6.2, 7.7, 8.6 and 11.3 μm are the UIR bands. The broad silicate feature peaking at $10 - 11$ μm can also be seen. The CGS3 spectrum (lower histogram) is plotted for comparison.

NASA. The United Kingdom Infrared Telescope is operated by the Joint Astronomy Centre on behalf of the U.K. Particle Physics and Astronomy Research Council.

REFERENCES

[1] Loreta E., *Astr. Nach.* **254** (1934) 151.
[2] O'Keefe J.A., *ApJ* **90** (1939) 254.
[3] Gillett F.C., Low F.J. and Stein W.A., *ApJ* **154** (1968) 677.
[4] Woolf N.J. and Ney E.P., *ApJ* **155** (1969) L181.
[5] Stein W.A., Gaustad J.E., Gillett F.C. and Knacke R., *ApJ* **155** (1969) L3.
[6] Friedemann C., *Physica* **41** (1969) 189.
[7] Gilman R.C., *ApJ* **155** (1969) L185.
[8] Bode M.F., Infrared Spectroscopy in Astronomy, edited by M.F. Bode (C.U.P., Cambridge, 1988) p. 317.
[9] Smith C.H., Aitken D.K. and Roche P.F., *MNRAS* **267** (1994) 225.
[10] Justtanont K. and Tielens A.G.G.M., *ApJ* **389** (1992) 400.
[11] Clayton G.C., Whitney B.A., Stanford S.A. and Drilling J.S., *ApJ* **397** (1992) 652.

[12] Woitke P., Goeres A. and Sedlmayr E., *A&A* **313** (1996) 217.
[13] Gehrz R.D., Jones T.J., Woodward C.E., *et al.*, *ApJ* **400** (1992) 671.
[14] Hackwell J.A., *A&A* **21** (1972) 239.
[15] Treffers R. and Cohen M., *ApJ* **188** (1974) 545.
[16] Ott U., *Nature* **364** (1993) 25.
[17] Bernatowicz T.J., From Stardust to Planetesimals, edited by Y.J. Pendleton and A.G.G.M. Tielens (ASP, San Francisco, 1997) p. 227.
[18] Speck A.K., Barlow M.J. and Skinner C.J., *MNRAS* **288** (1997) 431.
[19] Borghesi A., Bussoletti E., Colangeli L. and De Blasi C., *A&A* **153** (1985) 1.
[20] Papoular R., *A&A* **204** (1988) 138.
[21] Whittet D.C.B., Duley W.W. and Martin P.G., *MNRAS* **244** (1990) 427.
[22] Bernatowicz T.J. and Cowsik R., Astrophysical Implications of the Laboratory Study of Presolar Materials, edited by T.J. Bernatowicz and E. Zinner (AIP, Woodbury, 1997) p. 451.
[23] Jones B., Merrill K.M., Puetter R.C. and Willner S.P., *AJ* **83** (1978) 1437.
[24] Omont A., Astronomical Infrared Spectroscopy: Future Observational Directions, edited by S. Kwok (ASP, San Francisco, 1993) p. 87.
[25] Cox P., Astronomical Infrared Spectroscopy: Future Observational Directions, edited by S. Kwok (ASP, San Francisco, 1993) p. 163.
[26] Szczerba R., Omont A., Volk K., Cox P. and Kwok S., *A&A* **317** (1997) 859.
[27] Westbrook W.E., Willner S.P., Merrill K.M., *et al.*, *ApJ* **202** (1975) 407.
[28] Cox P., Gonzàlez-Alfonso E., Barlow M.J., *et al.*, *A&A* **315** (1996) L265.
[29] Waters L.B.F.M., Waelkens C., van Winckel H., *et al.*, *Nature* **391** (1998) 868.
[30] Barlow M.J., Dust and Molecules in Evolved Stars, edited by I. Cherchneff and T.J. Millar (Kluwer, Dordrecht, 1998) p. 15.
[31] Cernicharo J., Barlow M.J., Gonzàlez-Alfonso E., *et al.*, *A&A* **315** (1996) L201.
[32] Cohen M., Barlow M.J., Sylvester R.J., *et al.*, 1999, submitted.
[33] De Marco O., Barlow M.J. and Storey P.J., *MNRAS* **292** (1998) 86.
[34] Aitken D.K., Barlow M.J., Roche P.F. and Spenser P.M., *MNRAS* **192** (1980) 679.
[35] Waters L.B.F.M. , Beintema D.A., Zijlstra A.A., *et al.*, *A&A* **331** (1998) L61.
[36] Skinner C.J. and Whitmore B., *MNRAS* **235** (1988) 603.
[37] Sylvester R.J., Barlow M.J. and Skinner C.J., *MNRAS* **266** (1994) 640.
[38] Beck H.K.B., Gail H.-P., Henkel R. and Sedlmayr E., *A&A* **265** (1992) 626.
[39] Sylvester R.J., Skinner C.J. and Barlow M.J., *MNRAS* **301** (1998) 1083.

LECTURE 18

ISO Observations of Vega–Like Stars

C. Dominik*

Leiden Observatory, P.O. Box 9513,
2300 RA Leiden, The Netherlands

Abstract. We discuss the first results from ISO studies of Vega–like stars, which shows that the observed occurence of infrared excesses is in fact less than the earlier IRAS-based estimates of 30 – 50%. We also discuss new results on two cool discs around 55 Cnc and HD 207129.

1. INTRODUCTION

The discovery of infrared (IR) excess radiation associated with otherwise inauspicious main–sequence stars of all ages came as one of the big surprises of the IRAS mission [1–3]. Before this discovery it was assumed that the massive discs around T Tauri and Herbig AeBe stars would be entirely dissipated on a time scale of order 10^7 years, leaving insufficient dust in the system to produce such an excess. Even if material remained in planets and asteroids, the amount of dust produced by the collisional grinding of asteroids, in our own solar system, is so minor that it would be undetectable even from nearby stars with the instrumentation of IRAS and ISO. However, it is now clear that many stars preserve enough material in their disc over a time span of Gyrs that an easily observed IR excess is produced by these grains. In almost all systems studied so far, the grain lifetimes (determined mainly by mutually destructive collisions and by radiation forces) are much smaller than the age of the central star [3]. It is therefore now widely accepted that material is stored in larger bodies such as "asteroids" and "comets" [4]. These can easily survive for long times and *via*

* Presented on behalf of the HJHVEGA consortium.

collisional grinding and thermal evaporation and remain a long-lasting source of fresh dust which produces the observed infrared signatures. Therefore, these excesses are the first direct indicators that planetary material is present around stars other than our own sun. In the meantime, even if the presence of large planets in about 10 nearby systems is proven, studies of Vega–like stars will give very important and complementary insights into the nature of planetary discs.

Given these exciting results, a large number of ISO programs were designed to study these old debris discs in more detail. The ISO studies of Vega–like systems can be grouped into 3 categories.

1. "Classical" photometric searches for new candidates. These searches use the same method as the original detections: photometry at several wavelengths, in particular 60 μm. The different programs study interesting samples such as volume–limited (*e.g.* Becklin *et al.*, Habing *et al.*) or age sampling using clusters (*e.g.* Beckwith *et al.*).
2. Detailed photometric and mapping studies of bright objects where it may be possible to resolve the source with scans, multi–aperture photometry and mapping modes (*e.g.* Walker *et al.*, Backman and Stencil). First results of these observations include a map of αPsa [5].
3. Spectroscopic studies of the brightest sources. Unfortunately, SWS or LWS spectra are only feasible for Herbig AeBe stars which may well be precursors of β Pic like systems, but because of their youth form a class of their own. The fascinating ISO results on these stars are discussed elsewhere in this book. Even the disc of β Pictoris is just barely bright enough to allow for a spectral scan with SWS.

At the time of the Les Houches conference only a few of the above-mentioned proposals have produced published results, due to delays in both accurate calibration and the survey character of many programs. I will therefore in the following sections concentrate on our own program, the HJHVEGA volume limited sample.

2. THE HJHVEGA PROGRAM

The main goal of this program is to put the statistics of Vega–like discs associated with normal main sequence stars on much firmer statistical basis. In order to do this we selected a volume limited sample of spectral type A-K stars within 25 pc of the sun. In order to exclude secondary factors we excluded all close binary stars as well as variable stars. The selected 90 stars form a complete sample in the sense that we required detectability of the photosphere at 60 μm, equivalent to a minimum flux of 35 mJy. Thus we cover the whole range out to 25 pc only for A stars, but a smaller volume for later spectral types. We took photometric data with ISOPHOT, AOT P22 at 25, 60, and

Table I. — The frequency of Vega–like discs.

Spectral type	No. of stars total	with excess	Disc frequency
A0-A9	14	5	36%
F0-F9	28	6	21%
G0-G9	26	4	15%
K0-K9	22	3	14%
A0-K9	90	18	20%

160 μm. The observations at 25 μm where chopped. At 60 and 160 μm we used 3×3 raster mini maps with the C100 and C200 cameras, the recommended observation mode for faint sources.

2.1. Results of the general survey

Results obtained with the IRAS satellite indicated that a large fraction of main–sequence stars (between 30 and 50%) should have Vega–like discs, largely independent of the spectral type of the star [6, 7]. However, a recent re-evaluation of the IRAS data indicates that the frequency may be closer to 15% [8]. We have binned the stars in our sample into spectral class ranges and studied the occurrence of the IR excess. The results are summarized in Table I. We find an overall probability of about 20% for a star in our sample to have a Vega–like excess. A small trend is visible with spectral types: the disc frequency seems to decrease from A to K stars. Two effects are involved in this. First, discs of equal mass are much more easily detected around A stars than around K stars, because the IR emission depends upon the heating caused by the central star. Thus, for a given observational accuracy, we favor detection of discs around stars of early spectral type. Secondly, the trend observed in the spectral types can be interpreted as a trend with age: A and F stars in a volume–limited sample are on average much younger than the G and K stars. If the disc material is derived by collisional grinding from a population of planetesimals, it is quite possible that the reservoir of planetesimals diminishes with time. Recently, studies of the Kuiper Belt of the solar system indicate that the early Kuiper Belt was much more massive than it is today, resulting in a higher dust production rate [9, 10]. We are preparing a study of the IR excess vs. stellar ages in our sample.

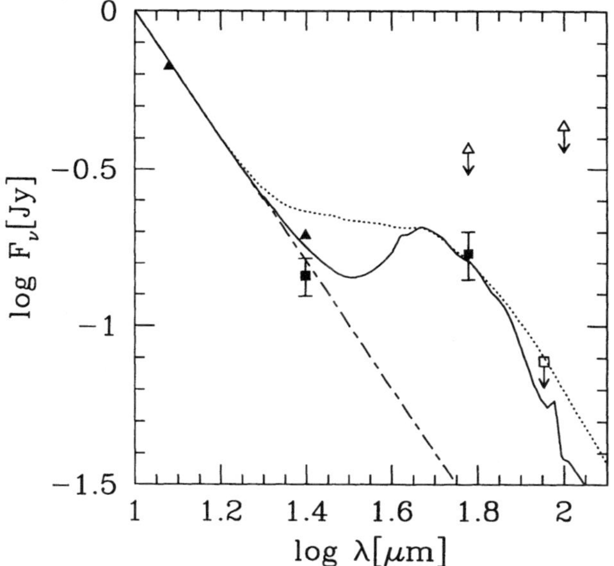

Fig. 1.— The measured fluxes from the star 55 Cnc. Triangles are IRAS observations, squares are from ISO. The solid line indicates the fit model described in the text. The broken line is an alternative model using ice–free dust.

2.2. 55 Cnc: Disc and planet?

It was realized very early on that the IR excess observed from main sequence stars is probably due to debris produced by a population of planetesimals in these systems. One of the important questions is therefore: does the presence of a large excess indicate the failure of a system to produce large planets which would presumably clear the disc of planetesimals? Or can an IR excess and planets coexist? One possibility is that the disc and planets occupy different regions. The dust seen is usually cold, about 100 K or less. Therefore it is located at relatively large distances from the star, of the order of 40 AU (in our solar system this would be outside the orbit of Neptune). There are several indications that some of the Vega–like stars might in fact have planets in their inner regions.

1. It can be shown that the space near the stars must be relatively free of dust. Dust near the star would be hot and produce an excess at near IR wavelengths, which is not seen [2]. Therefore, at least the inner regions of these discs have apparently been cleared.
2. The warp observed in the disc of β Pic [11] is presumably caused by a planet orbiting in an inclined orbit near the inner boundary of the disc [12].

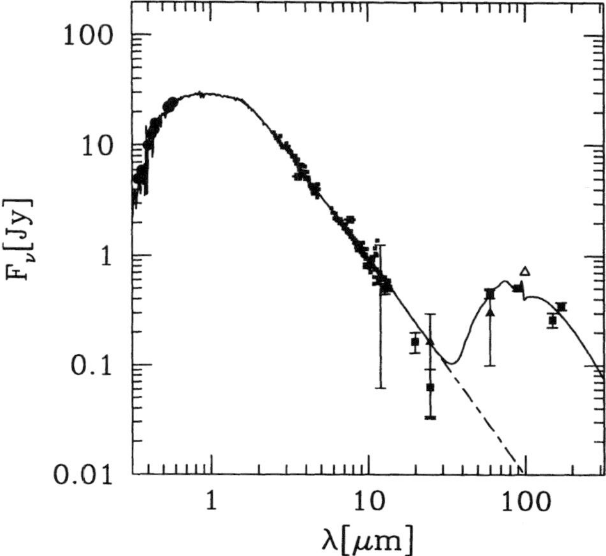

Fig. 2.— Observed fluxes from HD 207129. Circles indicate optical colors (Geneva), triangles are from IRAS, squares from ISO. The broken lines indicates the photospheric flux from the star. The solid line is the flux from star plus model disc.

3. Also, the asymmetry observed in the occurrence of variable circumstellar absorption features in the spectrum of β Pic and other stars [13] is interpreted as the influence of one or two planets [14].

In all 3 cases described above, the presence of planets is derived indirectly from the properties of the disc. However, in recent years we have witnessed the detection of the first extra–solar planets. One of the stars shown to host a Jupiter–sized companion in a short period orbit is 55 Cnc [15]. The planet is orbiting the star at a distance of 0.11 AU, its inferred mass is $M \sin i = 0.84\ M_{\rm Jup}$. We have observed 55 Cnc with ISO and detected a 60 μm excess (see Fig. 1) which can be interpreted as the presence of a disc of $4 \times 10^{-5}\ M_\oplus$ of small dust grains, orbiting the star at a distance of approximately 60 AU [16]. Thus the disc and the planet are clearly spatially separated.

2.3. HD 207129

The star HD 207129 was already identified in IRAS studies to show a Vega–like excess [17]. It was part of our sample. During the observations it was noticed that this star shows a very high far IR excess, pointing to a low temperature disc at approximately 50 K. This makes the disc the coldest studied so far. We show the measurements in Figure 2. The emission seen can be fitted with a disc containing $5 \times 10^{-8}\ M_\odot$ of dust grains. The dust disc has an inner hole of $100 - 300$ AU (depending upon the dust properties used) which is necessary

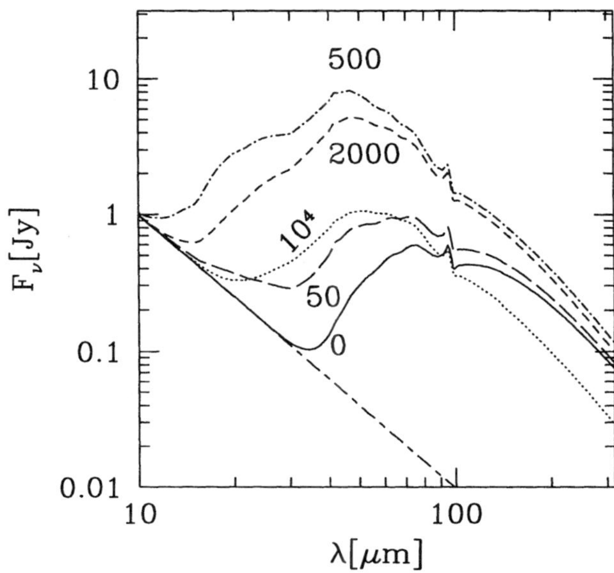

Fig. 3. — Time evolution of the IR excess from HD 207129 if only radiation forces acting on the dust grains are considered. The model starts from the model disc for todays excess. Times in Myrs are noted at the curves. Note how the effects of Poynting Robertson drag produces a strong near IR excess within approximately 100 Myrs.

to explain the lack of excess emission short of 40 μm. It is interesting to study the consequences of this result. Such an inner hole needs not only to be created in the first place, but it also needs to be maintained. To see this we have to consider the processes which remove dust grains from a disc. The main processes considered to be important for Vega–like discs are collisions, radiation forces and sublimation (see [3] for a review). Collisions in Vega–like systems are mostly destructive – *i.e.*, when two dust grains collide, the collision will result in breakup into smaller grains rather than aggregation. In this way the steady supply of small grains is kept up. Now, in the more luminous A and F stars, the radiation from the star is able to quickly expel the small grains from the system [18]. However, this is different for low luminosity stars like our sun [19] or HD 207129. Here, only a very small size range of particularly well absorbing grains can be removed from the disc by radiation pressure.

If radiation pressure fails to remove the particles, they stay in the disc and are pulled slowly into the star by the so–called Poynting-Robertson effect. For micron–sized dust located 100 AU away from a G2V star, this time scale is of the order of 10^7 yrs. Thus, if the disc has an inner hole to begin with, it will slowly be filled with dust on this timescale. This is illustrated by the results of a model calculation shown in Figure 3. In this calculation we have taken the disc model which fits the infrared excess observed today and calculated its dynamical evolution under the influences of radiation forces. As we can see, within 50 Myrs a large excess around 30 μm evolves, and this excess peaks after 500 Myrs. The subsequent decline of the excess is due to the drainage of dust from the outer disc, which is in this model not replenished by collisions.

The age of HD 207129 can be estimated at approximately 4 Gyrs from its distance and its position in the Hertzsprung-Russell diagram, *via* the use of theoretical isochrones. Thus, even if the disc had a 100 AU hole to begin with, the hole would in the time be filled with small dust grains and show a mid IR excess near 30 μm. We therefore have to conclude that an additional mechanism is required to keep this inner hole clean. Although the sweeping mechanism is still unclear, a possible explanation might be the presence of a planet near the inner edge of the disc.

3. CONCLUSIONS

The detection of Vega–like discs is still as difficult with ISO as it was with IRAS. The main improvement which ISO brings is a higher sensitivity which allows the detection of photospheric fluxes from more stars than was possible with IRAS. However, the difficulty remains that the detection of Vega–like discs requires absolute photometry with high accuracy in order to detect faint excesses only a little over and above the photospheric flux.

The new surveys indicate that ISO finds such excesses in 20% or less of the stars in the solar neighborhood – clearly less than the old studies with IRAS had suggested. This lower probability is in fact consistent with a new study of the IRAS catalogue using appropriate statistical tools.

The observations of 55 Cancri have shown that planets and Vega–like discs can coexist.

A second important advantage of ISO is that it can provide accurate photometry out to 200 μm, moving cooler discs into focus. The disc around HD 207129 is such a cold disc. The model fit of this excess requires a considerable inner hole in the disc. Since G type stars are unable to blow out their dust grains by radiation pressure (except for maybe a very small size range of particularly good absorbing grains), Poynting-Robertson drag would fill this hole on a time scale of 50 Myrs, giving rise to considerable mid IR excess which is not observed. Thus such systems highlight the need for other mechanisms to clear the inner discs. Planets may provide such a process - but more detailed studies are required.

REFERENCES

[1] Aumann H.H., Gillett F.C., Beichmann C.A., *et al.*, *ApJ* **278** (1984) L23.

[2] Gillett F.C., Light on Dark Matter, edited by F.P. Israel (Reidel, Dortrecht, 1986) p. 61.

[3] Backman D.E. and Paresce F., Protostars and Planets III, edited by E.H. Levy and J.I. Lunine (University of Arizona Press, Tuscon, 1993) p. 1253.
[4] Weissman P.R., *Science* **224** (1984) 987.
[5] Fajardo-Acosta S.B., Stencel R.E. and Backman D.E., *ApJ* **487** (1997) L151.
[6] Aumann H.H., *PASP* **97** (1985) 885.
[7] Aumann H.H., *ApJ* **96** (1988) 1414.
[8] Plets H., A systematic study of the occurrence of circumstellar dust around main-sequence stars and giants, Ph.D. thesis (University Leuven, Leuven, 1997) p. 1.
[9] Stern S.A. and Colwell J.E., *AJ* **114** (1997) 841.
[10] Stern S.A. and Colwell J.E., *ApJ* **490** (1997) 879.
[11] Burrows C.J., Krist J.E., Stapelfeldt K.R. and Team W.I.D., *Bull. Am. Astron. Soc.* **187** (1995) 3205.
[12] Mouillet D., Larwood J.D., Papaloizou J.C.B. and Lagrange A.M., *MNRAS* **292** (1997) 896.
[13] Beust H., Vidal-Madjar A., Ferlet R., Lagrange-Henri and A.M., *A&A* **241** (1991) 488.
[14] Levison H.F., Duncan M.J. and Wetherill G.W., *Nature* **372** (1994) 441.
[15] Butler R.P., Marcy G.W., Williams E., Hauser H. and Shirts P., *ApJ* **474** (1997) L115.
[16] Dominik C, Laureijs R.L., Jourdain de Muizon M. and Habing H.J., *A&A* **329** (1998) L53.
[17] Walker H.J. and Wolstencroft R.D., *PASP* **100** (1988) 1509.
[18] Artymowicz P., *ApJ* **355** (1988) L79.
[19] Burns J.A., Lamy P.L. and Soter S., *Icarus* **40** (1979) 1.

LECTURE 19

Dust Formation in Supernovae

P.O. Lagage[1], T. Douvion[1], J. Ballet[1], F. Boulanger[2], C.J. Cesarsky[1], D. Cesarsky[2] and A. Claret[1]

[1] *DSM/DAPNIA/Service d'Astrophysique, CEA/Saclay, 91191 Gif-sur-Yvette, France*
[2] *Institut d'Astrophysique Spatiale, Université d'Orsay, 91405 Orsay, France*

Abstract. In this paper we discuss the evidence for dust formation in supernovae as deduced from ISOCAM spectro-imaging observations.

1. INTRODUCTION

Dust grains constitute a major component of the Universe. In terms of the mass fraction, about half of the elements heavier than helium are in the form of dust in the interstellar medium. Dust plays a crucial role in many astrophysical processes, such as the formation of stars and planets, the catalytic formation of molecular hydrogen on grain surfaces, and the extinction of starlight.

Our current understanding of dust formation indicates that grains and grain cores are formed in circumstellar shells and are then ejected into the interstellar medium (ISM), where they are subject to major changes (*e.g.* [1] and references therein). On the one hand, grains can accrete ice and organic mantles when passing through interstellar clouds [2], but on the other hand, grains are destroyed when swept up by supernova (SN) shocks propagating into the ISM (*e.g.* [3] and references therein).

Evolved stars on the Asymptotic Giant Branch (AGB) are generally considered to be the main dust grain factories. This dust is known to form efficiently in the stellar atmosphere of these objects and to be blown out into the ISM by the radiation pressure from the central star (*e.g.* [4] and references therein). However, given the mass of condensible elements ejected during a supernova explosion, supernovae have the *potential* to be the major dust factories, especially in the formation of silicate dust (*e.g.* [5,6]).

The idea that *part* of the dust was made from SN material was introduced 30 years ago [7,8] and is now well supported by laboratory studies of meteorites. Indeed, SN are the only place where the nucleosynthesis can generate some of the isotopic anomalies observed in meteorites (*e.g.* [9] and references therein). The astrophysical evidence in favour of dust formation in SN ejecta is still scarce. Several strong indirect pieces of evidence have been deduced from the observations of SN1987a, the supernova which exploded in the Large Magellanic Cloud eleven years ago [10–14]. Additional evidence can be gathered by mapping the infrared (IR) emission from young supernova remnants (SNR). The four youngest SNR in our galaxy (Cassiopeia A, Kepler, Tycho and the Crab) have been observed with ISOCAM [15].

In this paper, we discuss only the results obtained from the observations of Cassiopeia A (Cas-A), the youngest known supernova remnant of our galaxy. Cas A exploded about 343 years ago at a distance of 3.4 kpc from us [16]. Section 2 deals with the first IR searches for emission from Cas-A and with the first ISO results. In Section 3, we present new ISOCAM results, which confirm the first claim that dust did form from material ejected during the explosion of Cas-A. Conclusions and perspectives are drawn in Section 4.

2. DUST IN CASSIOPEIA A

2.1. Dust detection with IRAS

The first attempts to detect dust in Cas A made using the Kuiper Airborne Observatory [17] and ground-based telescopes [18] failed. It is only with IRAS, the first satellite devoted to IR observations [19] that IR emission from Cas A was detected [20]. From the observed fluxes in the four IRAS bands at 12, 25, 60 and 100 μm, a dust temperature around 90 K was deduced.

However, due to the limited angular resolution of the IRAS observations, it was not possible to assess whether the emission originates from interstellar dust heated by the hot gas associated with the forward shock or from SN (or circumstellar) condensates heated by the hot gas associated with the reverse shock.

2.2. Dust origin as seen with ISO

With an angular resolution of 6 arcsec (compared to 30 arcsec for IRAS), ISOCAM, the camera on board the Infrared Space Observatory (ISO) [21], was well suited for follow-up studies of the IR emission from Cas A.

Initial observations revealed that the IR emission at the angular resolution of ISOCAM has a very patchy spatial structure comprised of knots ([22]; see

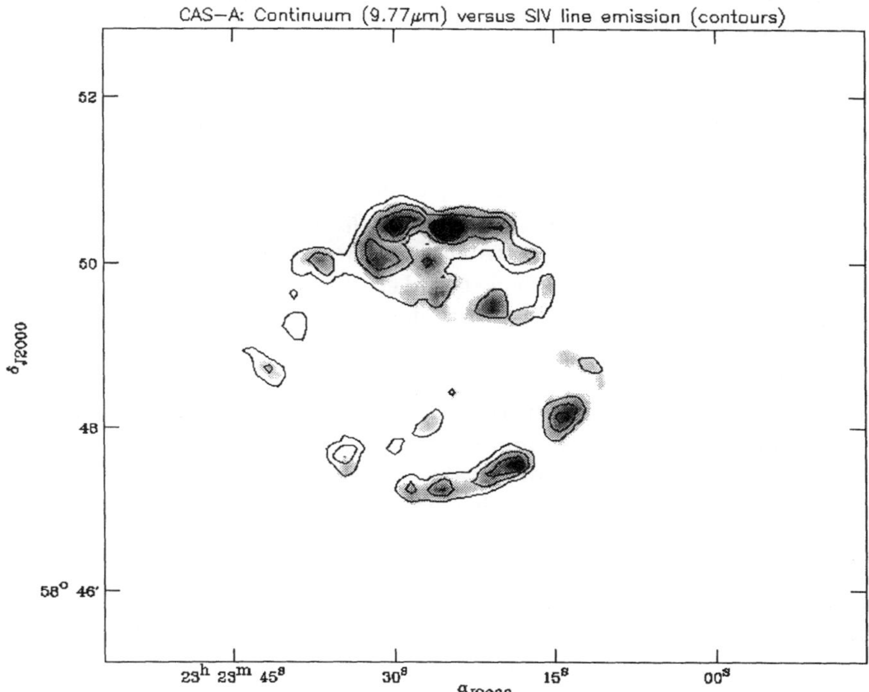

Fig. 1. — Cas-A as observed with ISOCAM in the narrow band circular variable filter (CVF) centred at 9.77 μm (continuum emission). The Pixel Field of View is 6 arcsec and the total field of view is 3×3 arcmin^2. A raster map, (5×4 frames with an overlap of half a frame, and a frame integration time of one minute), was made to map the whole remnant (4×4 arcmin^2). Overplotted on the image are the contours of the CVF image centred on the SIV ionic line at 10.5 μm. The flux is divided by 1.5 between each contour level.

also Fig. 1). The observations shown in Figure 1 were made with the narrow band circular variable filter (CVF) centred at 9.77 μm; the spectral resolution ($\lambda/\Delta\lambda$) was $\sim$50. At this wavelength, we do not expect line emission, but only continuum emission. This continuum emission is attributed to dust thermal emission, free-free emission and synchrotron emission being negligible at these wavelengths.

The dusty infrared knots are spatially well correlated with the so-called fast moving knots (FMK's) observed in the visible radiation. These gaseous knots, discovered more than 50 years ago, are known to consist of SN nuclear burning products [23–26]. While oxygen, sulfur and argon ionic lines are observed in abundance from these knots, no hydrogen or helium lines are detected. The detected lines are blue-shifted or red-shifted with velocities of several thousands of kilometres per second (*e.g.* [16]). The knots have a proper motion on the sky up to 1 arcsec per year.

The gaseous component of the knots was also observed with ISOCAM. Indeed various forbidden ionic lines of heavy elements, such as the ArII line at 7 μm, the ArIII line at 9 μm, the SIV line at 10.5 μm, the NeII line at 12.8 μm and the NeIII line at 15.5 μm, are observed within the wavelength coverage of ISOCAM ($3 - 17$ μm). The good spatial correlation between the gaseous knots consisting of SN ejecta and the dusty knots, both detected in the IR (Fig. 1) lead us to claim that the dust in the knots is made of ejecta material.

The temperature of the knots was found to be around 200 K, but how can the grains be heated to such temperatures? Dust is usually heated by the absorption of photons from an external source, generally a star. In the case of Cas A the main source of the dust heating is probably not photons but electrons. Indeed, the evaporating flow surrounding the knots consists of a hot plasma, in which the dust can be efficiently heated by collisions with the electrons of the plasma [27]. Given the expected temperature of the plasma (about $T_e \sim 5 \times 10^6$ K), the dust grains should have a size of $\sim$ 50 Å to be heated to 200 K [28]. This is the size expected from dust formation models developed within the framework of the supernova SN1987A observations [29]. Note that, according to the models, the dust formed soon (one year) after the explosion.

The mass of the brightest knot deduced from the IR is of the order of a few 10^{-7} solar masses. However, it should be stressed that most of the dust should be inside the knots and cold. This cold dust is not seen by ISOCAM. Only the dust in the evaporating layer is heated sufficiently to emit significantly in the wavelength range of ISOCAM (the thermal emission of a blackbody at a temperature between 300 K and 30 K has its peak of emission at a wavelength between 10 μm and 100 μm).

Part of the dust (the northernmost region in Fig. 1 of [22]) is not spatially correlated with FMK's. This dust is at a lower temperature ($\sim$100 K) than the dust associated with the knots. It is probably interstellar or circumstellar dust that has been heated by the blast wave resulting from the supernova explosion.

3. FURTHER ISOCAM EVIDENCE IN FAVOUR OF DUST FORMATION IN CAS-A

Three key arguments were used to claim that we have observed dust made from supernova ejecta material:

- the emission observed at 9.8 μm is continuum emission from dust;

- the dust is spatially well correlated with gaseous knots, detected through the ionic line emission from SIV at 10.5 μm;
- the gaseous knots seen in the IR are spatially well correlated with the gaseous knots seen in the visible. The gaseous knots seen in the visible are known to be made from SN material.

Additional ISOCAM observations were made to reinforce the three points mentioned above.

3.1. Full CVF spectrum

In order to better disentangle the continuum and line emission from the knots in Cas-A, we performed a complete CVF scan covering the $5-16.5$ μm wavelength range; the spatial resolution is 6 arcsec and the field of view is 3 arcmin, centred on the northern part of the remnant. We obtained 1024 spectra, which display significant differences, probably indicating that we are probing different layers of SN material [30]. The spectrum of the brightest knot is shown in Figure 2.

Line emission from ArII, ArIII, SIV, NeII and NeIII, and continuum emission are clearly present in the spectrum. The continuum can be fitted by a blackbody at a temperature of 220 K. The bump around 9.7 μm may indicate a silicate component to the dust.

Thus, these observations confirm the presence of dusty knots in Cas-A.

3.2. Spatial correlation between dusty and gaseous knots at the 1.5″ scale

The first ISOCAM observations were made with a Pixel Field of View (PFOV) of 6 arcsec in order to benefit from a large field (3×3 arcmin^2). In order to better assess the spatial correlation between continuum emission and line emission, follow-up observations were performed with the smallest ISOCAM PFOV: 1.5 arcsec. Observations of the gaseous ionic emission line were performed at 6.97 μm (ArII) and the continuum emission was taken at 9.77 μm. The results are shown in Figure 3.

As can be seen, the spatial correlation between line and continuum emission remains very good at this smaller scale. The signal to noise ratio is larger in the image of the line emission than in the image of the continuum emission. This can explain why some gaseous knots have no continuum counterpart. However, it is puzzling that some dusty knots do not seem to have a gaseous counterpart.

3.3. Observation of line doppler shifts

In order to ensure that we are observing fast moving knots, we have looked for doppler shifts of the ionic lines observed with ISO. The spectral resolution of ISOCAM CVF observations is low, but sufficient to observe the fastest knots, which can have velocities of up to 10 000 km s^{-1}.

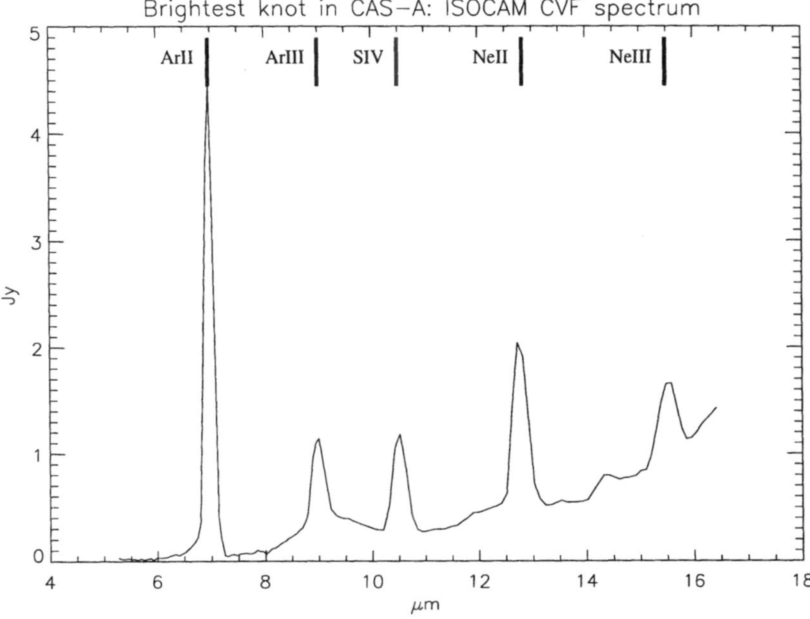

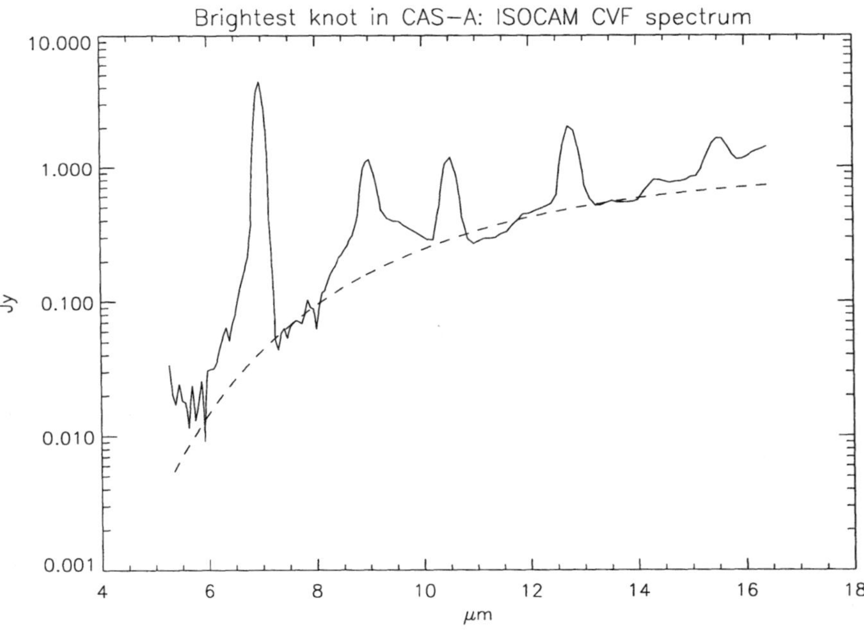

Fig. 2. — Flux of the brightest IR knots in Cas-A as a function of wavelength, at the spectral resolution of ISOCAM CVF observations ($\lambda/\Delta\lambda$ ~50). The flux has been integrated over a 3×3 pixel box (18×18 arcsec^2). Top curve: linear scale for the flux; Bottom curve: logarithmic scale for the flux in order to better show the underlying continuum emission. The dashed line represents the emission of a blackbody at a temperature of 220 K.

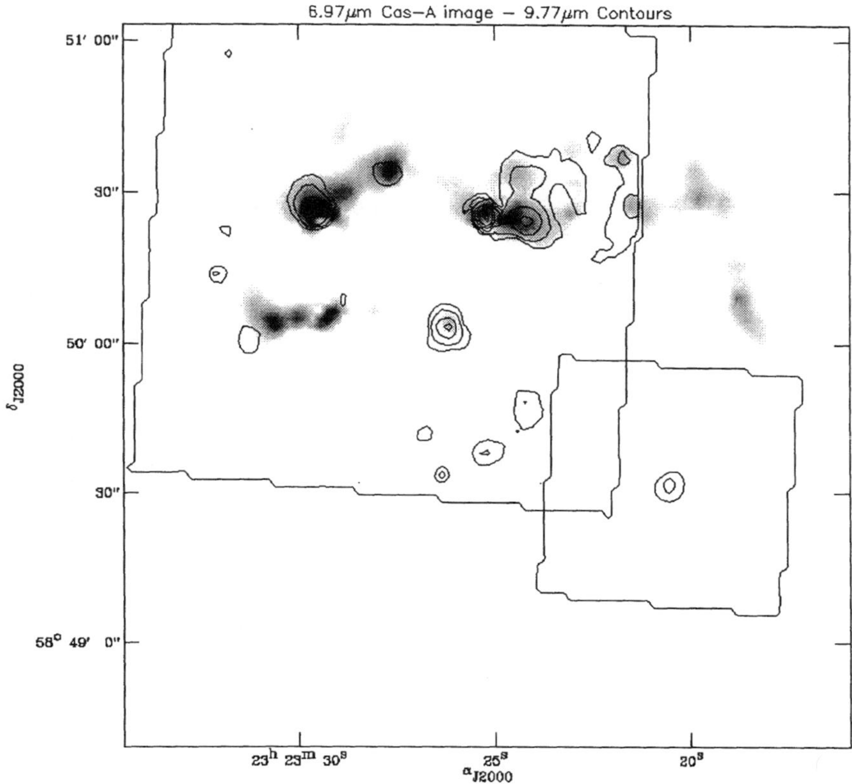

Fig. 3. — ISOCAM image of the northern part of the Cas-A SNR obtained with the CVF centred at 6.97 μm (ArII line) and a PFOV of 1.5 arcsec. Overplotted on the ArII image are the contours of two images obtained with the CVF centred at 9.77 μm (continuum emission) and a PFOV of 1.5 arcsec. The flux is divided by 2 between each of the contours. All the images were de-convolved using the point spread function algorithm developed by Pantin and Starck [31].

The doppler shift was indeed observed. This is illustrated in Figure 4 which is equivalent to Figure 3, but is shown at a wavelength red-shifted by 0.17 μm (6 000 km s^{-1}). New filamentary structures are detected. Note that the puzzling dusty knots which appear to be spatially uncorrelated with gaseous knots in Figure 3 now appear to correlate with the fast moving knots of the filaments in (Fig. 4).

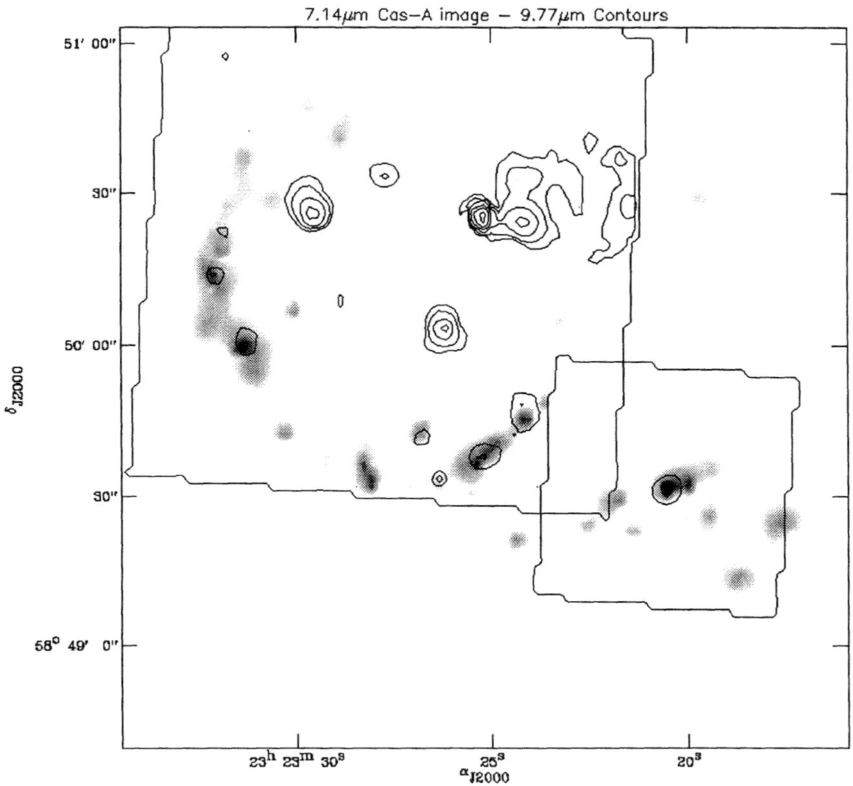

Fig. 4. — ISOCAM image of the northern part of the Cas-A SNR obtained with the CVF centred at 7.14 μm (red-shifted ArII line) and a PFOV of 1.5 arcsec. The image was obtained in the same way as for the image at 6.96 μm of Figure 3. Overplotted are the same contours as in Figure 3.

4. CONCLUSIONS AND PERSPECTIVES

As shown in this paper, ISOCAM observations of the Cas-A SNR have renewed the interest in supernovae as dust factories and triggered follow-up observations in the mm and sub-mm range. Such observations will allow the detection of the cold dust probably embedded in the dusty knots seen by ISOCAM. It will therefore be possible in the future to derive the efficiency of dust formation in SN.

Other results are expected to come from the data that we currently have at hand: SWS spectra of Cas-A, ISOCAM spectro imaging observations of the other young galactic supernova remnants (the Crab, Kepler).

REFERENCES

[1] Tielens A.G.G.M., *ApJ* **499** (1998) 267.

[2] Greenberg J.M., Comets, edited by L. Wilkening (Univ. Arizona Press, Tucson, 1982) p. 131.

[3] Jones A.P., Tielens A.G.G.M. and Hollenbach D.J., *ApJ* **469** (1996) 740.

[4] Habing H.J., *ARA&A* **7** (1996) 97.

[5] Jones A.P. and Tielens A.G.G.M., The cold Universe, edited by Th. Montmerle, Ch. Lada, I.F. Mirabel and J. Tran Thanh Van (Éditions Frontières, Gif-sur-Yvette, 1993) p. 35.

[6] Dwek E., *ApJ* **501** (1998) 643.

[7] Cernuchi F., Marsicano F. and Codina S., *Ann. d'Ap.* **30** (1967) 1039.

[8] Hoyle F. and Wickramasinghe N.C., *Nature* **226** (1970) 62.

[9] Zinner E., Cosmic Abundances, ASP Conf. Series 99, edited by S.S. Holt and G. Sonneborn (ASP, San Fransisco, 1996) p. 147.

[10] Moseley S.H., Dwek E., Glaccum W., Graham J.R. and Loewenstein R.F., *Nature* **340** (1989) 697.

[11] Suntzeff N.B. and Bouchet P., *AJ* **102** (1990) 650.

[12] Bouchet P., Danziger I.J. and Lucy L.B., *AJ* **102** (1991) 1135.

[13] Wooden D.H., Rank D.M., Bregman J.D., *et al.*, *ApJS* **88** (1993) 477.

[14] Colgan S.W.J., Haas M.R., Erickson E.F., Lord S.D. and Hollenbach D.J., *ApJ* **427** (1994) 874.

[15] Cesarsky C.J., Albergel A., Agnese P., *et al.*, *A&A* **315** (1996) L32.

[16] Reed J.E., Hester J.J., Fabian A.C. and Winkler P.F., *ApJ* **440** (1995) 706.

[17] Wright E.L., Harper D.A., Loewenstein R.F., Keene J. and Whitcomb S.E., *ApJ* **240** (1980) L157.

[18] Dinerstein H.L., Werner M.W., Capps R.W. and Dwek E., *ApJ* **255** (1982) 552.

[19] Neugebauer G., Habing H.J., Van Duinen R., *et al.*, *ApJ* **278** (1984) L1.

[20] Dwek E., Hauser M.G., Dinerstein H.L., Gillett F.C. and Rice W.L., *ApJ* **315** (1987) 571.

[21] Kessler M.F., Steinz J.A., Anderegg M.E., *et al.*, *A&A* **315** (1996) L27.

[22] Lagage P.O., Claret A., Ballet J., *et al.*, *A&A* **315** (1996) L273.

[23] Baade W.D. and Minkowski R., *ApJ* **119** (1954) 206.

[24] van den Bergh S., *ApJ* **219** (1971) 931.

[25] Kirshner R.P. and Chevalier R.A., *ApJ* **218** (1977) 142.

[26] Chevalier R.A. and Kirshner R.P., *ApJ* **219** (1978) 931.
[27] Dwek E. and Werner M.W., *ApJ* **248** (1981) 138.
[28] Dwek E., *ApJ* **302** (1986) 363.
[29] Kozasa T., Hasegawa H. and Nomoto K., *A&A* **249** (1991) 474.
[30] Douvion T., *et al.* (1998) in preparation.
[31] Pantin E. and Starck J.L., *A&AS* **118** (1996) 575.

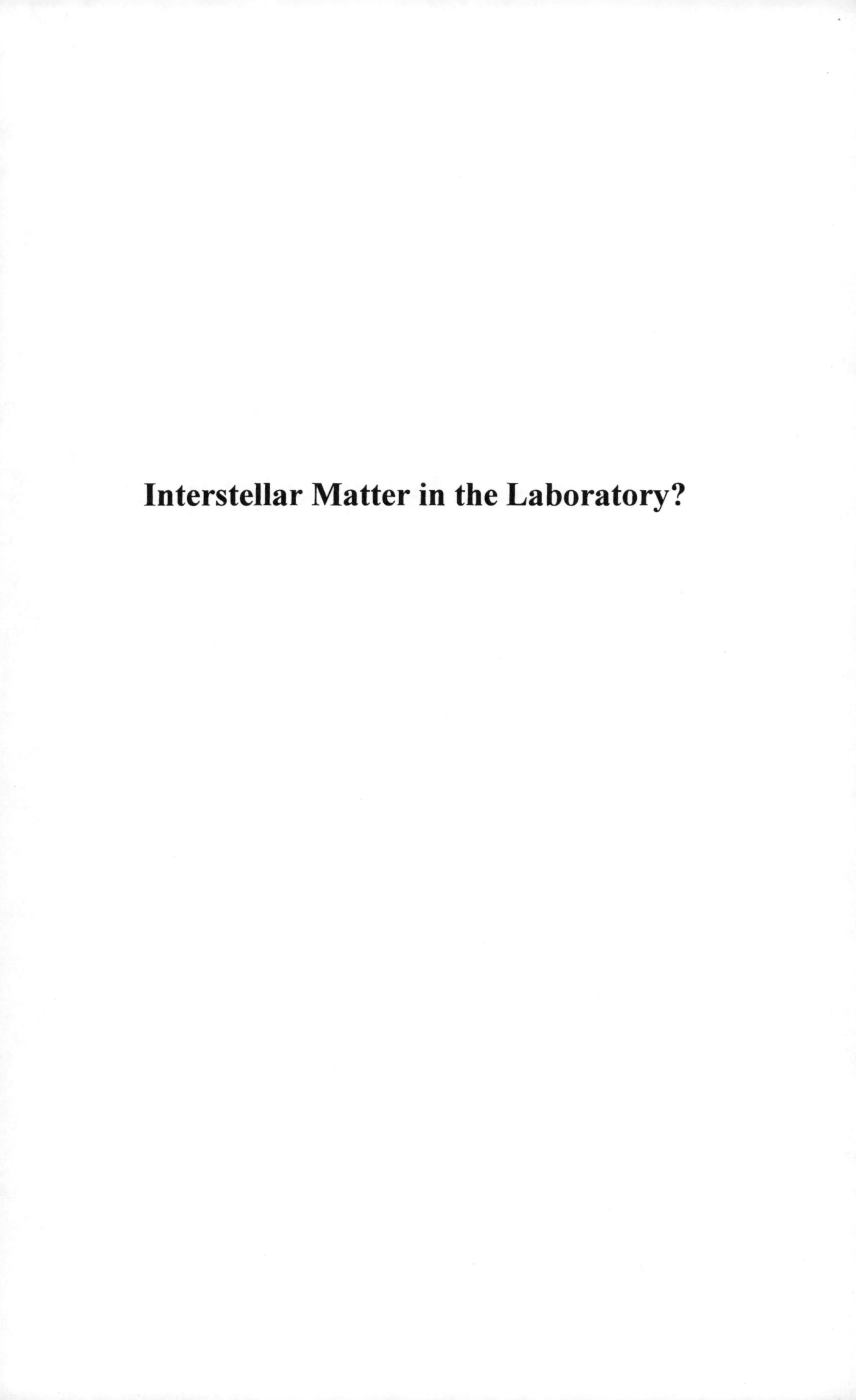

Interstellar Matter in the Laboratory?

LECTURE 20

Mg-Rich Olivine and Pyroxene Grains in Primitive Meteoritic Materials: Comparison with Crystalline Silicate Data from ISO

J.P. Bradley[1,2], T.P. Snow[3], D.E. Brownlee[4] and M.S. Hanner[5]

[1] *MVA Inc., 5500/200 Oakbrook Parkway, Norcross, GA 30093, U.S.A.*
[2] *School of Materials Science and Engineering, Georgia Institute of Technology, Atlanta, GA 30332-0245, U.S.A.*
[3] *Center for Astrophysics and Space Astronomy, University of Colorado, Boulder, CO 80309-0389, U.S.A.*
[4] *Department of Astronomy, University of Washington, Seattle, WA 98195, U.S.A.*
[5] *Jet Propulsion Laboratory, MS 183-501, California Institute of Technology, Pasadena, CA 91109, U.S.A.*

Abstract. The detection of crystalline olivine and pyroxene grains around young and old stars by the Infrared Space Observatory (ISO) provides an important clue in the search for presolar silicates in meteoritic materials. Although a variety of isotopically anomalous presolar circumstellar grains have been found in meteorites, not a single isotopically anomalous presolar circumstellar silicate grain has yet been identified. Olivine and pyroxenes are abundant in chondritic meteorites, micrometeorites, and interplanetary dust particles. The majority of these (Mg-Fe) olivines and pyroxenes appear to have formed by igneous melting and recrystallization in the solar system, but there are rare and exotic Mg-rich forsterite (and enstatite) grains, many of them exhibiting morphological, compositional, or crystallographic evidence of vapor phase growth, that appear to have survived since the earliest stages of solar system formation. Some "relict" forsterite grains in primitive meteorites were among the earliest high-temperature solids in the solar nebula and they exhibit ^{16}O

enrichments (to 50 parts per thousand) consistent with formation from or mass exchange with an (^{16}O-enriched) reservoir of grains and/or gas. ISO's discovery that crystalline circumstellar silicates are Mg-rich greatly simplifies distinguishing presolar silicates from the huge "background" of solar system Mg-Fe silicates in primitive meteoritic materials.

1. INTRODUCTION

Infrared spectroscopy provides a direct link between astronomical observations of dust in space and optical spectroscopy of meteoritic materials in the laboratory. The Infrared Space Observatory (ISO) was launched in 1995 with spectrometers covering the 2.5 – 240 μm wavelength range [1]. The short-wavelength and long-wavelength spectrometers (SWS and LWS) on board ISO provided coverage of the mid IR region where the ~10 and ~ 20 μm "silicate" features are observed and, for the first time, the 20 – 80 μm region where features due to crystalline silicate minerals occur [2]. Prior to ISO, it was widely believed that cosmic silicate grains were mostly amorphous (non-crystalline, *e.g.*, [3]). One of the most important findings of the ISO mission is that in addition to amorphous silicates, sub-μm Mg-rich forsterite and enstatite crystals are ubiquitous constituents of the dust shells of both young and old stars [4,5]. Another important finding is the remarkable similarity between the silicate mineralogy of comet Hale-Bopp's dust and dust around the isolated Herbig Ae/Be star HD100546 [5–7].

The ISO results are of importance in the search for presolar (circumstellar and interstellar) silicates in meteorites, micometeorites and interplanetary dust particles (IDPs). A variety of presolar circumstellar grains (diamond, silicon carbide, graphite, corundum, silicon nitride), recognizable by their non-solar isotopic compositions, have been recovered from carbonaceous chondrites [8,9]. However, not a single (isotopically anomalous) presolar circumstellar silicate grain has ever been positively identified in any meteoritic material. Failure to find presolar circumstellar silicates is one of the most perplexing issues confronting meteoriticists, especially because they are major constituents of the dust shells of evolved oxygen-rich stars [4,5]. The refractory nature of materials like diamond and silicon carbide undoubtedly favors their survival, but silicates like forsterite are also relatively refractory so at least some of them should have survived in primitive meteoritic materials.

The logical place to search for presolar circumstellar (and interstellar) silicates is in the most primitive meteoritic materials. Chondritic porous (CP) IDPs are believed to be the most chemically and isotopically primitive class of meteoritic materials [10–12], and some of them are probably of cometary origin [13]. Mg-rich forsterite and enstatite grains are found in IDPs, micrometeorites, and chondritic meteorites. There is evidence that at least some of these grains were among the first high-temperature solids in the solar nebula. In this paper, we discuss forsterite and enstatite grains in chondritic meteorites, polar micrometeorites, and IDPs. We explore their chemical and physical properties with the goal of determining grain formation mechanisms, and we evaluate whether any of them are presolar grains.

2. FORSTERITE AND ENSTATITE IN CHONDRITIC METEORITES

Chondritic meteorites are essentially lithified agglomerates of "cosmic detritus" that have been subjected to varying degrees of parent body processing since accretion into planetesimals [14,15]. There are at least nine classes of chondritic meteorites, of which the ordinary chondrites are by far the most common. Typical chondrites have been equilibrated due to long term metamorphic heating. Heating above 500 °C causes diffusion between grains and homogenization of mineral compositions. For example, all olivine grains in equilibrated L class chondrites contain about 17% by weight of iron. Only the rare unequilibrated chondrites preserve a range of olivine compositions and contain pure Mg olivine. The unequilibrated meteorites, including the unequilibrated ordinary chondrites (UOCs) and the carbonaceous chondrites (CCs), are considered to be the most primitive because their mineral compositions have not been severely altered by metamorphic heating inside a parent body. The carbonaceous chondrites are the most chemically primitive because they most closely match the composition of the Sun. CCs are believed to be samples of main-belt C-asteroids and UOCs may be from main-belt S-asteroids [16].

CCs and UOCs have a characteristic texture composed of millimeter-sized chondrules set in a fine-grained matrix [14]. The ratio of chondrules to matrix is variable from one meteorite to another. Chondrules are spherical or spheroidal objects $\sim$ 100 μm to 1 mm in diameter composed mostly of crystalline and glassy silicates. They exhibit a wide range of textures, mineral abundances, and bulk compositions [17]. Classic chondrules appear to have crystallized from a molten droplet and they are believed to have formed in regions of the solar nebula where clumps of solid material were heated close to or above their melting temperatures and then rapidly cooled prior to or during the accretion of planetesimals. The fine-grained matrix of CCs and UOCs consists of anhydrous silicates, sulfides, metal, carbonaceous material, as well as hydrous clay minerals (*e.g.*, serpentine) and other secondary minerals formed during parent body aqueous alteration [18].

All CCs and UOCs contain at least some olivine and pyroxene. Olivine has an orthorhombic crystal structure and the general formula $[Mg, Fe]_2SiO_4$ with complete solid solution between two end member compositions forsterite (Fo) Mg_2SiO_4 and fayalite (Fa) Fe_2SiO_4. Fe and Mg abundances are expressed in terms of the mole fractions of forsterite and fayalite respectively. For example, forsterite with stoichiometry $Mg_{1.96}Fe_{0.04}SiO_4$ contains 98% Mg_2SiO_4 and 2% Fe_2SiO_4 and can be referred to as either Fo98 or Fa2 olivine. Mg-Fe pyroxene has multiple crystal structures, the two most commonly observed are monoclinic clinopyroxene and orthorhombic orthopyroxene with the general formula [Mg, Fe]SiO_3 [19]. There is complete solid solution between the two end member compositions enstatite (En) $MgSiO_3$ and ferrosilite (Fs) $FeSiO_3$. Pyroxene with the stoichiometry $(Mg_{0.98}Fe_{.02})SiO_3$ can therefore be referred to as En98 or Fs2.

Olivine and pyroxene are among the most common minerals in chondrites. Their abundance is due, in part, to the fact that they are composed of Mg, Fe and Si, the most cosmically abundant rock forming elements (other than oxygen). Condensation theory predicts that pure Fe-free forsterite and enstatite plus Fe(Ni) metal should condense from the solar nebular at temperatures near 1300 K [20]. The high temperature olivines and pyroxenes are iron free and those containing iron cannot exist in equilibrium with nebular gas until lower gas temperatures. The presence of enstatite and forsterite condensates indicate formation at high temperature and lack of equilibration with cooler nebular gas. The atomic cosmic abundances of Mg, Si, and Fe are nearly identical and, considering only stoichiometry, it is interesting that a high temperature condensate composed only of pure enstatite and Fe metal could remove all of these elements from the gas phase. A lower temperature and fully oxidized condensate that could remove all of these elements from the gas phase would be pure Fo50 olivine, *i.e.*, [Mg, Fe]SiO_4.

Olivine and pyroxene are found both in chondrules and the fine-grained matrix of CCs and UOCs [18, 21–25]. In chondrules, they exhibit a variety of textures and Mg-Fe compositions, and grain sizes range from micrometers to millimeters. Most grains formed by melting and recrystallization of pre-existing solids. In the fine-grained matrices of some meteorites, olivine and pyroxene are rare and may have partially or completely altered to secondary hydrous silicates (*e.g.*, clays). In other meteorites, the fine-grained matrices are relatively anhydrous and contain abundant olivine and pyroxene grains. The origin of the olivine (and pyroxene) in the fine-grained matrices has been the subject of much debate. Some have proposed that they result from the breakdown of chondrules, while others propose that they were directly accreted into the matrix [26–31].

A small fraction of the olivine and pyroxene grains are Mg-rich forsterite ($0 < \mathrm{Fa} < 2$) and enstatite ($0 < \mathrm{Fs} < 2$). The average forsterite abundance is ~1% by weight with grains from millimeters in size down to $< 10\ \mu$m. Enstatite is less abundant (probably $< 0.1\%$ by weight) and grains $> 100\ \mu$m are extremely rare. A rare subset of these forsterites and enstatite grains ($1 - 100$ ppm of the bulk meteorites) are texturally distinct from other grains in terms of size, shape, and composition (Figs. 1 and 2). They are thought to represent grains that predate chondrule and matrix formation and may retain characteristics of solids from which chondrules formed [32, 33]. It has been proposed that these "relict" grains are vapor phase condensates which formed either in the solar nebula or in presolar environments [34–36].

Most relict grains are forsterite and they often exhibit unusual chemical zoning (from their cores to their outer rims) and elevated minor element

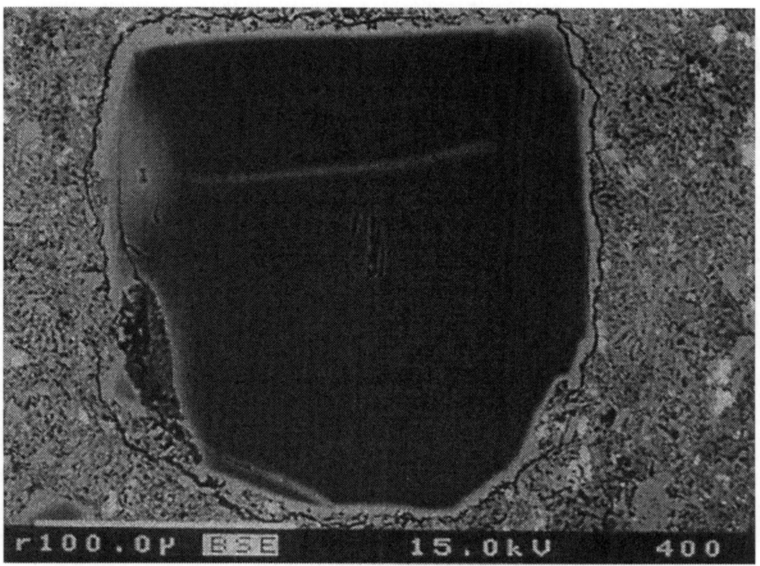

Fig. 1. — Backscattered electron image of a (polished thick-flat) relict forsterite (>Fo99) grain embedded in fine-grained matrix of the Allende (CV3) carbonaceous chondrite. The fine-grained matrix is dominated by Fo50 olivine and the brightness variations in the image are largely proportional to iron content. The thin bright rim on the grain is due either to Fe diffusion from the matrix or prior reaction with nebular gas. The horizontal bright streak is also Fe enrichment, probably from transport of Fe vapor along a fracture [27]. The same grain is seen in cathodoluminescence in Figure 2. Scale bar = 100 μm.

compositions [29,37]. In a study of these forsterites, Steele *et al.* [26] recognized a unique characteristic. In addition to containing unusually high levels of the trace elements Al, Ti, and Ca, they exhibit complex (red and blue) cathodoluminescence (CL) which correlates with the elevated trace element abundances (Fig. 2). Therefore, the CL provides a useful way of recognizing these grains. (Many minerals give off visible light, termed cathodoluminescence, when irradiated with electrons). For olivine and pyroxenes, CL is always associated with the Mg-rich end members (forsterite and enstatite) because appreciable levels of Fe quench CL in the visible (above $\sim$ 2% FeO no CL is observed). The observed CL colors range from bright blue ($\sim$ 450 nm) for high trace element levels to red ($\sim$ 800 nm) for low trace element levels [38]. "Blue" and "red" enstatites are found in chondrules and matrix. The CL allows recognition of complex textures for some grains (Fig. 2).

Two types of relict forsterites have been recognized from optical examination. One is described as dusty or cloudy due to high opacity resulting from high densities of μm-sized opaque inclusions, while the other type is clear and rarely contains inclusions. The clear forsterites exhibit a very characteristic blue cathodoluminescence and are often referred to as "blue" olivines (*e.g.*, Figs. 1 and 2). The FeO content of blue olivines ranges from 0.25 to 1.5 % by weight together with unusually high concentrations of Al, Ca, Ti, Sc and V and low concentrations of Mn and Ni (relative to most olivines). Dusty olivines have not been found within the fine-grained matrix of CCs although blue olivines have. However, blue olivines in the matrix may exhibit sharp euhedral outlines in luminescence (Fig. 2) while those in chondrules show embayments suggesting partial melting during chondrule formation.

There are grains in other classes of meteorites which may be relevant to the question of the survival of presolar silicate grains in meteorites. For example, another type of relict forsterite with similarities to those in CCs and UOCs occurs in enstatite chondrites. They are round forsterite grains which occur within enstatite blade-shape laths. Rambaldi and Wang [39] report that the cores of these olivines are Fe-rich relative to their rims. In cathodoluminescence, the round olivines are orange or blue [40]. The compositions of the blue olivines are similar to those of clear forsterites found in CCs and UOCs [23].

3. FORSTERITE AND ENSTATITE IN POLAR MICROMETEORITES

Thousands of micrometeorites, most of them measuring $50 - 500$ μm diameter, have been collected from ice meltwater in Greenland and Antarctica [41–44]. They are an important class of meteoritic materials because (a) they represent the bulk of the mass of particulate matter accreted by the Earth [45, 46], and (b) they fill an important size gap between < 50 μm diameter IDPs collected in the stratosphere and > 1 cm diameter meteorites. Millimeter-sized melted meteorites called deep-sea spheres are collected from the ocean floor [47, 48]. While some micrometeorites have been completely melted and recrystallized during atmospheric entry, others are not melted or have survived remarkably well without melting. The micrometeorite population appears to resemble some classes of CCs (CI, CM, and CR) [49–52].

Olivine and pyroxene grains are the most common anhydrous silicates in micrometeorites. While some of these grains appear to have formed by partial melting and dehydration of clays during atmospheric entry [51], others are clearly indigenous to the micrometeorites and their parent bodies. The compositions of the indigenous olivines and pyroxenes are similar but not identical to those observed in CCs and UOCs, but they appear to be less abundant in micrometeorites relative to CCs and UOCs [53, 54]. In addition, the pyroxene to olivine ratio is a factor of 10 times higher in micrometeorites than in related CCs [51, 54]. To date, no relict forsterites (or enstatites) have been identified

Fig. 2. — Cathodoluminescence (CL) image (same magnification and same field of view as in Fig. 1) showing oscillatory bright bands which correspond to fine-scale variations in minor refractory elements like Al, Ti, V, and Sc [37]. Such "blue" relict olivines (they luminesce at $\sim$ 450 nm) contain unusually high enrichments of trace refractory elements (and sometimes ^{16}O enrichments) which may result from grain growth by vapor phase condensation [23,27]. Images courtesy of I. Steele, U. Chicago.

in polar micrometeorites, although relict olivines with "blue" CL and elevated minor element abundances have been identified in deep sea spheres [37].

Mg-rich (forsteritic) olivine is a minor component in micrometeorites and those grains that have been identified are mostly between 5 μm and 100 μm in maximum dimension (smaller grains may exist although they have yet to be reported). In a study of 175 (melted and unmelted) Antarctic micrometeorites, Steele [53] identified 26 which contain olivine grains (5 μm diameter). Only two almost pure forsterite grains ($0 < \text{Fa} < 1$) were observed (most grains are $1 < \text{Fa} < 20$). None of the 48 grains analyzed, including 13 with $0 < \text{Fa} < 2$ exhibited observable cathodoluminescence. Flynn *et al.* [55] report $30 < \text{Fa} < 60$ olivines in Greenland and Antarctic micrometeorites. They observed pyroxene grains in 7 out of 17 particles although only one was low-Fe ($0 < \text{Fs} < 2$) enstatite. The size range of the pyroxenes is similar to that of the olivines. In a study of 13 unmelted chondritic micrometeorites from Antarctica, Kurat *et al.* [49] found that six were dominated by clays and the other seven consist primarily of olivine and pyroxene. The olivines range in composition from Fa2 to Fa37 with low levels of minor elements. All of the 7 analyzed olivines contain > 2 mole % Fa, and 2 of 8 pyroxenes are low-Fe ($0 < Fs < 2$) enstatite. Ten partially melted (scoriaceous) micometeorites were also analyzed. Four low-Fe forsterites and 4 enstatite grains were identified.

The most recent compilation (by C. Engrand, UCLA) of the compositions of olivine and (Mg-Fe) pyroxenes in micrometeorites is shown in Figure 3. Out of a total of 179 olivine grains, 19 are within the $0 < Fa < 1$ compositional range, 48 are within $0 < Fa < 2$, and 82 are within $0 < Fa < 5$. Thirty eight percent of the olivine grains contain less than 2 mole % Fa (Fig. 3). The pyroxenes are even more Mg-rich than the olivines. Out of a total of 133 pyroxene grains, 16 have $Fs < 1$ compositions, 43 are $Fs < 2$, and 90 are $Fa < 5$. Fourthly four percent of the pyroxenes contain less than 2 mole % $FeSiO_3$ (*i.e.*, $Fs < 2$). Note that 11% of the olivines contain contain <1 mole % Fe_2SiO_4 (*i.e.*, $Fa < 1$) and 12% of the pyroxenes contain < 1 mole % $FeSiO_3$ (*i.e.*, $Fs < 1$).

4. FORSTERITE AND ENSTATITE IN INTERPLANETARY DUST PARTICLES (IDPs)

IDPs are mostly $5-30$ μm diameter cosmic dust grains that are collected from the stratosphere using U2 aircraft [56]. Many of them contain high densities ($10^{10} - 10^{11}$ cm^{-2}) of implanted solar flare tracks and solar wind-irradiated surfaces which establish (a) their $\sim 10^4$ year lifetimes as discrete small bodies in solar orbit (as opposed to fragments of larger meteorites or micrometeorites) and, (b) that they were not severely pulse heated (above $\sim$ 650 °C) during atmospheric entry.

Based on the structure of their $\sim$ 10 μm "silicate feature" chondritic IDPs are divided into "pyroxene", "olivine", and "layer silicate" IR classes [57–59]. The "layer silicate" IR class are mineralogically and compositionally similar to the fine-grained matrices of CCs, and several have been linked directly to types CI and CM meteorites respectively and hence an asteroidal origin [60–62]. The status of the "olivine" IR class of chondritic IDPs is less secure. Although some particles contain solar flare tracks, indicating that they are indeed pristine objects, most are track-free with equilibrated silicate mineralogies suggestive of severe pulse heating during atmospheric entry. The olivine (and pyroxene) grains in three IDPs from the "olivine" IR class were studied using analytical electron microscopy [63]. As expected, olivine is the most abundant phase in all three samples and two of the three IDPs contain solar flare tracks. One IDP (Jedai) consists predominantly of a few large ($1-5$ μm diameter) and many small ($0.05-1$ μm diameter) anhedral olivine grains in fine-grained matrix. Another IDP (Essex) contains $0.05-0.1$ μm high-Ca (clino)pyroxene grains with (high-Fe) augitic compositions. The third IDP (Attila) contains abundant sub-μm olivine grains as well as some as large as several μm. The olivines in Jedai were uniformly forsteritic ($0 < Fs < 2$) and a single 0.1 μm enstatite (MgSiO3) grain was identified. The olivines in the other two IDPs were Fe-rich with Mg/(Fe+Mg) ratios of 0.72 and 0.39.

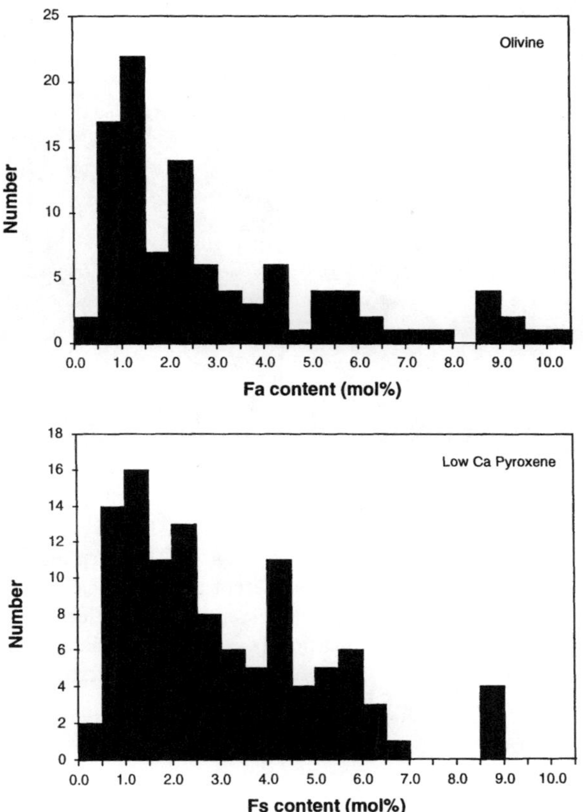

Fig. 3. — Compositions of olivines and pyroxenes in micrometeorites expressed in mole % fayalite-Fe_2SiO_4 (Fa) and ferrosilite-$FeSiO_3$ (Fs) respectively. (a) Olivine (179 grains) with 38% containing Fa < 2 (b) (Low-Ca) Mg-Fe pyroxene (133 grains) with 44% containing Fs < 2. (Data courtesy of C. Engrand, UCLA).

Most "pyroxene" IDPs belong to the chondritic porous (CP) subset of IDPs. These IDPs are potentially of highest relevance to the question of survival of presolar circumstellar and interstellar grains in meteoritic materials because they are the only known class of meteoritic materials that contain submicrometer forsterite, enstatite, and glassy silicate grains (GEMS, glass with embedded metal and sulfides) as major constituents (Figs. 4 and 5). Some of them have higher volatile element abundances than chondritic meteorites, suggesting

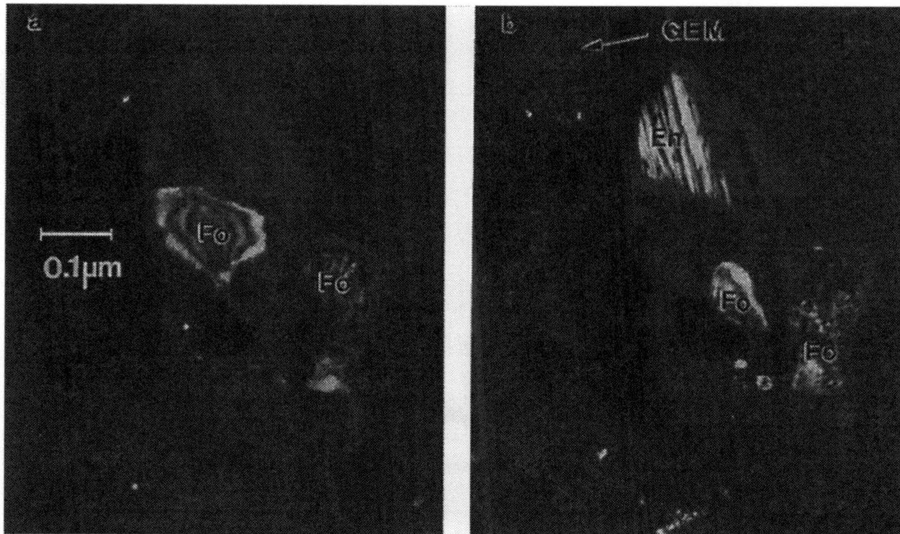

Fig. 4. — Darkfield electron micrographs (same field of view) of an electron transparent thin section of the chondritic porous (CP) interplanetary dust particle U219C2. Each image was acquired with a different electron beam/specimen incidence angle, such that a maximum number of (crystalline) grains could be observed in a high contrast (Bragg scattering) orientation (darkfield contrast depends on the crystallographic orientation of grain). The images show submicrometer forsterite (Fo) and enstatite (En) grains plus a GEM.

that they are even more chemically primitive than CCs, UOCs and micrometeorites [10, 11]. CP IDPs are typically highly porous with extremely fragile microstructures, indicating that they are also the most physically primitive meteoritic materials [64]. Such fragile microstructures have no counterpart among known classes of meteoritic materials.

An accumulating body of evidence indicates that some, and possibly all, CP IDPs are from comets (Fig. 6), where the fragile microstructure could have been supported by ice. One method of identifying cometary IDPs is based on the speed with which they enter the Earth's atmosphere. The entry speed can be estimated from the temperature at which they release Helium during stepwise heating in the laboratory [65]. Typical cometary IDPs enter at speeds > 16 km s^{-1}, whereas typical asteroidal IDPs enter at speeds of $12 - 14$ km s^{-1} [13]. The fine-grained texture and high carbon content [10] are also consistent with cometary origin. The identification of Mg-rich olivine and pyroxene in infrared spectra of comet Hale-Bopp has strengthened the link between CP IDPs and comet dust [6, 7, 66]. The En/Fo ratio inferred for Hale-Bopp is 2 to 12 [7]. This is similar to the ratio seen in chondritic porous IDPs but much higher than most other primitive meteoritic materials.

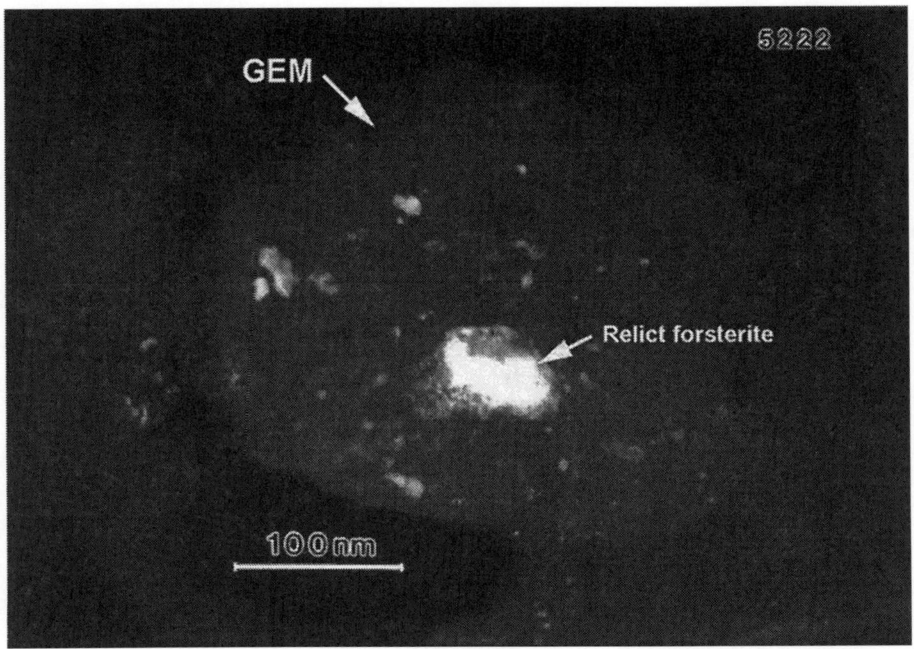

Fig. 5. — Darkfield electron micrograph of a GEM within chondritic porous IDP U222B28. The bright inclusion (lower center) is a relict forsterite grain embedded within the GEM. This GEM may be an interstellar "amorphous silicate" grain [67] and the embedded grain may be an example of the circumstellar forsterite detected by ISO. However, absolute proof of the presolar origins of these grains requires measurement of non-solar isotope abundances.

Of all sources of meteoritic materials comets are most likely to contain well-preserved pre-solar components (*e.g.*, [68]). Comets formed in the outer solar nebula (beyond the orbit of Saturn) where the temperature was < 150 K [69,70]. Unless there was strong radial mixing, more than models of the solar nebula predict [71,72], the material incorporated into comets was not subjected to strong heating in the solar nebula. Thus, if presolar forsterite and enstatite grains are indeed preserved in primitive (circumsolar) meteoritic materials, they are perhaps most likely to be found in IDPs. The highest D/H isotope ratio ever measured in a natural sample ($100 - 1000\times$ D/H in meteorites) was recently observed in a CP IDP, which suggests that some IDPs contain relatively pristine organic material formed by ion molecule reactions in cold interstellar molecular clouds [12,73].

Enstatite ($0 < \text{Fs} < 2$) and to a lesser extent forsterite ($0 < \text{Fa} < 2$) are common crystalline silicate grains in CP IDPs [58] (Fig. 5). In a study of 15 anhydrous chondritic IDPs from the "olivine" and "pyroxene" IR classes Zolensky

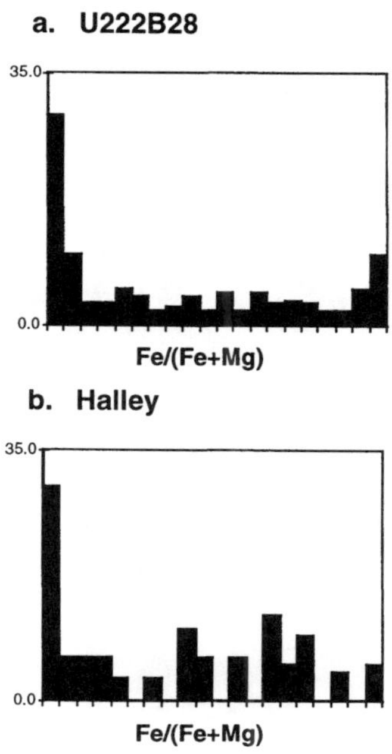

Fig. 6. — Histograms comparing the submicrometer-scale distribution of Fe/(Fe+Mg) in IDPs with comet Halley dust. The horizontal axis ranges from 0 to 1. (a) Chondritic porous "pyroxene" IDP U222B28 (266 analyses) which contains abundant Mg-rich silicates, mostly enstatite ($MgSiO_3$) and some forsterite (Mg_2SiO_4), resulting in a large number of Fe/(Fe+Mg) data points near zero [58]. (b) Halley (PUMA) mass spectrometry data (71 analyses) [74,75]. Only the anhydrous chondritic "pyroxene" IDPs show a submicrometer-scale distribution of Fe/(Fe+Mg) similar to comet Hallet's dust [76]. The Mg-rich silicate mineralogy of Halley is similar to that of comet Hale-Bopp as measured by ground based telescopes and ISO [6,7,77].

and Barrett [78] found that the range of compositions of olivine and (Mg, Fe) pyroxene grains is $52 < \mathrm{Fo} < 100$ and $46 < \mathrm{En} < 100$, with strong clustering around $90 < \mathrm{Fo} < 100$ and $90 < \mathrm{En} < 100$ compositions. Grain sizes range from $0.05 - 5$ μm, although most grains are between 0.05 and 0.5 μm in diameter. There is crystallographic evidence that some enstatite grains formed by direct gas-to-grain condensation from a nebula gas [19]. These crystals exhibit highly unusual whisker and ultrathin platelet morphologies and internal crystallographic defects (axial screw dislocations) which are typical of crystals grown from the vapor phase. However, such morphologies and defects have not been observed in the forsterite grains, possibly because forsterite has only one crystal structure (orthorhombic) whereas enstatite has several (*e.g.*, clinoenstatite and orthoenstatite). Both clinoenstatite and orthoenstatite are found in IDPs, often as complex intergrowths within a single crystal.

There is also compositional evidence that forsterite and enstatite grains in "pyroxene" IDPs grew by direct gas-to-grain condensation from a nebular gas [79]. Low Fe, Mn enriched (LIME) enstatite and forsterite grains in IDPs contain up to 5% by weight of MnO, in contrast to the majority of of pyroxenes and olivines in meteorites which contain $< 0.5\%$ MnO. It is likely that that the high Mn contents reflect high temperature condensation from a nebular gas. Forsterite (Mg_2SiO_4) is the first major silicate mineral to condense from a cooling gas of solar composition. Fe condenses out as metal but Mn is not stable as metal and condenses as (Mn_2SiO_4) in solid solution with forsterite. Whereas, Fe and Mn are coupled during mineral crystallization from a liquid melt (producing Fe- and Mn-rich olivine and pyroxenes), they are decoupled during vapor phase condensation producing low Fe- and Mn-enriched enstatite and forsterite. Similar LIME forsterites and enstatites were identified in the fine-grained matrix of CCs and a UOC suggesting a common origin for some of the grains in IDPs and meteorites [79].

"Pyroxene" IDPs also contain GEMS (Figs. 4 and 5). They are ubiquitous constituents of pyroxene-rich CP IDPs in general and cometary IDPs in particular where they are found admixed with enstatite and forsterite grains [13,80,81]. The properties of GEMS appear to have been shaped primarily by exposure to ionizing radiation and, since the exposure occurred prior to accretion of the (cometary) IDPs, GEMS must predate comets [67,80]. Many of the enstatite and forsterite grains contain amorphous rims suggesting that they too have been pre-accretionally exposed to ionizing radiation [82]. The properties of GEMS are consistent with those inferred for "amorphous silicate" grains which are ubiquitous throughout the interstellar medium [83,84].

5. DISCUSSION

Forsterite and enstatite grains from of the order of a tenth of a micrometer to hundreds of micrometers in diameter are present as minor constituents of meteorites and micrometeorites. Figure 7 illustrates a rare IDP that is composed almost entirely of forsterite. Sub-μm forsterite and enstatite grains are major constituents of the anhydrous CP "pyroxene" class of IDPs, some or all of which may be of cometary origin [64,80]. Are any of these grains examples of the circumstellar silicate crystals detected by ISO? If so, what kind of stars did they form around, how old are they, and how did they survive in the interstellar medium?

The only known way to unambiguously prove the presolar origin of (circumsolar) grains is to measure non-solar isotope abundances. The circumstellar silicon carbide and aluminum oxide grains found in meteorites have major isotopic anomalies in all their major elements (C, Si, Al, O) which reflect the isotopic compositions of their parent stars [8,9]. Therefore, well-preserved presolar circumstellar silicates should also exhibit major isotope anomalies. The oxygen isotopic compositions of some forsterite grains have been measured [85, 86]. Although they tend to be ^{16}O enriched relative to other meteoritic and terrestrial silicates [31, 87–89], the observed range of ^{16}O values is not significantly different from those measured in high-temperature refractory Ca-Al-rich inclusions (CAIs) which are also found in CCs and UOCs [90, 91]. CAIs were likely among the first high-temperature condensation products in the solar nebula, and their isotopic compositions are believed to reflect mixing and re-equilibration of a ^{16}O-enriched dust with ^{16}O-poor gas [92]. Although the oxygen isotopic compositions of the forsterites so far analyzed do not establish that they are presolar circumstellar grains, they at least reveal evidence of an isotopic "memory" of nucleosynthetic processes.

The expected isotopic compositions of interstellar silicate grains, *i.e.*, those that have experienced prolonged lifetimes in the interstellar medium (ISM) is difficult to predict. Typical interstellar grains might not, on average, have isotopic compositions significantly different from solar system material because theoretical calculations suggest that most grains are heavily processed by shocks and sputtering during their time in the interstellar medium [8, 93–96]. Presolar silicate grains in a protostellar cloud may consist of mixtures of interstellar grains, with a distribution of ISM exposure ages, whose chemical and isotopic compositions have been "homogenized" by processing in the ISM. The properties of CP IDPs (*i.e.*, chondritic mixtures of irradiated and sputtered GEMS, enstatite, forsterite, *etc.* [82, 97] are consistent with interstellar material, but proving it is difficult. Individual sub-μm grains within an IDP might be isotopically anomalous, but the entire IDP may have a bulk composition close to the solar system average. The O (and Mg and Si) isotopic compositions of several IDPs have been measured [85, 91, 98], and preliminary measurements of the Mg isotopic compositions of individual silicate grains (GEMS and an enstatite whisker) have been attempted [99]. No major anomalies outside of solar system values have been detected. Ultimately, it will be necessary to make measurements at the nanometer scale (*e.g.*, relict forsterite grains within GEMS (Fig. 5) to meaningfully measure the isotopic compositions of silicates in IDPs. In this regard, the latest generation, high spatial resolution "nanoSIMS" currently under development may play a key role in the future of IDP research.

ISO has provided exciting new insight about the nature of astronomical silicates. The SWS and LWS spectra reveal a common mineralogical link between circumstellar silicates, cometary silicates, and Mg-rich silicate grains found in primitive meteoritic materials. Perturbed (oxygen) isotope abundances as well as crystallographic and compositional evidence of vapor phase growth have been observed among the small number of Mg-rich meteoritic grains examined to date. Although these properties do not prove that the grains formed in presolar environments, (instead they may have formed in the solar nebula), they are consistent with the expected properties of well-preserved presolar circumstellar grains. In any case, ISO focuses renewed emphasis on the laboratory study of Mg-rich olivine ($90 < \mathrm{Fo} < 100$) and pyroxene ($90 < \mathrm{En} < 100$) in

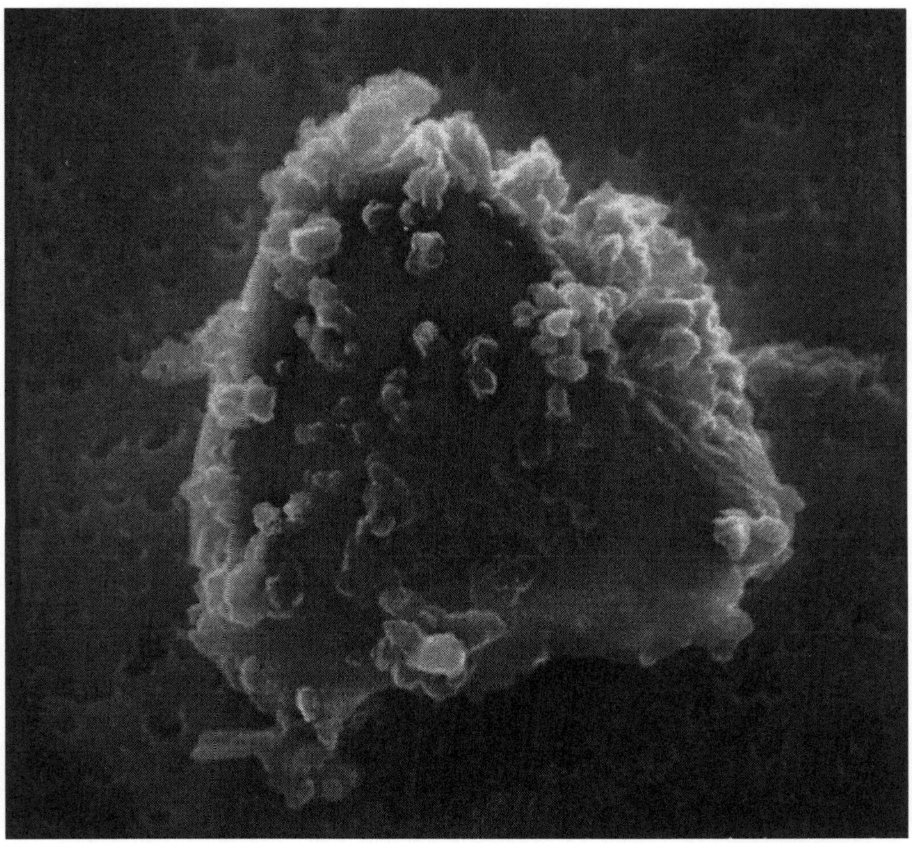

Fig. 7. — A scanning electron microscope image of a 10 μm interplanetary dust particle collected in the stratosphere. The particle is almost entirely composed of forsterite. The smooth main mass is a single optically transparent forsterite grain that is encrusted with sub-μm dark grains that are relatively iron rich and have nearly chondritic elemental composition. Presumably the dark grains are GEMS and they are residual bits of a fine-grained matrix in which the forsterite was formerly imbedded.

primitive meteoritic materials, especially those materials that are most likely of cometary origin.

ACKNOWLEDGMENTS

This research was supported by NASA grants NASW5035 and NAG 5-7450 (to JPB) and NAG5-7449 (T. Snow). We thank C. Engrand, G. Kurat, M. Maurette, and I. Steele for data and discussions.

REFERENCES

[1] Kessler M.S., Steinz J.A., Anderegg M.E., *et al.*, *A&A* **315** (1996) L27.
[2] Malfait K., Waelkens C., Waters L.B.F.M., *et al.*, *A&A* **332** (1998) L25.
[3] Sandford S.A., *Met. Planet. Sci.* **31** (1996) 449.
[4] Waters L.B.F.M., Molster F.J., de Jong T., *et al.*, *A&A* **315** (1996) L361.
[5] Waters L.B.F.M., Waelkens C., van Winckel H., *et al.*, *Nature* **391** (1998) 868.
[6] Hanner M.S., Gehrz R.D., Harker D.E., *et al.*, *Earth Moon Planets* (1998) in press.
[7] Wooden D.H., Harker D.E., Woddward C.E., *et al.* (1998) submitted.
[8] Zinner E., Astrophysical Implications of the Study of Presolar Materials, AIP Conf. Proc. 402, edited by T.J. Bernatowicz and E. Zinner (AIP, New York, 1997) p. 3.
[9] Bernatowicz T.J. and Walker R.M., *Phys. Today* **50** (1997) 26.
[10] Thomas, K.L., Blanford G.E., Keller L.E., Klock W. and McKay D.S., *Geochim. Cosochim. Acta* **57** (1993) 1551.
[11] Sutton S.R., Analysis of Interplanetary Dust M, AIP Conf. Proc. 310, edited by E. Zolensky, T.L. Wilson, F.J.M. Rietmeijer and G.J. Flynn (AIP, New York, 1994) p. 145.
[12] Messenger S. and Walker R.M., Astrophysical implications of the laboratory study of presolar materials, AIP Conf. Proc. 402, edited by T.J. Bernatowicz and E. Zinner (AIP, New York, 1997) p. 545.
[13] Brownlee D.E., Joswiak D.J., Schlutter D.J., *et al.*, *Lunar Planet Sci.* **26** (1995) 183.
[14] Dodd R.T., Meteorites (Cambridge University Press, New York, 1981) p. 368.
[15] Kerridge J.F. and Mathews M.S., Meteorites and the Early Solar System, edited by J.F. Kerridge and M.S. Mathews (University of Arizona Press, Tucson, 1988) p. 1252.
[16] Wetherill G.W. and Chapman C.R., Meteorites and the Early Solar System, edited by J.F. Kerridge and M.S. Mathews (University of Arizona Press, Tuscon, 1988) p. 35.
[17] Grossman J.N., Meteorites and the Early Solar System, edited by J.F. Kerridge and M.S. Mathews (University of Arizona Press, Tuscon, 1988) p. 680.
[18] Buseck P.R. and Hua X., *Ann. Rev. Earth Planet. Sci.* **21** (1993) 255.
[19] Bradley J.P., Brownlee D.E. and Veblen D.R., *Nature* **301** (1983) 473.
[20] Grossman L., *Geochim. Cosmochim. Acta* **36** (1972) 597.
[21] McSween H.Y. and Richardson S.M., *Geochim. Cosmochim. Acta* **41** (1977) 1145.
[22] Bunch T.E. and Chang S., *Geochim. Cosmochim. Acta* **44** (1980) 1543.
[23] Steele I.M., Meteorites and the Early Solar System, edited by J.F. Kerridge and M.S. Mathews (University Arizona Press, Tuscon, 1988) p. 808.
[24] Keller L.P. and Buseck P.R., *Geochim. Cosmochim. Acta* **54** (1990) 1155.

[25] Hua X. and Buseck P.R., *Geochim. Cosmochim. Acta* **59** (1995) 563.
[26] Steele I.M., Smith J.V. and Skiriu C., *Nature* **313** (1985) 294.
[27] Steele I.M., *Am. Min.* **71** (1986) 966.
[28] Steele I.M., *Geochim. Cosmochim. Acta* **50** (1986) 1379.
[29] Steele I.M., *Meteoritics* **25** (1990) 301.
[30] Jones R.H., *Geochim. Cosmochim. Acta* **56** (1992) 467.
[31] Jones R.H., Saxton J.M., Lyon I.C. and Turner G., *Lunar Planet. Sci.* **29** (1998) 1795.
[32] Nagahara H. and Kushiro I., *Mem. Natl. Inst. Polar Res.* **25** (1982) 66.
[33] Rambaldi E.R., *Nature* **293** (1981) 558.
[34] Fuchs L.H., Olsen E. and Jensen K.J., *Smithsonian Contrib. Earth Sci.* **10** (1973) 1.
[35] McSween H.Y., *Geochim. Cosmochim. Acta* **41** (1977) 411.
[36] Olsen E. and Grossman L., *Earth Planet. Sci. Lett.* **41** (1978) 111.
[37] Steele I.M., *Am. Min.* **80** (1995) 823.
[38] Benstock E.J., Buseck P.R. and Steele I.M., *Am. Min.* **82** (1997) 310.
[39] Rambaldi E.R. and Wang D., *Meteoritics* **17** (1982) 272.
[40] Leitch C.A. and Smith J.V., *Geochim. Cosmochim. Acta* **46** (1982) 2083.
[41] Maurette M., Jehanno C., Robin E. and Hammer C., *Nature* **328** (1987) 699.
[42] Maurette M., Olinger C., Michel-Levy M.C., *et al.*, *Nature* **351** (1991) 44.
[43] Maurette M., Kurat G., Perreau M. and Engrand C., *Microbeam Anal.* **2** (1993) 239.
[44] Maurette M. and Brownlee D.E., *Lunar Planet. Sci.* **20** (1989) 636.
[45] Hughes D.M. and Cosmic Dust, edited by J.A.M. McDonnell (Wiley, New York, 1978) p. 123.
[46] Dermott S.F., Jayaraman S., Xu Y.L., Gustavson B.E.S. and Liou J.C., *Nature* **369** (1994) 719.
[47] Brownlee D.E., *Ann. Rev. Earth Planet. Sci.* **13** (1985) 147.
[48] Steele I.M., Smith J.V. and Brownlee D.E., *Nature* **313** (1985) 297.
[49] Kurat G., Koeberl C., Presper T., Brandstdtter F. and Maurette M., *Geochim. Cosmochim. Acta* **58** (1994) 3879.
[50] Gounelle M., Maurette M., Engrand C. and Kurat G., *Met. Planet. Sci.* **33** (1998) A61.
[51] Maurette M. (1998) personal communication.
[52] Engrand C. and Maurette M., *Met. Planet. Sci.* **33** (1998) 565.
[53] Steele I.M., *Geochim. Cosmochim. Acta* **56** (1992) 2923.
[54] Kurat G., Brandstatter F., Presper T., Koeberl C. and Maurette M., *Russ. Geol. Geophys.* **34** (1993) 132.
[55] Flynn G.J., Sutton S.R. and Klock W., *Proc. NIPR Symp. Antarct. Meteorites* **6** (1993) 304.
[56] Sandford S.A., *Fund. Cosmic. Phys.* **12** (1987) 1.
[57] Sandford S.A. and Walker R.M., *ApJ* **291** (1985) 838.
[58] Bradley J.P., Germani M.S. and Brownlee D.E., *Earth Planet. Sci. Lett.* **93** (1989) 1.

[59] Germani M.S., Bradley J.P. and Brownlee D.E., *Earth Planet. Sci. Lett.* **101** (1990) 162.
[60] Bradley J.P. and Brownlee D.E., *Science* **251** (1992) 549.
[61] Keller L.P., Thomas K.L. and McKay D.S., *Geochim. Cosmochim. Acta* **56** (1992) 1409.
[62] Rietmeijer F.J.M., *Met. Planet. Sci.* **31** (1996) 278.
[63] Christoffersen R. and Buseck P.R., *Earth Planet. Sci. Lett.* **78** (1986) 53.
[64] Bradley J.P. and Brownlee D.E., *Science* **231** (1986) 1542.
[65] Nier A.O., Analysis of Interplanetary Dust, AIP Conf. Proc. 310, edited by M.E. Zolensky, T.L. Wilson, F.J.M. Rietmeijer and G.J. Flynn (AIP, New York, 1994) p. 115.
[66] Crovisier J., Leech K., Bockelee-Morvan D., *et al.*, *Science* **275** (1997) 1904.
[67] Bradley J.P., *Science* **265** (1994) 925.
[68] Greenberg J.M., Comets, edited by L.L. Wilkening (University of Arizona Press, Tucson, 1982) p. 131.
[69] Tscharnuter W.M. and Boss A.P., Protostars and Planets III, edited by E.H. Levy and J.I. Lunine (University Arizona Press, Tucson, 1993) p. 921.
[70] Boss A.P., *Proc. Lunar Planet. Sci. Conf.* **25** (1994) 149.
[71] Cuzzi J.N., Dobrovolskis A.R. and Champney J.M., *Icarus* **106** (1993) 102.
[72] Weidenschilling S.J., Meteorites and the Early Solar System, edited by J.F. Kerridge and M.S. Mathews (University of Arizona Press, Tuscon, 1988) p. 348.
[73] Messenger S., Clemett S.J., Keller L.P., *et al.*, *Meteoritics* **30** (1995) 546.
[74] Jessberger E.K., Christoforidis A. and Kissel J., *Nature* **332** (1988) 691.
[75] Brownlee D.E., Wheelock M.M., Temple S., Bradley J.P. and Kissel J., *Lunar Planet. Sci.* **18** (1987) 133.
[76] Bradley J.P., *Geochim. Cosmochim. Acta* **52** (1988) 889.
[77] Lawler M.E., Brownlee D.E., Temple S. and Wheelock M.M., *Icarus* **80** (1989) 225.
[78] Zolensky M.E. and Barrett R.A., *Meteoritics* **29** (1994) 616.
[79] Klock W., Thomas K.L., McKay D.S. and Palme H., *Nature* **339** (1989) 126.
[80] Joswiak D.J., Brownlee D.E., Bradley J.P., Schlutter D.J. and Pepin R.O., *Lunar Planet. Sci.* **28** (1996) 625.
[81] Bradley J.P., Humecki H.J. and Germani M.S., *ApJ* **394** (1992) 643.
[82] Bradley J.P., Brownlee D.E. and Snow T.P., From Stardust to Planetesimals, ASP Conf. Series 122, edited by Y.J. Pendleton and A.G.G.M. Tielens (ASP, San Francisco, 1997) p. 217.
[83] Flynn G.J., *Nature* **371** (1994) 287.
[84] Martin P.G., *ApJ* **445** (1995) L63.
[85] McKeegan K.D., PhD Thesis (Washington University, St. Louis, 1987) p. 1.
[86] Weinbruch S., Zinner E.K., El Goresy A., Steele I.M. and Palme H., *Geochim. Comochim. Acta* **57** (1993) 2649.

[87] Hervig R.L. and Steele I.M., *Lunar Planet. Sci.* **26** (1995) 589.
[88] Engrand C., McKeegan K.D. and Leshin L.A., *Met. Planet. Sci.* **32** (1997) A39.
[89] Leshin L.A., McKeegan K.D., Engrand C., *et al.*, *Met. Planet. Sci.* **33** (1998) A93.
[90] MacPherson G.J., Wark D.A. and Armstrong J.T., Meteorites and the early Solar System, edited by J. Kerridge and M.S. Mathews (University of Arizona Press, Tucson, 1988) p. 746.
[91] Greshake A., Hoppe P. and Bischoff A., *Met. Planet. Sci.* **31** (1996) 739.
[92] Clayton D., *QJRAS* **23** (1982) 174.
[93] Seab C.G. and Shull J.M., Interrelationships Among Circumstellar Interstellar and Interplanetary Dust, edited by J.A. Nuth and R.E. Stencel (NASA CP 2403, Washington DC, 1986) p. 37.
[94] Nuth J.A., Meteorites and the early Solar System, edited by J. Kerridge and M.S. Mathews (University of Arizona Press, Tucson, 1988) p. 457.
[95] Jones A.P., From Stardust to Planetesimals, ASP Conf. Series 122, edited by Y.J. Pendleton and A.G.G.M. Tielens (ASP, San Francisco, 1997) p. 97.
[96] Jones A.P., Tielens A.G.G.M., Hollenbach D.J. and McKee C.F., Astrophysical Implications of the Study of Presolar Materials, edited by T.J. Bernatowicz and E. Zinner (AIP Conf. Proc, New York, 402, 1997) p. 595.
[97] Bradley J.P., Snow T., Brownlee D.E. and Keller L.P., *Lunar Planet. Sci.* **29** (1998) 1737.
[98] Esat T.M., Brownlee D.E., Papanastassiou D.A. and Wasserburg G.J., *Science* **206** (190) 1979.
[99] Bradley J.P. and Ireland T., Physics Chemistry and Dynamics of Interplanetary, Dust, ASP Conf. Series 122, edited by Y.J. Pendleton and A.G.G.M. Tielens (ASP, San Francisco, 1996) p. 217.

Imprimé en France. JOUVE, 18 rue Saint-Denis, 75001 PARIS
Dépôt légal : avril 1999